Mycotoxins and Fungal Toxins: Current Status and Future Perspectives

Mycotoxins and Fungal Toxins: Current Status and Future Perspectives

Guest Editors

Jianhua Wang
Josefa Tolosa

MDPI

Basel • Beijing • Wuhan • Barcelona • Belgrade • Novi Sad • Cluj • Manchester

Guest Editors
Jianhua Wang
Institute for Agro-Food Standards and Testing Technology
Shanghai Academy of Agricultural Sciences
Shanghai
China

Josefa Tolosa
Preventive Medicine and Public Health, Food Science, Toxicology and Forensic Medicine
University of Valencia
Valencia
Spain

Editorial Office
MDPI AG
Grosspeteranlage 5
4052 Basel, Switzerland

This is a reprint of the Special Issue, published open access by the journal *Toxins* (ISSN 2072-6651), freely accessible at: https://www.mdpi.com/journal/toxins/special_issues/3NDGM73EAY.

For citation purposes, cite each article independently as indicated on the article page online and as indicated below:

Lastname, A.A.; Lastname, B.B. Article Title. *Journal Name* **Year**, *Volume Number*, Page Range.

ISBN 978-3-7258-5101-0 (Hbk)
ISBN 978-3-7258-5102-7 (PDF)
https://doi.org/10.3390/books978-3-7258-5102-7

Cover image courtesy of Jianhua Wang

Contents

About the Editors

Jianhua Wang

Dr. Wang is currently a Research Professor at the Shanghai Academy of Agricultural Sciences. He has long been engaged in molecular biology research of foodborne pathogens, focusing on: (1) the isolation and identification of mycotoxin-producing fungi and the construction/screening of mutant strains; (2) the functional genomics of pathogens utilizing gene-editing technologies; (3) the discovery and functional characterization of toxin biosynthesis-related genes through integrated approaches including chromatography-mass spectrometry, transcriptomic analysis, and comparative genomics; and (4) the elucidation of the molecular mechanisms and regulatory networks underlying fungal toxin biosynthesis, providing theoretical foundations and technical support for mycotoxin control strategies. His relevant research has been published in journals such as *Trends in Food Science & Technology, Journal of Fungi, Toxins, Food Control, Fungal Genetics and Biology, PNAS, Journal of Phytopathology, Plant Disease, International Journal of Molecular Sciences,* and other international journals.

Josefa Tolosa

Dr. Tolosa is currently Associate Professor of Toxicology at the Department of Preventive Medicine and Public Health, Food Science, Toxicology and Forensic Medicine at Universitat de València (UV). Her research activity focuses on food chemistry and food toxicology, and more concretely on the determination of mycotoxins in food samples and the risk assessment and determination of toxicity by means of alternative in silico methods. Throughout her research career, she has been focused mainly on two major lines of research. The first of these is the determination and identification of contaminants and metabolites or degradation products present in food, mainly of animal origin, and also in raw materials and feed intended for animal feeding. To this end, she has worked with advanced methodologies such as high-resolution liquid chromatography and diverse high-resolution mass spectrometry detectors, such as time of flight (TOF) and the Orbitrap. Meanwhile, her research also focuses on the evaluation of the toxicokinetics and toxicodynamics of different contaminants, mainly mycotoxins, using computational models based on artificial intelligence (AI) and machine learning. Dr. Tolosa has participated in several competitive national and international projects. Recently, she participated as a member of the national project (PID2020-115871RB-I00) "Combined exposure to mycotoxins and pesticides: Risk assessment, safety and mitigation strategies" funded by the Ministry of Science and Innovation. The main objective of this project was to analyze the interaction between mycotoxins and pesticides and their combined toxic effects to develop strategies to mitigate their impact on human and animal health. More than 30 publications have been derived from this project, and 50 presentations of results have been made at national and international congresses.

toxins

MDPI

Editorial

Mycotoxins and Fungal Toxins: Current Status and Future Perspectives

Jianhua Wang [1,*], Josefa Tolosa [2], Wenyu Wang [1] and Xianli Yang [1,*]

1 Institute for Agro-Food Standards and Testing Technology, Shanghai Academy of Agricultural Sciences, 1000 Jinqi Road, Shanghai 201403, China; m230100125@st.shou.edu.cn

2 Laboratory of Toxicology, Department of Preventive Medicine and Public Health, Food Sciences, Toxicology and Forensic Medicine, University of Valencia, 46100 Buriassot, Valencia, Spain; josefa.tolosa@uv.es

* Correspondence: wangjianhua@saas.sh.cn (J.W.); yangxianli@saas.sh.cn (X.Y.)

1. Introduction

Many toxigenic fungi are devastating pathogens of crop, fruit, and vegetable diseases worldwide. Serious yield losses in grains and fruits occur annually due to fungal diseases in the field and during storage. In addition to reducing agricultural product yields, devastating fungal diseases can also lead to the contamination of consumption goods with toxic secondary metabolites which pose a threat to human and animal health. Food/feed safety is by far the greatest concern as infected grains are often contaminated with various mycotoxins, such as trichothecenes, aflatoxins, and patulin, and these toxic secondary metabolites can also co-occur. In recent years, the impact of mycotoxins on food safety and human health has aroused considerable public concern. Toxigenic fungi and mycotoxins therefore pose a significant threat to the global food supply and food/feed security.

2. There Are Various Types of Mycotoxins Which Are Key Biological Hazards in the Field of Food and Feed Safety

Mycotoxins have attracted global attention because of their wide-ranging appearance in various grains, nuts, fruit juices, and related products. It was estimated by Eskola et al. [1] that approximately 25% of the world's food/feed commodities are contaminated by mycotoxins annually. On the other hand, multiple toxic effects (emetic, anorexic, immunosuppressive, and even death in cases of severe exposure) induced by these secondary metabolites toward humans and animals have also aroused considerable public concern. Furthermore, several mycotoxins have been shown to behave as virulence factors in plants, allowing fungi to colonize and spread in host tissues.

There is currently a wide range of identified fungal toxins, the most common being aflatoxins, trichothecenes, zearalenone (ZEN), fumonisins, ochratoxin, and penicillin. Members of the *Fusarium* genus represent a large group of toxigenic fungi, which have been reported to produce at least eight kinds of mycotoxins with toxicological significance, including trichothecenes, ZEN, fumonisins, fusarin C, moniliformin, enniatins, beauvericin, and fusaproliferin. Toxigenic fungi in the section of *Aspergillus flavi* synthesize four main molecules, aflatoxin B1, B2, G1, and G2, all of which can yield other derivatives when metabolized by animals after their ingestion [2,3].

Furthermore, the identification of novel fungal toxins in recent years has presented new and significant threats to food and feed safety, such as masked/modified mycotoxins and novel toxins. Modifications of trichothecenes, for example, glycosylation on the C-3 of deoxynivalenol (DON), are more common in nature; this phenomenon has primarily

https://doi.org/10.3390/toxins17040176

been found in many plants [4]. Most recently, two HT-2-alpha-glucosides produced by *Fusarium sporotrichilides* were identified on a rice medium [5]. Different modified forms of mycotoxins have been found in nature. For example, two kinds of sulfate conjugation of DON, DON-15S and DON-3S, were detected in wheat treated with DON in 2015 [6]. It is worth noting that most mycotoxin modifiers are also easily reduced to the non-modified form in animal intestines or even during food processing [4,6–8]. In addition, recently, several novel mycotoxins were identified, for example, NX-2 and NX-3 produced by *Fusarium graminearum* [9,10]. It must be pointed out that among these toxic secondary metabolites, most are currently not regulated in cereals and related food and feed by governments; the toxicity data investigated should provide a robust framework for setting their maximum residue limit values and protecting the health of human beings and animals. There is still a long way to go in the toxicological assessment of different mycotoxins and the development of related limit standards.

3. Significant Differences Were Observed in the Species Composition and Genetic Diversity of Toxigenic Fungi and Their Mycotoxins

The biosynthesis of different mycotoxins is complex, being highly influenced by the genetic basis. One fungus has the capability to produce multiple types of mycotoxins in infected matrices. On the other hand, certain toxins can be produced by different pathogenic fungi. For example, the *Fusarium graminearum* clade, *Fusarium culmorum*, and *Fusarium cerealis* are the main sources of DON and ZEN contamination in wheat, maize, and other cereal grains. Meanwhile, fumonisins can be produced by *Fusarium verticillioides* and *Fusarium proliferatum*. Moreover, several subsets of strain-specific trichothecene genotypes have been identified both in the type A and type B trichothecene-producing fusaria. For example, according to the trichothecene production profiles, *Fusarium graminearum* clade strains were subdivided into three different genotypes/chemotypes: the 3ADON genotype/chemotype that can produce DON and 3ADON; the 15ADON genotype/chemotype that can produce DON and 15ADON; and the NIV genotype/chemotype that can produce NIV and 4ANIV [11–15]. *Fusarium graminearum* clade strains typically produce one of three strain-specific profiles of type B trichothecenes, whereas no *Fusarium graminearum* clade strain was reported to produce type A trichothecene prior to 2015 [9,10]. Similarly, in *Fusarium goolgardi* (a type A trichothecene-producing species), two different trichothecene genotypes, DAS and T-2, were identified by Roach et al. in 2015 [16].

Toxigenic fungi are not distributed evenly around the climatic areas of the world, and such ecological differences may contribute to establishing specific regional grain contamination. According to surveys, it is clear that the geographic distributions of mycotoxin-producing fungi are significantly influenced by several elements, such as climatic conditions (temperature, sunshine, humidity, etc.), agricultural practices (soil cultivation, nitrogen fertilization, fungicides, crop rotations, etc.), and even trade [17–19]. Due to the different mycotoxins produced by these fungi, the regional differences in these pathogens certainly lead to different risks to crop production and food safety. The *Fusarium graminearum* clade has been intensively studied since its chemotype significantly differs from that of other *Fusarium* species. In America, 15ADON producers are a prevalent population, while 3ADON producers are predominant in China [20–22]. A recent report by Senatore et al. [23] indicated that members of the *Fusarium tricinctum* species complex are replacing *F. graminearum* in European countries, for example, Italy.

4. The Genetic and Biochemical Approaches and Molecular Mechanisms of Mycotoxin Biosynthesis Are Complex

At the moment, the molecular regulatory mechanisms of the biosynthesis of several fungal toxins are rather evident, but there are still many unanswered questions. Containing the genes required for mycotoxin biosynthesis in the genome of a certain strain is the basic genetic basis for its ability to produce specific toxins. It is clear that most secondary metabolite genes are located in a cluster. In *Fusarium* spp., the core genes responsible for trichothecene biosynthesis are located in an approximately 25 kb cluster on chromosome 2 [24]. This cluster includes 15 genes, and its activation is mainly regulated by *Tri6* and *Tri10* [25]. As recently reviewed by our group, the evolutionary process of the core trichothecene biosynthesis gene cluster and specific *Tri* genes is complex in fusaria [26]. As an example, the results of Roach et al. [16] indicated that a single-nucleotide mutation/deletion occurred in the *Tri1* and *Tri16* genes, which gave rise to the differentiation of the DAS and T-2 genotypes in *Fusarium goolgardi*. Thus, all these apparent genetic differences among mycotoxin-producing fungi highlight the need for monitoring and more phenotypic characterizations of *Fusarium* species. Understanding the evolutionary pathways and molecular mechanisms of *Fusarium* and its mycotoxins will benefit the toxic potential and prediction and identification of unknown mycotoxins.

The biosynthesis of mycotoxins is significantly influenced both by the genetic basis and external environmental conditions. Plenty of genes involved in mycotoxin biosynthesis have been identified and functionally characterized in different fungi, especially in *Aspergillus* spp. and *Fusarium* spp. Among *Fusarium* spp., for example, more than 252 genes have been identified to be involved in DON biosynthesis in *Fusarium graminearum*, as of 2019 [12]. The synthesis of mycotoxins, as secondary metabolites, has been proved to be influenced by other complex mechanisms triggered in response to environmental stimuli, including pH, light, nutrient sources, and other stresses, which may activate different cell signaling pathways resulting in the modulation of the expression of genes involved in toxin production [27–31]. Revealing the connection between gene clusters and environmental stimuli in depth may help to define efficient strategies to decrease mycotoxin production.

Hence, this Special Issue, titled "Mycotoxins and Fungal Toxins: Current Status and Future Perspectives" (https://www.mdpi.com/journal/toxins/special_issues/3NDGM73EAY, accessed on 5 March 2025), was formulated to compile current research and future perspectives on all aspects of toxigenic fungi and mycotoxins. The topic was created on 12 August 2021 with the support of the MDPI management team and officially closed on 30 November 2023. In this Special Issue, a total of eleven articles were published by a diverse group of scientists from different countries. Experts from the fields of mycotoxin toxicology, mycotoxin detection and prevention, and mycotoxin biosynthesis molecular mechanisms contributed their newest findings in the form of research articles or reviews. These findings collectively contribute to the ongoing efforts concerning mycotoxins and offer novel insights into future research work.

Author Contributions: Writing—original draft preparation, J.W., W.W. and X.Y.; writing—review and editing, J.W., J.T. and X.Y.; supervision, J.W.; project administration, J.W.; funding acquisition, J.W. All authors have read and agreed to the published version of the manuscript.

Funding: This research was funded by the Natural Science Foundation of Shanghai Municipality (23ZR1455700), the National Natural Science Foundation of China (32472461), and the National Agricultural Product Quality and Safety Risk Assessment Project of China (GJFP20240102).

Acknowledgments: The editors express their gratitude to the authors who made significant contributions to this Special Issue. The success of this Special Issue cannot be separated from the efforts of the peer reviewers. We thank the peer reviewers for their professional input through their valuable evaluations and comments, enhancing the quality of the research. The editors also extend their thanks to the MDPI management team and staff for their continuous contributions and support.

Conflicts of Interest: The authors declare no conflicts of interest.

List of Contributions:

1. Xu, J.; Sun, L.; He, M.; Zhang, S.; Gao, J.; Wu, C.; Zhang, D.; Dai, J. Resveratrol protects against zearalenone-induced mitochondrial defects during porcine oocyte maturation via PINK1/Parkin-mediated mitophagy. *Toxins* **2022**, *14*, 641.
2. Cardoso, P.; Boleti, A.; Silva, P.; Mukoyama, L.; Guindo, A.; Moraes, L.; Oliveira, C.; Macedo, M.; Carvalho, C.; de Castro, A.; et al. Evaluation of a novel synthetic peptide derived from cytolytic mycotoxin candidalysin. *Toxins* **2022**, *14*, 696.
3. Gong, A.; Lei, Y.; He, W.; Liao, Y.; Ma, L.; Zhang, T.; Zhang, J. The inhibitory effect of *Pseudomonas stutzeri* YM6 on *Aspergillus flavus* growth and aflatoxins production by the production of volatile dimethyl trisulfide. *Toxins* **2022**, *14*, 788.
4. Hassan, M.; Abo-Elmaaty, A.; Zaglool, A.; Mohamed, S.; Abou-Zeid, S.; Farag, M.; Alagawany, M.; Di Cerbo, A.; Azzam, M.; Alhotan, R.; et al. *Origanum vulgare* essential oil modulates the AFB1-induced oxidative damages, nephropathy, and altered inflammatory responses in growing rabbits. *Toxins* **2023**, *15*, 69.
5. Wang, J.; Huang, Q.; Guo, W.; Guo, D.; Han, Z.; Nie, D. Fe_3O_4@COF(TAPT–DHTA) nanocomposites as magnetic solid-phase extraction adsorbents for simultaneous determination of 9 mycotoxins in fruits by UHPLC–MS/MS. *Toxins* **2023**, *15*, 117.
6. Tang, H.; Han, W.; Fei, S.; Li, Y.; Huang, J.; Dong, M.; Wang, L.; Wang, W.; Zhang, Y. Development of acid hydrolysis-based UPLC–MS/MS method for determination of *Alternaria* toxins and its application in the occurrence assessment in solanaceous vegetables and their products. *Toxins* **2023**, *15*, 201.
7. Dong, T.; Qiao, S.; Xu, J.; Shi, J.; Qiu, J.; Ma, G. Effect of abiotic conditions on growth, mycotoxin production, and gene expression by *Fusarium fujikuroi* species complex strains from maize. *Toxins* **2023**, *15*, 260.
8. Tolosa, J.; Serrano Candelas, E.; Vallés Pardo, J.; Goya, A.; Moncho, S.; Gozalbes, R.; Palomino Schätzlein, M. MicotoXilico: An interactive database to predict mutagenicity, genotoxicity, and carcinogenicity of mycotoxins. *Toxins* **2023**, *15*, 355.
9. Vila-López, M.; Pallarés, N.; Ferrer, E.; Tolosa, J. Mycotoxin determination and cccurrence in pseudo-cereals intended for food and feed: A review. *Toxins* **2023**, *15*, 379.
10. Wang, J.; Zhang, M.; Yang, J.; Yang, X.; Zhang, J.; Zhao, Z. Type A trichothecene metabolic profile differentiation, mechanisms, biosynthetic pathways, and evolution in *Fusarium* species—A mini review. *Toxins* **2023**, *15*, 446.
11. He, W.; Wang, J.; Han, M.; Wang, L.; Li, L.; Zhang, J.; Chen, S.; Guo, J.; Zhai, X.; Yang, J. Potential toxicity and mechanisms of T-2 and HT-2 individually or in combination on the intestinal barrier function of porcine small intestinal epithelial cells. *Toxins* **2023**, *15*, 682.

References

1. Eskola, M.; Kos, G.; Elliott, C.T.; Hajšlová, J.; Mayar, S.; Krska, R. Worldwide contamination of food-crops with mycotoxins: Validity of the widely cited 'FAO estimate' of 25%. *Crit. Rev. Food Sci. Nutr.* **2020**, *60*, 2773–2789. [PubMed]
2. Caceres, I.; Khoury, A.; Khoury, R.; Lorber, S.; Oswald, I.P.; Khoury, A.; Atoui, A.; Puel, O.; Bailly, J. Aflatoxin biosynthesis and genetic regulation: A review. *Toxins* **2020**, *12*, 150. [CrossRef] [PubMed]
3. Navale, V.; Vamkudoth, K.R.; Ajmera, S.; Dhuri, V. *Aspergillus* derived mycotoxins in food and the environment: Prevalence, detection, and toxicity. *Toxicol. Rep.* **2021**, *8*, 1008–1030. [PubMed]
4. Khaneghah, A.M.; Martins, L.M.; von Hertwig, A.M.; Bertoldo, R.; Sant'Ana, A.S. Deoxynivalenol ant its masked forms: Characteristics, incidence, control and fate during wheat and wheat based products processing-a review. *Trends Food Sci. Technol.* **2018**, *71*, 13–24.
5. Svoboda, T.; Labuda, R.; Sulyok, M.; Krska, R.; Bacher, M.; Berthiller, F.; Adam, G. *Fusarium sporotrichioides* produces two HT-2-α-glucosides on rice. *Toxins* **2024**, *16*, 99. [CrossRef]

6. Warth, B.; Fruhmann, P.; Wiesenberger, G.; Kluger, B.; Sarkanj, B.; Lemmens, M.; Hametner, C.; Fröhlich, J.; Adam, G.; Krska, R.; et al. Deoxynivalenol-sulfates: Identification and quantification of novel conjugated (masked) mycotoxins in wheat. *Anal. Bioanal. Chem.* **2015**, *407*, 1033–1039.
7. Deng, Y.; You, L.; Nepovimova, E.; Wang, X.; Musilek, K.; Wu, Q.; Wu, W.; Kuca, K. Biomarkers of deoxynivalenol (DON) and its modified form DON-3-glucoside (DON-3G) in humans. *Trends Food Sci. Technol.* **2021**, *110*, 551–558.
8. Zhang, H.; Wu, L.; Li, W.; Zhang, Y.; Li, J.; Hu, X.; Sun, L.; Du, W.; Wang, B. Conversion of deoxynivalenol-3-glucoside to deoxynivalenol during Chinese steamed bread processing. *Toxins* **2020**, *12*, 225. [CrossRef]
9. Varga, E.; Wiesenberger, G.; Hametner, C.; Ward, T.J.; Dong, Y.; Schöfeck, D.; McCormick, S.; Broz, K.; Stückler, R.; Schuhmacher, R.; et al. New tricks of an old enemy: Isolates of *Fusarium graminearum* produce a type A trichothecene mycotoxin. *Environ. Microbiol.* **2015**, *17*, 2588–2600.
10. Chen, L.; Yang, J.; Wang, H.; Yang, X.; Zhang, C.; Zhao, Z.; Wang, J. NX toxins: New threat posed by *Fusarium graminearum* species complex. *Trends Food Sci. Technol.* **2022**, *119*, 179–191.
11. Miller, J.D.; Greenhalgh, R.; Wang, Y.; Lu, M. Trichothecene chemotypes of three *Fusarium* species. *Mycologia* **1991**, *83*, 121–130. [CrossRef]
12. Chen, Y.; Kistler, H.C.; Ma, Z. *Fusarium graminearum* trichothecene mycotoxins: Biosynthesis, regulation, and management. *Annu. Rev. Phytopathol.* **2019**, *57*, 15–39. [PubMed]
13. Wang, J.; Li, H.; Qu, B.; Zhang, J.; Huang, T.; Chen, F.; Liao, Y. Development of a generic PCR detection of 3-acetyldeoxynivalenol-, 15-acetyldeoxynivalenol- and nivalenol-chemotypes of *Fusarium graminearum* clade. *Int. J. Mol. Sci.* **2008**, *9*, 2495–2504. [CrossRef] [PubMed]
14. Wang, J.H.; Ndoye, M.; Zhang, J.B.; Li, H.P.; Liao, Y.C. Population structure and genetic diversity of the *Fusarium graminearum* species complex. *Toxins* **2011**, *3*, 1020–1037. [CrossRef] [PubMed]
15. Ward, T.J.; Bielawski, J.P.; Kistler, H.C.; Sullivan, E.; O'Donnell, K. Ancestral polymorphism and adaptive evolution in the trichothecene mycotoxin gene cluster of phytopathogenic *Fusarium*. *Proc. Natl. Acad. Sci. USA* **2002**, *99*, 9278–9283. [CrossRef]
16. Rocha, L.O.; Laurence, M.H.; Proctor, R.H.; McCormick, S.P.; Summerell, B.A.; Liew, E.C.Y. Variation in type A trichothecene production and trichothecene biosynthetic genes in *Fusarium goolgardi* from natural ecosystems of Australia. *Toxins* **2015**, *7*, 4577–4594. [CrossRef]
17. Bottalico, A.; Perrone, G. Toxigenic *Fusarium* species and mycotoxins associated with head blight in small-grain cereals in Europe. *Eur. J. Plant Pathol.* **2002**, *108*, 611–624. [CrossRef]
18. Kelly, A.C.; Clear, R.M.; O'Donnell, K.; McCormicK, S.; Turkington, T.K.; Tekauz, A.; Gilbert, J.; Kistler, H.C.; Busman, M.; Ward, T.J. Diversity of fusarium head blight populations and trichothecene toxin types reveals regional differences in pathogen composition and temporal dynamics. *Fungal Genet. Biol.* **2015**, *82*, 22–31.
19. Kelly, A.; Proctor, R.H.; Belzile, F.; Chulze, S.N.; Clear, R.M.; Cowger, C.; Elmer, W.; Lee, T.; Obanor, F.; Waalwijk, C.; et al. The geographic distribution and complex evolutionary history of the NX-2 trichothecene chemotype from *Fusarium graminearum*. *Fungal Genet. Biol.* **2016**, *95*, 39–48. [CrossRef]
20. Goswami, R.S.; Kistler, H.C. Heading for disaster: *Fusarium graminearum* on cereal crops. *Mol. Plant Pathol.* **2004**, *5*, 515–525. [CrossRef]
21. Ward, T.J.; Clear, R.M.; Rooney, A.P.; O'Donnell, K.; Gaba, D.; Patrick, S.; Starkey, D.E.; Gilbert, J.; Geiser, D.M.; Nowicki, T.W. An adaptive evolutionary shift in fusarium head blight pathogen populations is driving the rapid spread of more toxigenic *Fusarium graminearum* in North America. *Fungal Genet. Biol.* **2008**, *45*, 473–484. [PubMed]
22. Zhang, J.; Li, H.; Dang, F.; Qu, B.; Xu, Y.; Zhao, C.; Liao, Y. Determination of the trichothecene mycotoxin chemotypes and associated geographical distribution and phylogenetic species of the *Fusarium graminearum* clade from China. *Mycol. Res.* **2007**, *111*, 967–975. [PubMed]
23. Senatore, M.T.; Ward, T.J.; Cappelletti, E.; Beccari, G.; McCormick, S.P.; Busman, M.; Laraba, I.; O'Donnell, K.; Prodi, A. Species diversity and mycotoxin production by members of the *Fusarium tricinctum* species complex associated with fusarium head blight of wheat and barley in Italy. *Int. J. Food Microbiol.* **2021**, *358*, 109298.
24. Kimura, M.; Tokai, T.; Takahashi-Ando, N.; Ohsato, S.; Fujimura, M. Molecular and genetic studies of *Fusarium* trichothecene biosynthesis: Pathways, genes, and evolution. *Biosci. Biotechnol. Biochem.* **2007**, *71*, 2105–2123. [PubMed]
25. Jiang, C.; Zhang, C.; Wu, C.; Sun, P.; Hou, R.; Liu, H.; Wang, C.; Xu, J.-R. *Tri6* and *Tri10* play different roles in the regulation of deoxynivalenol (DON) production by cAMP signaling in *Fusarium graminearum*. *Environ. Microbiol.* **2016**, *18*, 3689–3701.
26. Wang, J.; Zhang, M.; Yang, J.; Yang, X.; Zhang, J.; Zhao, Z. Type A trichothecene metabolic profile differentiation, mechanisms, biosynthetic pathways, and evolution in *Fusarium* species—A mini review. *Toxins* **2023**, *15*, 446. [CrossRef]
27. Klich, M.A. Environmental and developmental factors influencing aflatoxin production by *Aspergillus flavus* and *Aspergillus parasiticus*. *Mycoscience* **2007**, *48*, 71–80.
28. Montibus, M.; Pinson-Gadais, L.; Richard-Forget, F.; Barreau, C.; Ponts, N. Coupling of transcriptional response to oxidative stress and secondary metabolism regulation in filamentous fungi. *Crit. Rev. Microbiol.* **2013**, *41*, 295–308.

29. Georgianna, D.R.; Payne, G.A. Genetic regulation of aflatoxin biosynthesis: From gene to genome. *Fungal Genet. Biol.* **2009**, *46*, 113–125.
30. Affeldt, K.J.; Carrig, J.; Amare, M.; Keller, N.P. Global survey of canonical *Aspergillus flavus* G protein-coupled receptors. *Am. Soc. Microbiol.* **2014**, *5*, e01501–e01514.
31. Yu, J.-H.; Keller, N. Regulation of secondary metabolism in filamentous fungi. *Annu. Rev. Phytopathol.* **2005**, *43*, 437–458.

toxins

MDPI

Article

Resveratrol Protects against Zearalenone-Induced Mitochondrial Defects during Porcine Oocyte Maturation via PINK1/Parkin-Mediated Mitophagy

Jiehuan Xu [1,2,3], Lingwei Sun [1,2], Mengqian He [1,2], Shushan Zhang [1,2,3,4], Jun Gao [1,4], Caifeng Wu [1,4], Defu Zhang [1,2,3,4,*] and Jianjun Dai [1,2,3,4,*]

1 Institute of Animal Husbandry and Veterinary Science, Shanghai Academy of Agricultural Sciences, Shanghai 201106, China
2 Shanghai Municipal Key Laboratory of Agri-Genetics and Breeding, Shanghai 201106, China
3 Shanghai Engineering Research Center of Breeding Pig, Shanghai 201106, China
4 Key Laboratory of Livestock and Poultry Resources (Pig) Evaluation and Utilization, Ministry of Agriculture and Rural Affairs, Shanghai 201106, China
* Correspondence: zhangdefu@saas.sh.cn (D.Z.); daijianjun@saas.sh.cn (J.D.)

Abstract: Mitochondria hold redox homeostasis and energy metabolism as a crucial factor during oocyte maturation, while the exposure of estrogenic mycotoxin zearalenone causes developmental incapacity in porcine oocyte. This study aimed to reveal a potential resistance of phytoalexin resveratrol against zearalenone during porcine oocyte maturation and whether its mechanism was related with PTEN-induced kinase 1 (PINK1)/Parkin-mediated mitophagy. Porcine oocytes were exposed to 20 μM zearalenone with or without 2 μM resveratrol during in vitro maturation. As for the results, zearalenone impaired ultrastructure of mitochondria, causing mitochondrial depolarization, oxidative stress, apoptosis and embryonic developmental incapacity, in which mitophagy was induced in response to mitochondrial dysfunction. Phytoalexin resveratrol enhanced mitophagy through PINK1/Parkin in zearalenone-exposed oocytes, manifesting as enhanced mitophagy flux, upregulated PINK1, Parkin, microtubule-associated protein light-chain 3 beta-II (LC3B-II) and downregulated substrates mitofusin 2 (MFN2), voltage-dependent anion channels 1 (VDAC1) and p62 expressions. Resveratrol redressed zearalenone-induced mitochondrial depolarization, oxidative stress and apoptosis, and accelerated mitochondrial DNA copy during maturation, which improved embryonic development. This study offered an antitoxin solution during porcine oocyte maturation and revealed the involvement of PINK1/Parkin-mediated mitophagy, in which resveratrol mitigated zearalenone-induced embryonic developmental incapacity.

Keywords: zearalenone; mitophagy; resveratrol; PINK1/Parkin; porcine oocyte

Key Contribution: The application of phytoalexin resveratrol redresses oocyte developmental incapacity against zearalenone contamination. PINK1/Parkin-mediated mitophagy is involved in the antitoxin effect of resveratrol.

1. Introduction

Grains are major ingredients in feed formulation in intensive livestock farms, while around 70% of cereal feeds is polluted by mycotoxins [1]. Zearalenone is a non-steroidal estrogenic mycotoxin produced by multiple species of the *Fusarium* genus that is frequently detected in cereals [2]. Based on a ten-year global survey from more than 74,000 samples in 100 countries, zearalenone is one of the top-three mycotoxins in animal feedstuffs [3]. Serious studies have reported that zearalenone and its metabolites have estrogen-like activities in human and domestic animals that compete with the estrogen receptor [4]. As a result, excessive estrogen syndrome and a series of pathological changes in the reproductive organs occur in domestic animals through ingestion of zearalenone-polluted fodder [5], causing economic loss to husbandry and the culling of breeding stocks.

As a main livestock and medical model, pigs have been considered more sensitive to zearalenone toxicity [6]. High-dose zearalenone intake causes permanent pathologic alterations in pigs, leading to edema and prolapse of the vagina, infertility, pseudopregnancy and embryo lethal resorption [7]. The toxicity of zearalenone on porcine oocytes is presumed to relate to oxidative stress, apoptosis, autophagy and epigenetic modification variation [8]. Similar research on porcine blastocysts also claimed that zearalenone exposure promoted DNA damage, apoptosis and autophagy [9]. Other studies demonstrated that zearalenone exposure impaired mitochondrial membrane potential ($\Delta\Psi m$) [10], causing oxidative stress through the mitochondrial apoptosis pathway in pigs [11]. These studies hinted the state of mitochondria as a presumable toxicology of zearalenone, while in this study, further disclosure of their interrelation and potential rescue solution of in vitro maturation (IVM) oocytes will be evaluated.

Each oocyte equips a large supply of mitochondria to meet the energy requirement of maturation and embryonic development. Mitochondria are a crucial factor determining the developmental competence of oocytes, as they control intracellular Ca^{2+} homeostasis and produce ATP for continuous transcription and translation [12]. During mitochondrial renewal or mitochondrial stress, mitophagy, oocytes selectively degrading impaired or supernumerary mitochondria through autophagy, are involved in maintaining mitochondrial homeostasis and oocyte survival [13].

As an evolutionarily conserved intracellular function, autophagy is considered a stress-responsive autonomous process. Cells selectively dispose of redundant, potentially harmful or dysfunctional cytoplasmic entities via autophagy to maintain intracellular homeostasis [14]. Mitophagy, selectively eliminating mitochondria via autophagy, is critical for the quality and quantity control of mitochondria. Impaired mitochondria are targeted by the autophagic system to form mitophagosomes and subsequently delivered to lysosomes for degradation. A primary signaling pathway to regulate mitophagy is through ubiquitin-dependent PTEN-induced kinase 1 (PINK1)/Parkin. PINK1, a Ser/Thr kinase, stabilizes on depolarized mitochondria and phosphorylates mitofusin (MFN), voltage-dependent anion channels 1 (VDAC1) and other substrates on the outer-mitochondrial membrane (OMM) [15,16]. E3 ubiquitin ligase Parkin is then recruited by phosphorylated substrates from cytoplasm to depolarized mitochondria and catalyzes the transferring of ubiquitin to substrates on OMM [17,18]. These polyubiquitinated substrates are recognized by p62 and interact with microtubule-associated protein light chain 3 (LC3) through LC3-interacting region motif and further cause the bilayer membrane structure to encapsulate damaged mitochondria. The mitophagosome is then formed. In order to degrade target mitochondria as its cargo inside, fashioned mitophagosome subsequently fused with lysosome, forming mitophagolysosome. Target mitochondria are degraded into matrix substances in the acidic environment of lysosomes and recycled by cell.

Due to cytoplasmic immaturity, the developmental capacity of IVM oocytes is considered lower than ovulatory oocytes [19]. Therefore, numerous antioxidants have been evaluated to rescue IVM oocytes. Resveratrol, a natural phytoalexin, has been widely studied for its resistant properties against oxidation, carcinogenicity and inflammation. A previous study demonstrated multiple effects of resveratrol towards oocytes, such as enhancing the clearance of mitochondrial damage [20], protecting oocytes form oxidative stress and apoptosis, promoting oocyte maturation and hindering postovulatory aging [21]. Recent studies revealed that these beneficial effects are closely related to autophagy/mitophagy indued by resveratrol through multiple signaling pathways. For example, mitophagy and enhanced mitochondrial protein FOXO3a induced by resveratrol was considered as a potential mechanism against postovulatory oocyte aging [22]. In consideration of the antitoxin effect of resveratrol and its close relation with mitophagy, this study presumed a hypothesis that resveratrol could rescue zearalenone-induced maturation impairment of porcine oocyte through PINK1/Parkin-mediated mitophagy. Resveratrol was concurrently co-incubated with zearalenone during porcine oocyte maturation for the hypothetical resistance against zearalenone.

The purpose of this study was to (1) elucidate whether resveratrol could alleviate zearalenone-indued oxidative stress, apoptosis and developmental incapacity during porcine oocyte maturation through mitophagy and (2) clarify whether the PINK1/Parkin signaling pathway was involved in it.

2. Results

2.1. Resveratrol Alleviated Zearalenone-Induced Embryonic Developmental Failures

To clarify the effect of estrogen-like mycotoxin zearalenone and the hypothetical antitoxin effects of resveratrol against zearalenone during porcine oocyte maturation, parthenogenetic activation (PA) and in vitro zygote culture were performed to determine the cleavage and blastosphere rates as embryonic developmental potential (Figure 1A–C). The results revealed that zearalenone hindered the development of PA embryo, demonstrated as significantly decreased cleavage rate (90.67% ± 3.24% vs. 67.74% ± 3.96%, $p < 0.05$) and blastosphere rate (18.64% ± 1.98% vs. 7.28% ± 1.50%, $p < 0.05$). Compared to zearalenone-exposed oocytes, the cleavage rate (67.74% ± 3.96% vs. 82.78% ± 2.31%, $p < 0.05$) and blastosphere rate (7.28% ± 1.50% vs. 13.20% ± 2.23%) raised up when resveratrol co-incubated with zearalenone during oocyte maturation.

2.2. Resveratrol Alleviated Zearalenone-Induced Oxidative Stress and Apoptosis during Oocyte Maturation

To reveal the mechanism of embryonic developmental failures induced by zearalenone and the antitoxin effect of resveratrol, oocyte ROS levels (Figure 1D,E) and apoptosis rates (Figure 1F,G) were, respectively, determined. Compared to maturated oocytes, zearalenone significantly upregulated the ROS generation (1.00 ± 0.08 vs. 9.23 ± 0.67, $p < 0.05$) and apoptosis rate (6.67% ±1.93% vs. 66.02% ± 3.31%, $p < 0.05$) in porcine oocytes, while, compared to zearalenone-exposed oocytes, resveratrol significantly inhibited the ROS generation (9.23 ± 0.67 vs. 5.89 ± 0.60, $p < 0.05$) and apoptosis (66.02% ± 3.31% vs. 51.25% ± 3.08%, $p < 0.05$) when it co-incubated with zearalenone during oocyte maturation.

2.3. Resveratrol Alleviated Zearalenone-Induced Mitochondrial Defects during Oocyte Maturation

To explore the effect of zearalenone towards oocyte mitochondrial state, ultrastructure observation and determinations of $\Delta\Psi m$ and mitochondrial DNA (mtDNA) copy numbers were performed. As for the results, numerous mitochondria in regular morphology with mitochondrial crista were equably distributed in the cytoplasm in maturated porcine oocytes. After zearalenone exposure, the destruction of mitochondrial internal structure, including vacuolated mitochondria and vague or disabled cristae (white arrows), was observed in oocytes. Typical structure of mitophagosome (red arrow) was observed in zearalenone-exposed oocytes (Figure 2A), which indicated the activation of mitophagy.

In the meantime, the ultrastructural damage caused by zearalenone led to the variation in $\Delta\Psi m$ (Figure 2B,C) and relative mtDNA copy number (Figure 2D), which were determined by fluorescence probe JC-1 and realtime-PCR. As for the results, compared to maturated oocytes, the $\Delta\Psi m$ (1.56 ± 0.08 vs. 0.66 ± 0.08, $p < 0.05$) and relative mtDNA copy number (1.00 ± 0.11 vs. 0.76 ± 0.08) were both decreased in zearalenone-exposed oocytes. The results above revealed that zearalenone exposure led to mitochondrial structural damage, depolarization and decreased mtDNA copy number and also hinted the involvement of mitophagy in zearalenone-induced mitochondrial defects during porcine oocyte maturation.

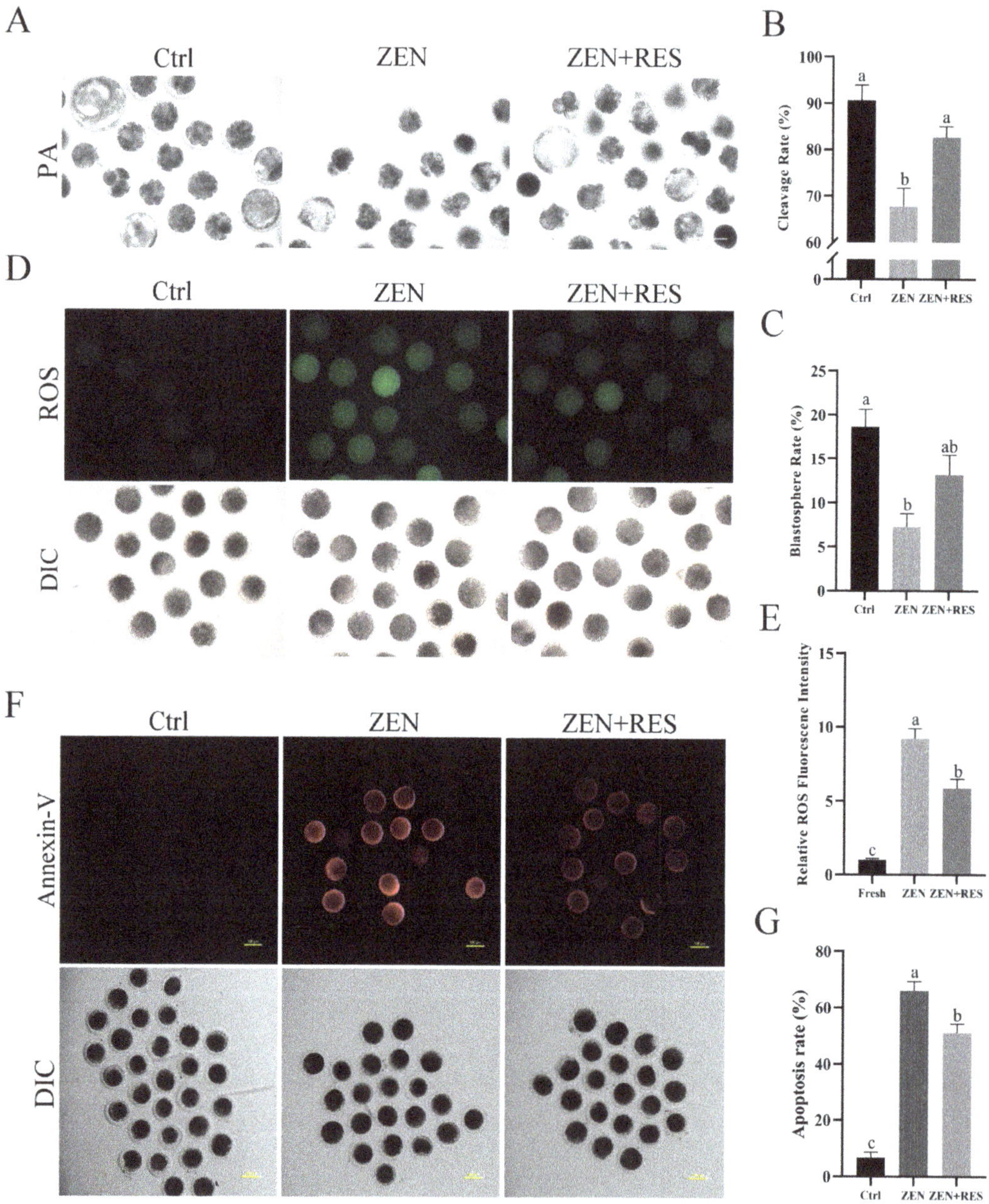

Figure 1. Resveratrol alleviated zearalenone-induced oocyte oxidative stress, early apoptosis and loss of embryonic developmental potential. (**A**) Morphological observation of PA embryos in 7 d. Bar = 100 μm. (**B**,**C**) Statistical analysis on the ratio of cleavage and blastosphere number. (**D**) Determination of ROS generation. Bar = 100 μm. (**E**) Statistical analysis of ROS generation. (**F**) Fluorescence observation of oocyte apoptosis. Bar = 100 μm. (**G**) Statistical analysis of apoptosis rates. Typical images of each different treatment towards oocytes were present. Independent replications were performed three times in each experiment. Different lowercase letters on the statistical graphs indicated significant differences between treatments (p-value < 0.05).

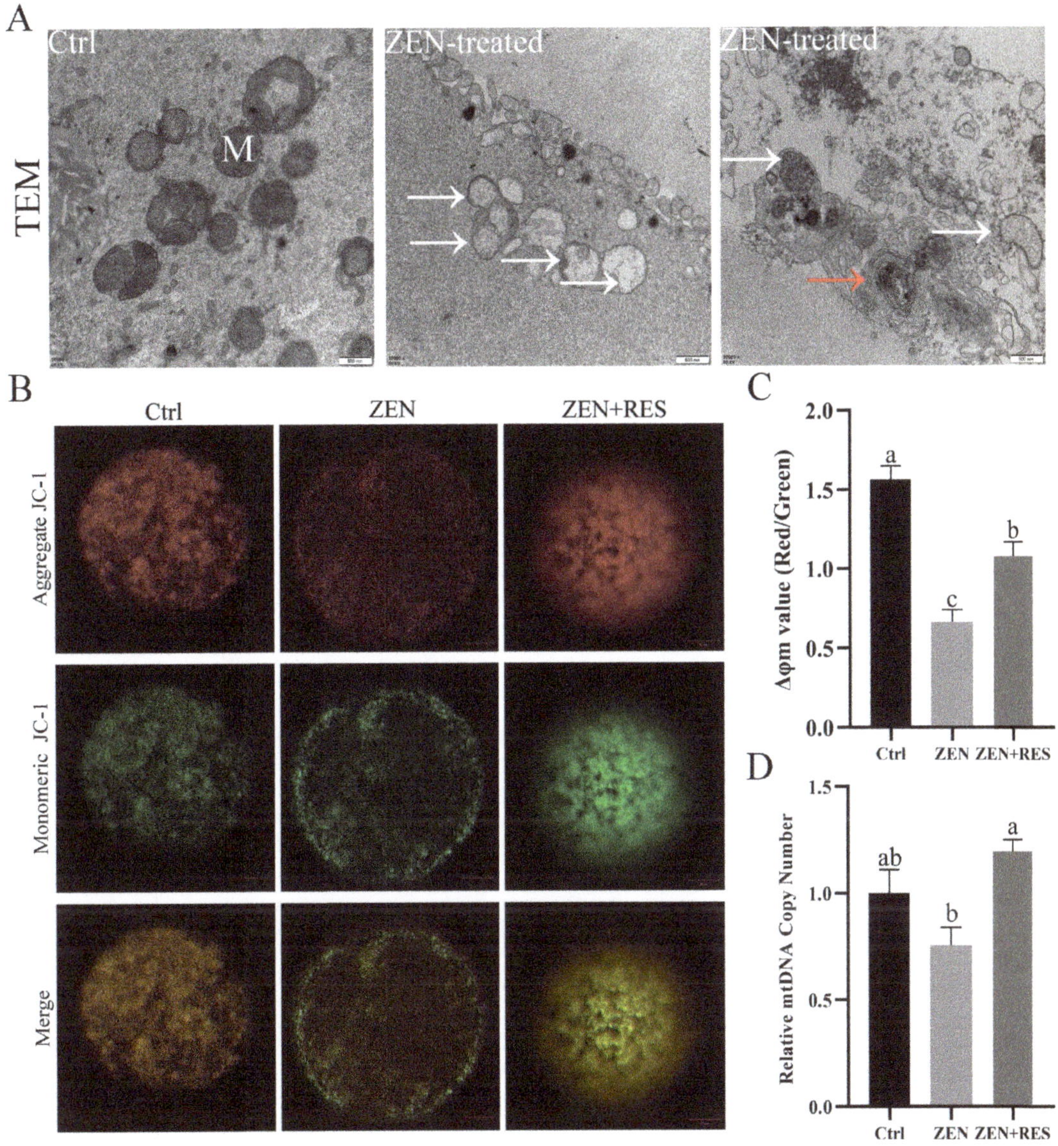

Figure 2. Resveratrol alleviated zearalenone-induced mitochondrial dysfunction during porcine oocyte maturation. (**A**). Mitochondrial ultrastructure of zearalenone-exposed porcine oocytes was observed by transmission electron microscopy. M: mitochondria. White arrows: damaged mitochondria with vague or disabled cristae. Red arrow: mitophagosome. Bar = 500 nm. (**B**) Detection of ΔΨm. Red fluorescence represented high ΔΨm while green fluorescence represented low ΔΨm. Bar = 20 μm. (**C**) Statistical analysis of ΔΨm value (Red/Green). (**D**) Comparison on relative mtDNA copy numbers. Relative gene expression of NADH dehydrogenase subunit 1 (ND1) was determined by real-time PCR. Typical ultrastructural images or intracellular fluorescence images were given representing different oocyte treatments. Independent replications were performed three times in each treatment. Different lowercase letters on the statistical graphs indicated significant differences between treatments (p-value < 0.05).

To explore the effect of resveratrol on mitochondrial defects against zearalenone, ΔΨm and relative mtDNA copy numbers were determined when resveratrol was co-incubating with zearalenone during porcine oocyte maturation (Figure 2B–D). As for the results, resveratrol significantly upregulated the ΔΨm (0.66 ± 0.08 vs. 1.08 ± 0.09, $p < 0.05$) and relative mtDNA copy number (0.76 ± 0.08 vs. 1.20 ± 0.05, $p < 0.05$) in zearalenone-exposed oocytes. The results reveled that resveratrol alleviated zearalenone-induced mitochondrial defects in porcine oocytes during maturation.

2.4. Resveratrol Enhanced Mitophagy Flux during the Maturation of Zearalenone-Exposed Oocytes

To reveal the mechanism of resveratrol alleviating zearalenone-induced mitochondrial dysfunction and the effect of resveratrol on mitophagy flux in zearalenone-exposed oocytes, the formation and degradation of mitophagosomes were, respectively, determined. Representing the formation of mitophagosomes, mitochondrial outer-membrane marker translocase of outer-mitochondrial membrane 20 (TOMM20) and autophagic vacuole marker microtubule-associated protein 1 light-chain 3 beta (LC3B) were stained using immunofluorescence and colocalized under a laser scanning confocal microscope (LSCM) (Carl Zeiss, Jena, Germany) (Figure 3). After oocyte exposing to zearalenone during maturation, the fraction of TOMM20 overlapping LC3B significantly raised up (0.30 ± 0.04 vs. 0.63 ± 0.06, $p < 0.05$). Compared to it, resveratrol further upregulated the formation of mitophagosomes in zearalenone-exposed oocytes, demonstrating a significantly enhanced fraction of TOMM20 overlapping LC3B (0.63 ± 0.06 vs. 0.88 ± 0.05, $p < 0.05$).

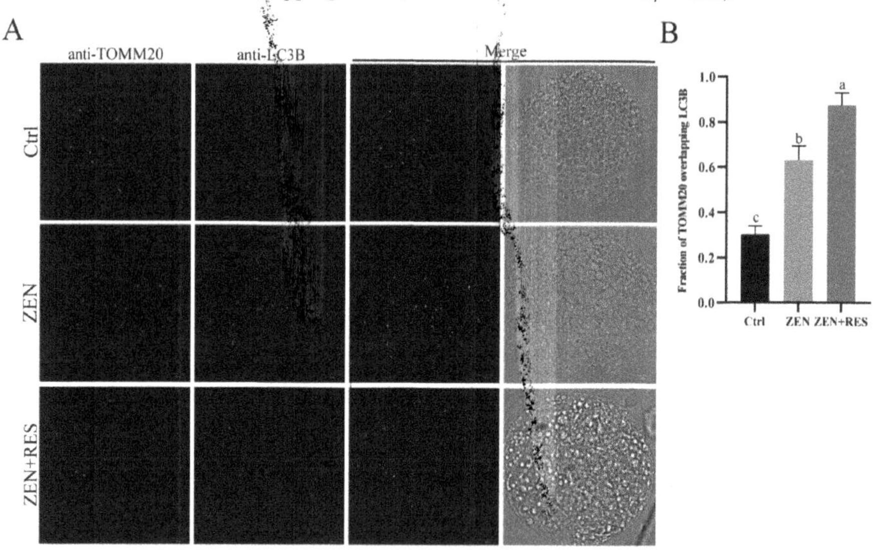

Figure 3. Fluorescence colocalization of mitophagosomes in porcine oocytes. (**A**) Fluorescence colocalization between VDAC1 and LC3B. Bar = 20 μm. (**B**) Statistical analysis of the fraction of VDAC1 overlapping LC3B. Typical images of each different treatment towards oocytes were present. Independent replications were performed three times. Different lowercase letters on the statistical graphs indicated significant differences between treatments (p-value < 0.05).

Representing the formation of mitophagolysosomes, the fluorescence colocalization of mitochondria and lysosomes was stained by fluorescence probes Mito-tracker green and Lyso-tracker red and observed under LSCM (Figure 4). The fluorescence dots of mitochondria and lysosomes equally distributed in maturated oocytes, showing a loose bond between mitochondria and lysosomes. Compared to it, the colocalization coefficient between mitochondria and lysosomes significantly upregulated in zearalenone-exposed oocytes (0.13 ± 0.02 vs. 0.33 ± 0.04, $p < 0.05$), which revealed the enhanced formation of mitophagolysosomes induced by zearalenone. Compared to zearalenone-exposed oocytes, the colocalization coefficient was further upregulated (0.33 ± 0.04 vs. 0.54 ± 0.06, $p < 0.05$) when resveratrol co-incubated with zearalenone during oocyte maturation.

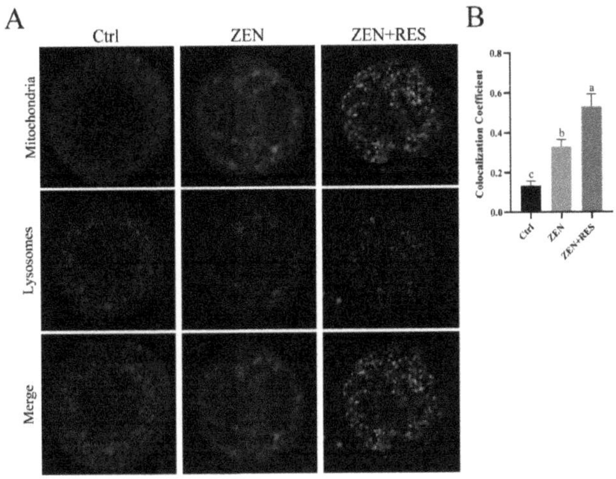

Figure 4. Fluorescence colocalization of mitophagolysosomes in porcine oocytes. (**A**) The fluorescence of mitochondria (represented in green) and lysosomes (represented in red) was inspected by LSCM. Bar = 20 μm. Typical images of each different treatment towards oocytes were present. Independent replications were performed three times. (**B**) Statistical analysis of colocalization coefficient. Different lowercase letters on the statistical graphs indicated significant differences between treatments (p-value < 0.05).

The results above revealed that mitophagy was activated in zearalenone-exposed porcine oocytes. Based on it, antitoxin resveratrol further enhanced mitophagy flux in zearalenone-exposed oocytes.

2.5. *Resveratrol Enhanced Mitophagy through PINK1/Parkin Signaling Pathway in Zearalenone-Exposed Oocytes*

To reveal the mechanism of resveratrol altering mitophagy flux in zearalenone-exposed oocytes, the fluorescence aggregation of Parkin and the protein expressions of PINK1/Parkin, a primary signaling pathway of mitophagy, were determined by immunofluorescence and Western blotting. The results showed that, compared to zearalenone-exposed oocytes, resveratrol significantly enhanced fluorescence aggregation of Parkin during oocyte maturation (1.56 ± 0.16 vs. 2.31 ± 0.14, $p < 0.05$) (Figure 5A,B).

Meanwhile, the results of the Western blot assay revealed that mitophagy was induced by zearalenone in porcine oocyte during maturation (Figure 5B,C), demonstrated as significantly upregulated autophagy marker LC3B-II (0.44 ± 0.03 vs. 0.56 ± 0.03, $p < 0.05$), and p62 degradation (1.06± 0.06 vs. 0.72 ± 0.04, $p < 0.05$), comparing zearalenone-exposed oocytes to maturated oocytes. Meanwhile, compared to zearalenone-exposed oocytes, resveratrol upregulated the protein expressions of the PINK1/Parkin signaling pathway, demonstrated as significantly upregulated PINK1 (0.48 ± 0.04 vs. 0.64 ± 0.03, $p < 0.05$), Parkin (1.01 ± 0.04 vs. 1.34 ± 0.06, $p < 0.05$), significant degradation of ubiquitinated substrates VDAC1 (0.73 ± 0.03 vs. 0.49 ± 0.02, $p < 0.05$) and MFN2 (1.02 ± 0.04 vs. 0.36 ± 0.05, $p < 0.05$) and significantly upregulated downstream autophagy marker LC3B-II (0.56 ± 0.03 vs. 0.77 ± 0.04, $p < 0.05$) and p62 degradation (0.73 ± 0.04 vs. 0.40 ± 0.05, $p < 0.05$).

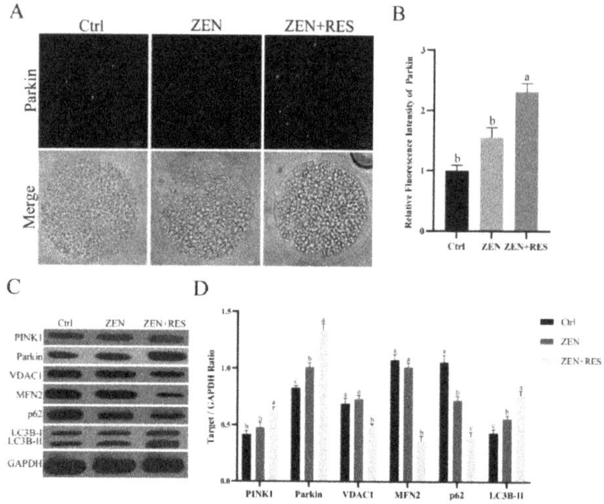

Figure 5. Resveratrol enhanced PINK1/Parkin signaling pathway in zearalenone-exposed porcine oocytes. (**A**) Fluorescence localization of Parkin. Bar = 20 μm. Typical intracellular fluorescent distributions were given representing different oocyte treatments. (**B**) Statistical analysis of relative fluorescence intensity of Parkin. (**C**) Protein expressions of PINK1/Parkin-mediated mitophagy determined by Western blotting. (**D**) Statistical analysis of protein expressions. Independent replications were performed three times in each experiment. Different lowercase letters on the statistical graphs indicated significant differences between treatments (p-value < 0.05).

3. Discussion

As a non-steroidal estrogenic mycotoxin, zearalenone is ingested by domestic animals through grains and cereal products and generally causes hypofertility and considerable financial loss in animal husbandry. A previous study on porcine oocytes attributed the toxicity of zearalenone as impairing the function and distribution of organelles, interdicting the extrusion of polar body and disturbing the expansion of cumulus granulosa cells [10]. In vivo research showed that 1.04 mg/kg zearalenone environmental intake induced follicular atresia in pigs, inhibiting proliferation of porcine granulosa cells and inducing its apoptosis and necrosis [23]. However, the dose of zearalenone enriched to porcine follicle fluid was still required to be clarified. On the other side, in vitro dose test on porcine cumulus–oocyte complexes revealed that 10 μM to 30 μM zearalenone incubation led to a significant decline in oocyte maturation rate [8] and it further decreased to less than 5% when the concentration of zearalenone up to 50 μM [24]. A similar study on mouse oocytes also claimed that zearalenone disturbed G2/M transition [25]. In spite of this evidence, however, oxidative-stress-mediated apoptosis was also considered as a plausible mechanism of zearalenone toxicity [9,26], which hinted a potential toxicity effect towards mitochondria and the application of antioxidants. In consideration of these studies, we cultured cumulus–oocyte complexes supplied with 20 μM zearalenone to explore its effect on the mitochondrial function of porcine oocytes and the potential rescue mechanism. Therefore, in this study, we first investigated the outcome from oocyte meiosis to PA embryo, verifying the fate of oocytes that zearalenone exposure during IVM led to the failure of cleavage and blastosphere formation, whereas the application of resveratrol gained higher embryonic developmental capacity of PA oocytes (Figure 1A–C), which led to further investigations of the antitoxin effects and their mechanism.

In this study, zearalenone was confirmed to disrupt mitochondrial homeostasis, demonstrating structural destruction, depolarization and DNA damage in mitochondria (Figure 2). Dysfunctional mitochondria are less competent to counteract ROS production and the supraphysiological levels of ROS in zearalenone-exposed oocyte led to oxidative stress (Figure 1D,E). Mitochondria are not only the primary endogenous source of ROS synthesis and release, but also the main target organelle of ROS effects. ROS increases the permeability of the inner-mitochondrial membrane to solutes through activating the mitochondrial permeability transition pore [27], which induces mitochondrial depolarization. Therefore, in this study, zearalenone-induced supraphysiological ROS and structural damage of mitochondria caused a vicious loop between oxidative stress and aggravated mitochondrial dysfunction, in which apoptosis was activated. In addition, zearalenone was previously proved to induce apoptosis and/or autophagy in oocytes [10], spermatogonia [28] and embryo [29], whereas, in this study, we provided evidence that mitophagy was activated in response to zearalenone-induced mitochondrial defects and oxidative stress in porcine oocyte, demonstrated as enhanced mitophagy flux (Figures 3 and 4).

Several studies attempted to mitigate zearalenone-induced impairment and oxidative stress in vivo or in vitro. In vivo research claimed that resveratrol was efficacious in reducing DNA lesions and the modulation of antioxidant enzymes caused by zearalenone intake in rats [30]. In vitro comparison in aged mice and humans revealed that about 1.0 μm was a more appropriate concentration of resveratrol than 0.1 μm and 10 μm in IVM medium that induced oocyte maturation and benefitted mitochondrial quality [31]. Meanwhile, another comparison study further revealed that 2 μM was the optimum concentration of resveratrol in a range of 0 to 4 μM to gain increased first polar body extrusion rate and hinder postovulatory aging [21]. A similar study in pigs claimed that reductant Vitamin C prevented hormonal disorders and vulval deformities caused by zearalenone intake [32]. In vitro research claimed that melatonin ameliorated oxidative stress, aberrant mitochondria distribution and DNA damage in zearalenone-exposed porcine embryo [29]. In the meantime, we confirmed that resveratrol relieved mitochondrial dysfunction (Figure 2) and prevented oxidative-stress-induced apoptosis (Figure 1) in zearalenone-exposed oocytes through PINK1/Parkin-mediated mitophagy (Figures 3–5).

The progress of apoptosis as well as excessive autophagy generally lead to cell death, while a certain extent of autophagy contributes to maintaining cellular metabolism and environmental homeostasis. In this study, we considered mitophagy as a self-healing mechanism in response of zearalenone infringement towards mitochondria, of which the positive effects of resveratrol-induced autophagy/mitophagy were demonstrated in multiple studies. Resveratrol was verified to enhance SIRT1-mediated autophagy in oocytes against aging [33], alleviating the disorder of mitochondrial biogenesis against benzo(a)pyrene toxicity by promoting mitophagy [34]. A similar effect was also confirmed in flavonoid quercetin, which improved IVM outcomes in porcine by promoting mitophagy, improving mitochondrial function and reducing oxidative stress [35].

Parkin, an E3 ubiquitin ligase belonging to the RING-between-RING family, was involved in the inducement of multiple nerve diseases, including Parkinson's [36], Alzheimer's [37] and Huntington disease [38], of which the pathogenesis was closely related with mitochondrial dysregulation. Moreover, it brought about widespread attention whether the deficiency of Parkin-mediated mitophagy might be a potential pathogenesis of these diseases [39]. Nevertheless, the implementation of PINK1/Parkin-mediated mitophagy was ubiquitin dependent. A noteworthy fact was that MFN1 and MFN2, the ubiquitinated substrates of PINK1/Parkin, were recognized as key regulators of mitochondrial fusion in mammals. Although mitochondria were considered morphologically static organelles, they continuously change their shape in response to a variety of cellular signals, which were known as mitochondrial dynamics [40]. Damaged mitochondria may lose their inner-membrane potential, causing accumulation of toxic ROS, then contaminating other mitochondria through mitochondrial fusion. In this study, the enhanced degradation of MFN2 through the ubiquitin–proteasome proteolytic system was determined when resveratrol co-incubated with zearalenone during IVM, which revealed a synchronization between enhanced clearance through mitophagy and inhibited fusion of damaged mitochondria as the antitoxin effect of resveratrol against zearalenone in porcine oocytes.

This in vitro study gave the evidence of phytoalexin resveratrol application showing reproductive toxicity resistance in pigs. However, each oocyte was suspended in follicular fluid of a follicle in ovaries, so the concentration of resveratrol from in vitro to in vivo needs to consider the absorptivity of RES from blood to follicular fluid and its accumulation effect. The bioavailability of oral RES still needs further in vivo studies, since many factors affect its absorption, such as different species, the amount of fat in the diet and drug-delivery methods (oral vs. intraperitoneal injection) et al. [41].

Nevertheless, it occurred with high frequency when livestock underwent a co-exposure between zearalenone and deoxynivalenol. The ecology of deoxynivalenol production often mirrors that of zearalenone, since it is produced by the same fungi and usually detected in crops at the same time. A similar reproductive toxic effect to zearalenone was also detected, of which the exposure of deoxynivalenol towards oocyte hindered the normal progression of meiosis by disrupted meiotic spindle. The inducement of autophagy and apoptosis was also detected in oocytes after deoxynivalenol exposure [42]. Moreover, their co-exposure toward oocytes was confirmed to alter DNA methylation levels [43], which caused the loose embryo developmental capacity [44]. Therefore, whether resveratrol would show a similar resistant effect against deoxynivalenol or their joint toxicity in oocytes through mitophagy remains a worthy research issue.

4. Conclusions

We attributed the oxidative stress and apoptosis of porcine oocytes to functional defects in mitochondria induced by zearalenone during IVM, which further led to embryonic developmental incapacity. Phytoalexin resveratrol alleviated zearalenone-induced mitochondrial dysfunction, oxidative stress and apoptosis by upregulating PINK1/Parkin-mediated mitophagy, which improved embryonic developmental potential. This study proposed a feasible protocol for porcine oocytes to resist reproductive toxicity of estrogenic mycotoxin zearalenone with the application of phytoalexin resveratrol during oocyte maturation.

5. Materials and Methods

5.1. Chemicals

Chemicals applied in this study were purchased from Sigma-Aldrich (St Louis, MO, USA) unless otherwise specified.

5.2. Oocyte Maturation

Prepubertal gilts were slaughtered in a local slaughter house. Their ovaries were subsequently separated form enterocoelia, preserved in 0.9% (*w/v*) saline at 39 °C and transported to laboratory within 1 h. A 10 mL syringe (combining with an 18-gauge needle) was used to aspirate porcine follicular fluid (PFF) from antral follicles, which contained cumulus–oocyte complexes (COCs). After being naturally precipitated for 30 min at 39 °C, the superstratum was centrifuged at 3000 r/min for 15 min and filtered to gain PFF, while the sediment was inspected under microscopy to gain COCs. Every 55 COCs were cultured in a well of four-well dishes, with each well consisting of 500 μL in vitro maturation medium [45] and covered with 200 μL mineral oil. 20 μM zearalenone (Pribolab Pte. Ltd., Singapore) with or without 2 μM resveratrol added into maturation medium. The COCs were cultured at 39 °C in a humidified atmosphere of 5% CO_2 for 44 h and transferred into 0.1% hyaluronidase to stripped oocyte form COCs.

5.3. Embryonic Developmental Capacity

PA was performed on cumulus-denuded oocytes after 44 h maturation so as to verify their embryonic developmental capacity. After being washed with PA medium [46] 3 times, cumulus-denuded oocytes were transferred to a microslide 0.5 mm fusion chamber (Model 450, BTX, Holliston, MA, USA) and went through a 1.2 kV/cm direct current pulse in 60 ms by BTX2001 (BTX, Holliston, MA, USA). About 150 electric activated oocytes were cultured in porcine zygote medium-3 [47] at 39 °C in a humidified atmosphere of 5%

CO_2. Number of cleavages was observed and counted at 2 d. Number of blastospheres was counted at 7 d.

5.4. Oocyte ROS Level

Next, 30 oocytes were stained by 50 μM 2′,7′-dichlorofluorescin diacetate (DCFH-DA) (S0033, Beyotime, Shanghai, China) for 30 min at 39 °C in a humidified atmosphere of 5% CO_2, then washed 3 times with PBS and inspected by a fluorescence microscope (Olympus, Tokyo City, Japan). Relative fluorescence intensity was analyzed by image J.

5.5. Oocyte Apoptosis

Then, 30 oocytes were stained by 2.5% Annexin V-mCherry (C1069, Beyotime, China) for 30 min at 39 °C in a humidified atmosphere of 5% CO_2, then washed 3 times with PBS and inspected under fluorescence microscope. Oocytes observed with red fluorescence were considered as apoptotic oocytes.

5.6. Intracellular Ultrastructure Observation

The procedures referred to the previous study [46]. In brief, 500 oocytes were fixed in 2.5% glutaraldehyde, embedded in 4% agar, fixed with 1% osmium tetroxide, fully dehydrated with a series of increased concentration of ethanol, replaced with propylene oxide, transferred in Epon-812 and polymerized in polymerization reactor. Ultrathin sections were sliced to form semithin sections, stained with uranyl acetate-lead citrate and observed under transmission electron microscopy.

5.7. Mitochondrial Membrane Potential

Following this, 30 oocytes were stained with 5,5′,6,6′-tetrachloro-1,1′,3,3′-tetraethyl-imidacarbocyanine iodide (JC-1) (C2006, Beyotime, China) for 30 min in a humidified atmosphere of 5% CO_2 to determine the $\Delta\Psi m$. After staining, oocytes were washed with PBS 3 times, inspected under an LSCM and visualized using software ZEN (Carl Zeiss, Germany). When the $\Delta\Psi m$ was low, JC-1 was kept as monomer which could be detected in green fluorescence, while it formed into aggregation in mitochondria with high $\Delta\Psi m$ and could be detected as red fluorescence. The ratio of red (JC-1 aggregation) to green (JC-1 monomer) fluorescence intensity represented the $\Delta\Psi m$.

5.8. Relative mtDNA Copy Number

Next, 100 oocytes were pooled as one sample for total RNA isolation using TaKaRa MiniBEST Universal RNA Extraction Kit (9767, Takara, Kusatsu, Shiga, Japan). PrimeScript RT Master Mix (RR036A, Takara, Japan) and TB Green® Premix Ex Taq™ II (Tli RNaseH Plus, Takara, Japan) were used for reverse transcription and real-time PCR following manufacturers' instruction. ND1, specific primer for coding region of mitochondria DNA, was designed on behalf of mitochondrial DNA copy numbers. Glyceraldehyde-3-phosphate dehydrogenase (GAPDH) was designed as reference gene. The primer sequences were listed as follows. ND1, Forward: TCCTACTGGCCGTAGCATTCCT; Reword: TTGAGGATGTGGCTGGTCGTAG. GAPDH, Forward: CGATGGTGAAGGTCGGAGTG; Reword: TGCCGTGGGTGGAATCATAC. The results were calculated by $2^{-\Delta\Delta Ct}$ method.

5.9. Immunofluorescence

Oocytes were fixed in 4% paraformaldehyde for 30 min, washed 3 times with 0.1% Tween-20 and transferred to 0.1% Triton X-100 overnight for permeabilization. Permeabilized oocytes were washed 3 times and then blocked with 1% BSA for 1 h.

Blocked oocytes were successively incubated with the primary and secondary antibodies Rabbit anti-Parkin antibody (ab233434, Abcam, Boston, MA, USA) and Goat anti-rabbit IgG H&L (Alexa Fluor® 647) (ab150079, Abcam, USA), washed with PBS 3 times and the fluorescence aggregation of Parkin was observed with LSCM.

For colocalization of LC3B and TOMM20, blocked oocytes were incubated with Rabbit Anti-LC3B (ab192890, Abcam, USA) and Mouse Anti-TOMM20 (ab283317, Abcam, USA) for 1 h, then incubated with secondary antibodies Goat anti-mouse IgG H&L (Alexa Fluor® 488) (ab150113, Abcam, USA) and Goat anti-rabbit IgG H&L (Alexa Fluor® 647) (ab150079, Abcam, USA), washed with PBS 3 times and observed with LSCM. The overlapping coefficient was analyzed using Image J as the fraction of VDAC1 overlapping LC3B.

5.10. Fluorescent Colocalization of Mitophagolysosomes

Then, 30 oocytes were co-stained with Mito-Tracker Green (C1048, Beyotime, China) and Lyso-Tracker Red (C1046, Beyotime, China) following the manufacturer's instructions. Oocytes were then washed 3 times with PBS and observed under LSCM. The colocalization coefficient between the fluorescence of mitochondria and lysosomes was analyzed as the Pearson's coefficient using Image J software, which represented mitophagolysosomes.

5.11. Protein Quantification by Western Blot Assay

For protein extraction, 100 oocytes were gathered and transferred in protein lysis buffer as a sample, with three duplicates for each group and the determination of protein concentration was performed using a BCA Protein Assay Kit (02912E, Cwbiotech, Beijing, China). Western blotting assay was subsequently performed following manufacturers' protocols to ensure the levels of protein expression. The information of primary and secondary antibodies is listed as follows: Rabbit Anti-PINK1 (ab23707, Abcam, USA), Rabbit anti-Parkin (ab233434, Abcam, USA), Rabbit Anti-VDAC1/Porin (ab15895, Abcam, USA), Rabbit anti-Mitofusin 2 (ab124773, Abcam, USA), Rabbit Anti-LC3B, Rabbit Anti-SQSTM1/p62 (ab233207, Abcam, USA), Rabbit Anti-GAPDH (ab9484, Abcam, USA) and the secondary antibody Goat anti-rabbit IgG (H+L) (HRP) (111-035-003, Jackson, West Grove, PA, USA). The expression of GAPDH was determined as a loading control.

5.12. Statistical Analysis

At least three independent replicates were performed in each experiment. Image J was used to determine fluorescence intensity, colocalization coefficient and Western blot quantification. ANOVAs with Duncan multiple comparisons were executed for data comparisons in SPSS Statistics 22 (IBM, Armonk, NY, USA). The results were provided as the mean $\pm$ SEM. Different lowercase letters on the statistical graph indicate significant differences between treatments (p-value < 0.05).

Author Contributions: Conceptualization, D.Z.; Formal analysis, S.Z.; Funding acquisition, D.Z. and J.D.; Investigation, J.X. and L.S.; Methodology, J.D.; Project administration, J.D.; Resources, C.W.; Software, M.H.; Supervision, D.Z.; Validation, J.G.; Visualization, M.H.; Writing—original draft, J.X. and L.S.; Writing—review and editing, J.X. All authors have read and agreed to the published version of the manuscript.

Funding: This research was funded by the National Key Research and Development Plan, grant number 2021YFD1200301; Chongqing Technology Innovation and Application Development Project, grant number CSTC2021-JSCX-DXWTBX0004; and Ningbo Science and Technology Innovation 2025 Major Project, grant number 2019B10023.

Institutional Review Board Statement: The animal study protocol was approved by the Animal Ethics Committee of SHANGHAI ACADEMY OF AGRICULTURAL SCIENCES with the Permit Code: SAASPZ0522057, Approval Date: 1 April 2022. The slaughter and sampling were in accordance with the 'Guidelines on Ethical Treatment of Experimental Animals' (2006) No.398 set by the Ministry of Science and Technology, China.

Informed Consent Statement: Not applicable.

Data Availability Statement: Not applicable.

Acknowledgments: We acknowledge the funding support for this study and the valuable contributions from the MDPI staff.

Conflicts of Interest: The authors declare no conflict of interest.

References

1. Streit, E.; Schwab, C.; Sulyok, M.; Naehrer, K.; Krska, R.; Schatzmayr, G. Multi-mycotoxin screening reveals the occurrence of 139 different secondary metabolites in feed and feed ingredients. *Toxins* **2013**, *5*, 504–523. [CrossRef]
2. Metzler, M.; Pfeiffer, E.; Hildebrand, A. Zearalenone and its metabolites as endocrine disrupting chemicals. *World Mycotoxin J.* **2010**, *3*, 385–401. [CrossRef]
3. Gruber-Dorninger, C.; Jenkins, T.; Schatzmayr, G. Global mycotoxin occurrence in feed: A ten-year survey. *Toxins* **2019**, *11*, 375. [CrossRef]
4. Zinedine, A.; Soriano, J.M.; Molto, J.C.; Manes, J. Review on the toxicity, occurrence, metabolism, detoxification, regulations and intake of zearalenone: An oestrogenic mycotoxin. *Food Chem. Toxicol.* **2007**, *45*, 1–18. [CrossRef]
5. Xu, Y.; Sun, M.H.; Li, X.H.; Ju, J.Q.; Chen, L.Y.; Sun, Y.R.; Sun, S.C. Modified hydrated sodium calcium aluminosilicate-supplemented diet protects porcine oocyte quality from zearalenone toxicity. *Environ. Mol. Mutagenesis* **2021**, *62*, 124–132. [CrossRef]
6. EFSA, K.H.; Alexander, J. Risks for animal health related to the presence of zearalenone and its modified forms in feed. *J. EFSA J.* **2017**, *15*, e04851.
7. Liu, J.; Applegate, T. Zearalenone (ZEN) in livestock and poultry: Dose, toxicokinetics, toxicity and estrogenicity. *Toxins* **2020**, *12*, 377. [CrossRef]
8. Han, J.; Wang, T.; Fu, L.; Shi, L.-Y.; Zhu, C.-C.; Liu, J.; Zhang, Y.; Cui, X.-S.; Kim, N.-H.; Sun, S.-C. Altered oxidative stress, apoptosis/autophagy, and epigenetic modifications in Zearalenone-treated porcine oocytes. *Toxicol. Res.* **2015**, *4*, 1184–1194. [CrossRef]
9. Yao, X.; Jiang, H.; Gao, Q.; Li, Y.-H.; Xu, Y.N.; Kim, N.-H. Melatonin alleviates defects induced by zearalenone during porcine embryo development. *Theriogenology* **2020**, *151*, 66–73. [CrossRef]
10. Wang, Y.; Xing, C.-H.; Chen, S.; Sun, S.-C. Zearalenone exposure impairs organelle function during porcine oocyte meiotic maturation. *Theriogenology* **2022**, *177*, 22–28. [CrossRef]
11. Wang, J.; Li, M.; Zhang, W.; Gu, A.; Dong, J.; Li, J.; Shan, A. Protective effect of n-acetylcysteine against oxidative stress induced by zearalenone via mitochondrial apoptosis pathway in SIEC02 cells. *Toxins* **2018**, *10*, 407. [CrossRef]
12. Boyman, L.; Karbowski, M.; Lederer, W.J. Regulation of mitochondrial ATP production: Ca2+ signaling and quality control. *Trends Mol. Med.* **2020**, *26*, 21–39. [CrossRef]
13. Shen, Q.; Liu, Y.; Li, H.; Zhang, L. Effect of mitophagy in oocytes and granulosa cells on oocyte quality. *Biol. Reprod.* **2021**, *104*, 294–304. [CrossRef]
14. Levine, B.; Kroemer, G. Biological functions of autophagy genes: A disease perspective. *Cell* **2019**, *176*, 11–42. [CrossRef]
15. Geisler, S.; Holmström, K.M.; Skujat, D.; Fiesel, F.C.; Rothfuss, O.C.; Kahle, P.J.; Springer, W. PINK1/Parkin-mediated mitophagy is dependent on VDAC1 and p62/SQSTM1. *Nat. Cell Biol.* **2010**, *12*, 119–131. [CrossRef]
16. Chan, N.C.; Salazar, A.M.; Pham, A.H.; Sweredoski, M.J.; Kolawa, N.J.; Graham, R.L.; Hess, S.; Chan, D.C. Broad activation of the ubiquitin–proteasome system by Parkin is critical for mitophagy. *Hum. Mol. Genet.* **2011**, *20*, 1726–1737. [CrossRef]
17. Narendra, D.; Walker, J.E.; Youle, R. Mitochondrial quality control mediated by PINK1 and Parkin: Links to parkinsonism. *Cold Spring Harb. Perspect. Biol.* **2012**, *4*, a011338. [CrossRef]
18. Exner, N.; Lutz, A.K.; Haass, C.; Winklhofer, K.F. Mitochondrial dysfunction in Parkinson's disease: Molecular mechanisms and pathophysiological consequences. *EMBO J.* **2012**, *31*, 3038–3062. [CrossRef]
19. De Vos, M.; Grynberg, M.; Ho, T.M.; Yuan, Y.; Albertini, D.F.; Gilchrist, R.B. Perspectives on the development and future of oocyte IVM in clinical practice. *J. Assist. Reprod. Genet.* **2021**, *38*, 1265–1280. [CrossRef]
20. Hara, T.; Kin, A.; Aoki, S.; Nakamura, S.; Shirasuna, K.; Kuwayama, T.; Iwata, H. Resveratrol enhances the clearance of mitochondrial damage by vitrification and improves the development of vitrified-warmed bovine embryos. *PLoS ONE* **2018**, *13*, e0204571. [CrossRef]
21. Abbasi, B.; Dong, Y.; Rui, R. Resveratrol hinders postovulatory aging by modulating oxidative stress in porcine oocytes. *Molecules* **2021**, *26*, 6346. [CrossRef]
22. Zhou, J.; Xue, Z.; He, H.-N.; Liu, X.; Yin, S.-Y.; Wu, D.-Y.; Zhang, X.; Schatten, H.; Miao, Y.-L. Resveratrol delays postovulatory aging of mouse oocytes through activating mitophagy. *Aging* **2019**, *11*, 11504. [CrossRef]
23. Dai, M.; Jiang, S.; Yuan, X.; Yang, W.; Yang, Z.; Huang, L. Effects of zearalenone-diet on expression of ghrelin and PCNA genes in ovaries of post-weaning piglets. *Anim. Reprod. Sci.* **2016**, *168*, 126–137. [CrossRef]
24. Lu, Y.; Zhang, Y.; Liu, J.-Q.; Zou, P.; Jia, L.; Su, Y.-T.; Sun, Y.-R.; Sun, S.-C. Comparison of the toxic effects of different mycotoxins on porcine and mouse oocyte meiosis. *PeerJ* **2018**, *6*, e5111. [CrossRef]
25. Ji, Y.-M.; Zhang, K.-H.; Pan, Z.-N.; Ju, J.-Q.; Zhang, H.-L.; Liu, J.-C.; Wang, Y.; Sun, S.-C. High-dose zearalenone exposure disturbs G2/M transition during mouse oocyte maturation. *Reprod. Toxicol.* **2022**, *110*, 172–179. [CrossRef]
26. Feng, Y.-Q.; Zhao, A.-H.; Wang, J.-J.; Tian, Y.; Yan, Z.-H.; Dri, M.; Shen, W.; De Felici, M.; Li, L. Oxidative stress as a plausible mechanism for zearalenone to induce genome toxicity. *Gene* **2022**, *829*, 146511. [CrossRef]
27. Wang, Z.-H.; Clark, C.; Geisbrecht, E.R. Drosophila clueless is involved in Parkin-dependent mitophagy by promoting VCP-mediated Marf degradation. *Hum. Mol. Genet.* **2016**, *25*, 1946–1964. [CrossRef]

28. Lee, R.; Kim, D.-W.; Lee, W.-Y.; Park, H.-J. Zearalenone Induces Apoptosis and Autophagy in a Spermatogonia Cell Line. *Toxins* **2022**, *14*, 148. [CrossRef]
29. Xu, Y.; Zhang, K.-H.; Sun, M.-H.; Lan, M.; Wan, X.; Zhang, Y.; Sun, S.-C. Protective effects of melatonin against zearalenone toxicity on porcine embryos in vitro. *Front. Pharmacol.* **2019**, *10*, 327. [CrossRef]
30. Virk, P.; Al-Mukhaizeem, N.A.R.; Morebah, S.H.B.; Fouad, D.; Elobeid, M. Protective effect of resveratrol against toxicity induced by the mycotoxin, zearalenone in a rat model. *Food Chem. Toxicol.* **2020**, *146*, 111840. [CrossRef]
31. Liu, M.-J.; Sun, A.-G.; Zhao, S.-G.; Liu, H.; Ma, S.-Y.; Li, M.; Huai, Y.-X.; Zhao, H.; Liu, H.-B. Resveratrol improves in vitro maturation of oocytes in aged mice and humans. *Fertil. Steril.* **2018**, *109*, 900–907. [CrossRef]
32. Su, Y.; Sun, Y.; Ju, D.; Chang, S.; Shi, B.; Shan, A. The detoxification effect of vitamin C on zearalenone toxicity in piglets. *Ecotoxicol. Environ. Saf.* **2018**, *158*, 284–292. [CrossRef] [PubMed]
33. Sugiyama, M.; Kawahara-Miki, R.; Kawana, H.; Shirasuna, K.; Kuwayama, T.; Iwata, H. Resveratrol-induced mitochondrial synthesis and autophagy in oocytes derived from early antral follicles of aged cows. *J. Reprod. Dev.* **2015**, *61*, 251–259. [CrossRef]
34. Chen, K.-G.; Kang, R.-R.; Sun, Q.; Liu, C.; Ma, Z.; Liu, K.; Deng, Y.; Liu, W.; Xu, B. Resveratrol ameliorates disorders of mitochondrial biogenesis and mitophagy in rats continuously exposed to benzo (a) pyrene from embryonic development through adolescence. *Toxicology* **2020**, *442*, 152532. [CrossRef]
35. Cao, Y.; Zhao, H.; Wang, Z.; Zhang, C.; Bian, Y.; Liu, X.; Zhang, C.; Zhang, X.; Zhao, Y. Quercetin promotes in vitro maturation of oocytes from humans and aged mice. *Cell Death Dis.* **2020**, *11*, 1–15. [CrossRef]
36. Imai, Y. PINK1-Parkin signaling in Parkinson's disease: Lessons from Drosophila. *Neurosci. Res.* **2020**, *159*, 40–46. [CrossRef]
37. Quinn, P.M.; Moreira, P.I.; Ambrósio, A.F.; Alves, C.H. PINK1/PARKIN signalling in neurodegeneration and neuroinflammation. *Acta Neuropathol. Commun.* **2020**, *8*, 1–20. [CrossRef]
38. Aladdin, A.; Király, R.; Boto, P.; Regdon, Z.; Tar, K. Juvenile Huntington's Disease Skin Fibroblasts Respond with Elevated Parkin Level and Increased Proteasome Activity as a Potential Mechanism to Counterbalance the Pathological Consequences of Mutant Huntingtin Protein. *Int. J. Mol. Sci.* **2019**, *20*, 5338. [CrossRef]
39. Roverato, N.D.; Sailer, C.; Catone, N.; Aichem, A.; Stengel, F.; Groettrup, M. Parkin is an E3 ligase for the ubiquitin-like modifier FAT10, which inhibits Parkin activation and mitophagy. *Cell Rep.* **2021**, *34*, 108857. [CrossRef]
40. Zungu, M.; Schisler, J.; Willis, M.S. All the little pieces. –Regulation of mitochondrial fusion and fission by ubiquitin and small ubiquitin-like modifer and their potential relevance in the heart.–. *Circ. J.* **2011**, *75*, 2513–2521. [CrossRef]
41. Smoliga, J.M.; Blanchard, O. Enhancing the delivery of resveratrol in humans: If low bioavailability is the problem, what is the solution? *Molecules* **2014**, *19*, 17154–17172. [CrossRef] [PubMed]
42. Han, J.; Wang, Q.-C.; Zhu, C.-C.; Liu, J.; Zhang, Y.; Cui, X.-S.; Kim, N.-H.; Sun, S.-C. Deoxynivalenol exposure induces autophagy/apoptosis and epigenetic modification changes during porcine oocyte maturation. *Toxicol. Appl. Pharmacol.* **2016**, *300*, 70–76. [CrossRef] [PubMed]
43. Zhu, C.-C.; Hou, Y.-J.; Han, J.; Liu, H.-L.; Cui, X.-S.; Kim, N.-H.; Sun, S.-C. Effect of mycotoxin-containing diets on epigenetic modifications of mouse oocytes by fluorescence microscopy analysis. *Microsc. Microanal.* **2014**, *20*, 1158–1166. [CrossRef] [PubMed]
44. Malekinejad, H.; Schoevers, E.J.; Daemen, I.J.; Zijlstra, C.; Colenbrander, B.; Fink-Gremmels, J.; Roelen, B.A. Exposure of oocytes to the Fusarium toxins zearalenone and deoxynivalenol causes aneuploidy and abnormal embryo development in pigs. *Biol. Reprod.* **2007**, *77*, 840–847. [CrossRef]
45. Xu, J.; Zhang, D.; Ju, S.; Sun, L.; Zhang, S.; Wu, C.; Rui, R.; Dai, J. Mitophagy is involved in the mitochondrial dysfunction of vitrified porcine oocytes. *Mol. Reprod. Dev.* **2021**, *88*, 427–436. [CrossRef]
46. Xu, J.; Sun, L.; Wu, C.; Zhang, S.; Ju, S.; Rui, R.; Zhang, D.; Dai, J. Involvement of PINK1/Parkin-mediated mitophagy in mitochondrial functional disruption under oxidative stress in vitrified porcine oocytes. *Theriogenology* **2021**, *174*, 160–168. [CrossRef]
47. Yoshioka, K.; Suzuki, C.; Tanaka, A.; Anas, I.M.-K.; Iwamura, S. Birth of piglets derived from porcine zygotes cultured in a chemically defined medium. *Biol. Reprod.* **2002**, *66*, 112–119. [CrossRef]

toxins

MDPI

Article

Evaluation of a Novel Synthetic Peptide Derived from Cytolytic Mycotoxin Candidalysin

Pedro Henrique de Oliveira Cardoso [1], Ana Paula de Araújo Boleti [1], Patrícia Souza e Silva [1], Lincoln Takashi Hota Mukoyama [1], Alexya Sandim Guindo [1], Luiz Filipe Ramalho Nunes de Moraes [1], Caio Fernando Ramalho de Oliveira [2], Maria Ligia Rodrigues Macedo [2], Cristiano Marcelo Espínola Carvalho [1], Alinne Pereira de Castro [1] and Ludovico Migliolo [1,*]

[1] S-Inova Biotech, Programa de Pós-graduação em Biotecnologia, Universidade Católica Dom Bosco, Campo Grande 79117-900, Mato Grosso do Sul, Brazil

[2] Laboratório de Purificação de Proteínas e suas Funções Biológicas, Unidade de Tecnologia de Alimentos e da Saúde Pública, Universidade Federal de Mato Grosso do Sul, Campo Grande 79070-900, Mato Grosso do Sul, Brazil

* Correspondence: ludovico@ucdb.br; Tel.: +55-67-33123473

Abstract: The importance of neuroinflammation in neurology is becoming increasingly apparent. In addition to neuroinflammatory diseases such as multiple sclerosis, the role of neuroinflammation has been identified in many non-inflammatory neurological disorders such as stroke, epilepsy, and cancer. The immune response within the brain involves the presence of CNS resident cells; mainly glial cells, such as microglia, the CNS resident macrophages. We evaluated the peptide Ca-MAP1 bioinspired on the *C. albicans* immature cytolytic toxin candidalysin to develop a less hemolytic peptide with anti-neuroinflammatory, antibacterial, and cytotoxic activity against tumor cells. *In silico* and *in vitro* studies were performed at various concentrations. Ca-MAP1 exhibits low hemolytic activity at lower concentrations and was not cytotoxic to MRC-5 and BV-2 cells. Ca-MAP1 showed activity against *Acinetobacter baumannii*, *Escherichia coli* ATCC, *E. coli* KPC, *Klebsiella pneumoniae* ATCC, *Pseudomonas aeruginosa*, and *Staphylococcus aureus* ATCC. Furthermore, Ca-MAP1 exhibits anti-neuroinflammatory activity in the BV-2 microglia model, with 93.78% inhibition of nitrate production at 18.1 μM. Ca-MAP1 presents cytotoxic activity against tumor cell line NCI-H292 at 36.3 μM, with an IC_{50} of 38.4 μM. Ca-MAP1 demonstrates results that qualify it to be evaluated in the next steps to promote the control of infections and provide an alternative antitumor therapy.

Keywords: *Candida albicans*; drug design; peptide toxin; neuroinflammation; multiactivity peptide

Key Contribution: In this work, we described a novel synthetic peptide with biological activity against bacteria, LPS-mediated neuroinflammation, and cancer cell lines.

1. Introduction

Neuroinflammation is defined as the response of central nervous system (CNS) cells to infections, as well as to the infiltration of the brain and spinal cord by cells of the innate and adaptive immune system [1]. A prolonged or very intense inflammatory response in the CNS causes neuronal death by increasing levels of pro-inflammatory cytokines, proteases, glutamate, free radicals, and excessive activation of glial cells [1,2].

The origin of neuroinflammation can be multifactorial and includes infections, accumulation of toxic metabolites in obese or aging individuals, chemotherapy-related cognitive impairment, and peptides or proteins, as seen in various neurodegenerative diseases, Alzheimer's disease, Parkinson's disease, amyotrophic lateral sclerosis, ischemic injury, and multiple sclerosis [3,4].

The membrane component of Gram-negative bacteria, such as lipopolysaccharide (LPS), is a trigger of the inflammatory process that can cause an exacerbated reaction and induce the production of reactive molecules that directly damage cells by activating cascades of reactions that might cause sepsis and death; it is also related to neuroinflammation that can lead to the onset of neurodegenerative diseases. Forms of treatment that deal with both bacteria and LPS are highly desirable from a therapeutic viewpoint [5–7].

The search for a new form of treatment has led to the search for molecules called antimicrobial peptides (AMPs), especially those with a cationic helical structure, that act as mediators in cases of routine infections, hospital sepsis, or bacterial resistance by inhibiting the growth or destroying the membranes of invading cells [8,9]. Fungi are microorganisms that share many ecological niches with bacteria and end up competing with each other; this competition causes them to develop molecules, such as AMP, that can eliminate their competitors [10–12]. New molecules from organisms such as fungi are an alternative in the search for the control and/or combat of pathogenic organisms.

The pathogenic fungi *Candida albicans*, which grows in the mouth, digestive tract, and in the external portion of the reproductive system of most of the world population, can compete with human pathogenic bacteria and is capable of synthesizing a cationic cytolytic peptide toxin called candidalysin. Candidalysin in lower concentrations (1.5–15 μM) has already induced immunomodulatory effects such as DNA binding of c-Fos, G-CSF, and GM-CSF. Additionally, concentrations of 70 μM are already able to induce cell damage and damage-associated cytokines such as IL-1 α and IL-6 [13–15].

The cytolytic peptide toxin candidalysin is critical for the pathogenesis of the *C. albicans* fungi, thus showing the potential to be used in drug design to modulate its activity, because fungi can eliminate or inhibit other microorganisms when changing to its filamentous form and causing candidiasis [13–15]. The present study proposes to evaluate the anti-neuroinflammatory activity of synthetic peptides based on the protein sequence originating from the candidalysin of the fungus *C. albicans* as a new alternative molecule with potential against neurodegenerative diseases and bacterial infections, and also against cancer.

2. Results

2.1. Rational Design of the Ca-MAP1 Peptide

The generation of the derivate peptide was designed based on the parental peptide immature candidalysin ($_{NH2}$-SIIGIIMGIIMGILGNIPQVIQIIMSIVKAFKGNKR-$_{COOH}$), an important cytolytic toxin from *Candida albicans* [15], so that it should present at least 70% similarity to the parental peptide, preserving its physicochemical properties.

To perform the rational design of the derivates, a segment with half the number of amino acid residues of the parental peptide candidalysin primary sequence was chosen (Figure 1); the criteria to choose the segment were the presence of characteristics close to the antimicrobial peptides described in the literature, such as the charge from +2 to +4, percentage of hydrophobic residues close to 45 to 55%, and amphipathicity, thus being subjected to only four amino acid modifications to preserve its identity [16,17].

The derivate was defined as Ca-MAP1 (*Candida albicans*-Multiactivity Peptide 1), generated from the final segment of the candidalysin $_{17}$IQIIMSIVKAFKGNKR$_{32}$ (Table 1), undergoing modifications of the amino acids at positions 2, 3, 12, and 15, replaced by residues Lys, Gly, Leu, and Leu, giving rise to the peptide Ca-MAP1 ($_{NH2}$-IKGIMSIVSKAFLGNLR-$_{COOH}$). To demonstrate the sequence development, Table 1 shows the alignment of the paternal peptide candidalysin, and the peptide Ca-MAP1 furthermore shows the number of amino acid residues.

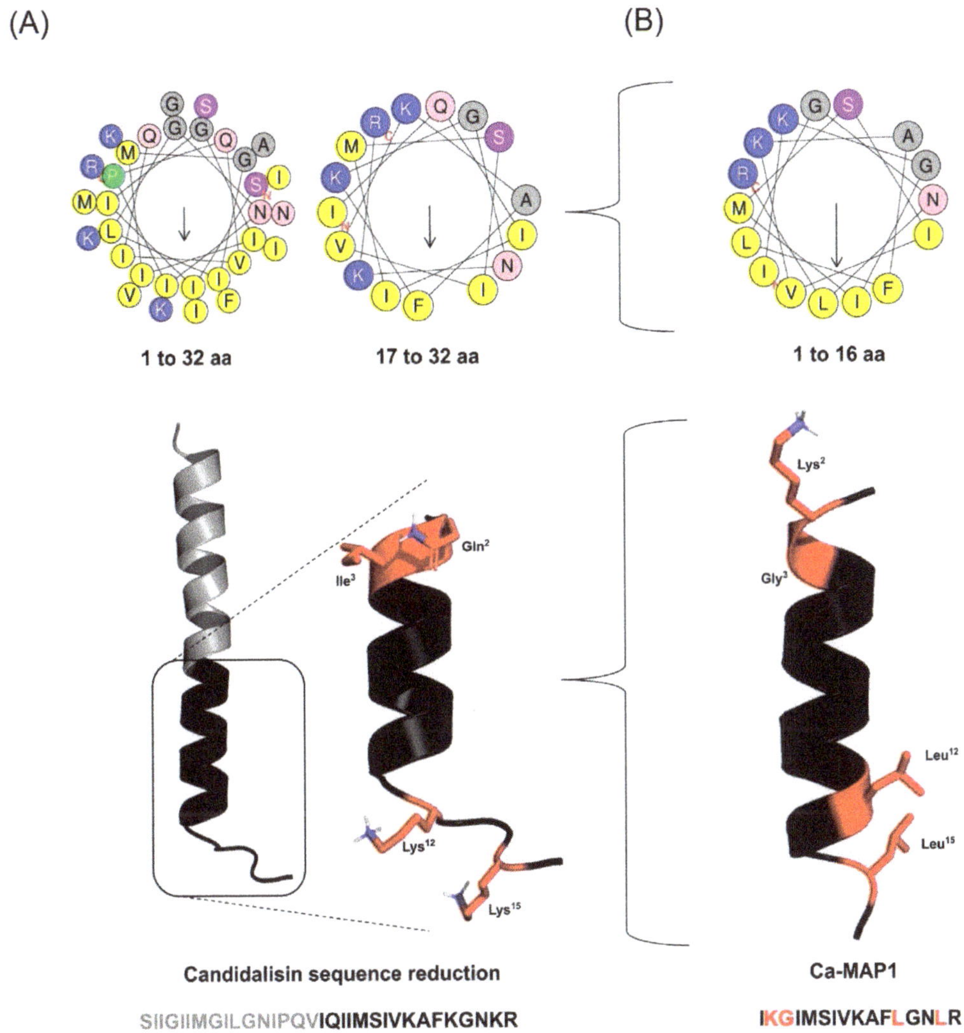

Figure 1. The strategy of rational design of the Ca-MAP1 peptide derivate from candidalysin: (**A**) Three-dimensional structure of candidalysin and reduced sequence portion of candidalysin with AMP physical-chemical characteristics with primary sequence and helical wheel projections; (**B**) Ca-MAP1 three-dimensional structure with primary sequence and helical wheel projection. The black amino acid residues in the three-dimensional structure are conserved, and the red is the changed amino acids. Legend: Positively charged amino acid residues are blue, negatively charged are red, hydrophobic aliphatic or aromatic are yellow/gray, and uncharged polar ones are pink/purple. The arrows within the diagram represent the hydrophobic moment.

The percentage of structural homology of the Ca-MAP1 peptide with its respective candidalysin portion is 75% of identity, which represents the fully conserved amino acid residue. It has a 6.25% strong similarity, indicating the modification of amino acids with similar physical-chemical properties; 18.75% are differences demonstrating a complete change of amino acid residue.

Table 1. Alignment between primary sequences of candidalysin and Ca-MAP1, showing in bold the portion used from the parental cytolytic toxin candidalysin as a framework for the design of Ca-MAP1, and the number of amino acid residues.

Name	Alignment	N° of Residues
Candidalysin	SIIGIIMGILGNIPQV**IQIIMSIVKAFKGNKR**	32
Ca-MAP1	----------------IKGIMSIVKAFLGNLR	16

The candidalysin sequence ($_{NH2}$-SIIGIIMGILGNIPQVIQIIMSIVKAFKGNKR-$_{COOH}$) has 32 amino acid residues, a net +4 charge, and hydrophobicity of 56.25%. The synthetic Ca-MAP1 ($_{NH2}$-IKGIMSIVKAFLGNLR-$_{COOH}$) has only 16 amino acid residues (50% of candidalysin length) and a reduced charge compared to the native +3 (Net charge of +3 is the most common net charge found in AMP deposited in APD [17]) and equal hydrophobicity of 56.25%. The physicochemical properties of these peptides were predicted using the Antimicrobial Peptide Database and are on display in Table 2 (https://aps.unmc.edu/prediction, accessed on 15 July 2022) and Heliquest (https://heliquest.ipmc.cnrs.fr/cgi-bin/ComputParams.py, accessed on 15 July 2022) [17,18].

Table 2. The *in silico* calculated physicochemical properties of the parental peptide candidalysin and its derivate peptide Ca-MAP1. Legend: <Z>, net charge; <H>, hydrophobicity; <μH>, hydrophobic moment; Da, Daltons.

Name	<Z>	Apolar (%)	<H>	<μH>	Mass (Da)
Candidalysin	+4	56.25	0.679	0.408	3464.05
Ca-MAP1	+3	56.25	0.607	0.631	1759.05

2.2. Hemolytic Activity

After the synthesis of Ca-MAP1, the hemolytic activity of the Ca-MAP1 peptide was tested with murine-derived erythrocytes (Figure 2) and caused 91.09% hemolysis at a concentration of 72.7 μM., 71.95% at a concentration of 36.3 μM, 32.25% at a concentration of 18.1 μM, 5% at a concentration of 9.1 μM, and no hemolysis at concentrations below 4.5 μM.

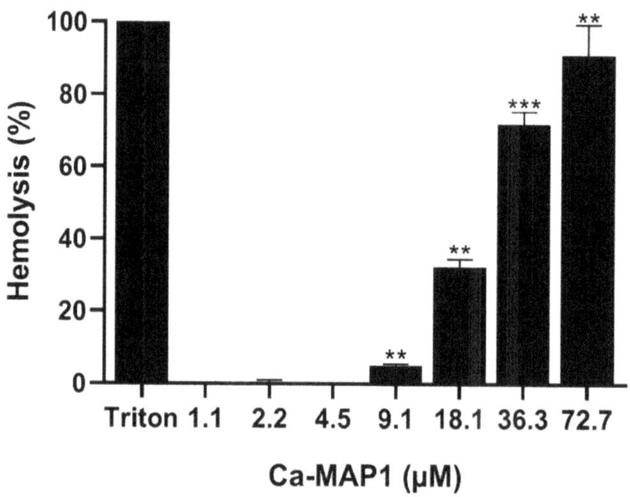

Figure 2. The hemolytic percentage from murine erythrocytes challenged with the Ca-MAP1 peptide in different micromolar concentrations. All experiments were performed in triplicate. Values of $p \leq 0.01$ and $p \leq 0.0001$ represent ** and ***, respectively. The p values were compared with cells treated with control.

2.3. Antibacterial Activity

Table 3 shows the minimum inhibitory concentration and minimum bactericidal concentration values of the Ca-MAP1 peptide against various resistant and sensitive Gram-negative and -positive bacteria. *E. coli*, *A. baumannii*, and *S. aureus* showed minimum inhibitory concentrations (MIC) of 9.1, 18.1, and 36.3 μM, respectively, and the minimum bactericidal concentration (MBC) resulted in the same concentration values. Ca-MAP1 did not demonstrate an inhibitory or bactericidal effect on the Gram-negative bacterium *P. aeruginosa* at the addressed concentrations, while ciprofloxacin presents activity with 6 μM to the same Gram-negative bacteria.

Table 3. Results of the minimal inhibitory concentration and minimal bactericidal concentration assays of Ca-MAP1 peptide and ciprofloxacin against Gram-negative and -positive bacteria. All experiments were performed in triplicate.

Bacteria	**Ca-MAP1 (μM)**		**Ciprofloxacin (μM)**	
	MIC	MBC	MIC	MBC
Gram-negative				
Acinetobacter baumannii	18.1	>72.7	96.5	>386.3
Escherichia coli	9.1	9.1	96.5	>386.3
Escherichia coli (KPC + 001812446)	9.1	9.1	386.3	96.5
Klebsiella pneumoniae (ATCC + 13883)	18.1	18.1	6	193.1
Pseudomonas aeruginosa	>72.7	>72.7	6	6
Gram-positive				
Staphylococcus aureus	36.3	36.3	6	6

The antibacterial activity of the Ca-MAP1 presents a more effective activity with a lower concentration than the commercial antibacterial ciprofloxacin because all the ciprofloxacin concentrations of MIC/MBC were higher than the Ca-MAP1 concentrations in the cases of *A. baumannii* and *E. coli* ATCC and KPC.

To investigate the time of action of the Ca-MAP1 peptide, a membrane permeability assay with the dye Sytox Green was performed with *E. coli* KPC bacteria treated with the peptide at a concentration 30 times higher than the minimum inhibitory concentration [19,20]. The analyses show that in 15 min 84% of bacterial membrane permeabilization already occurs, and in 20 min Ca-MAP1 causes 98% permeabilization, as shown in Figure 3.

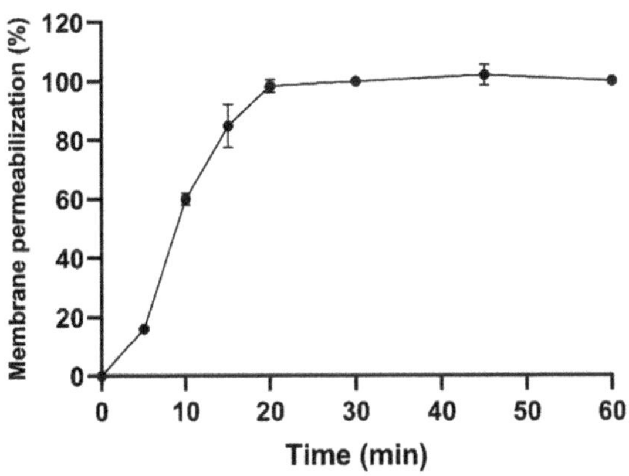

Figure 3. Effect on the permeabilization of E. coli bacterial membrane by Ca-MAP1 peptide over time, assed with Sytox Green dye assay. Three independent experiments were performed, in triplicate.

2.4. Anti-Neuroinflammatory Activity

To analyze the activity of Ca-MAP1 peptide in preventing LPS stimulation in BV-2 microglia cells, nitrite quantification was performed by Griess reaction methodology, as shown in Figure 4 [21].

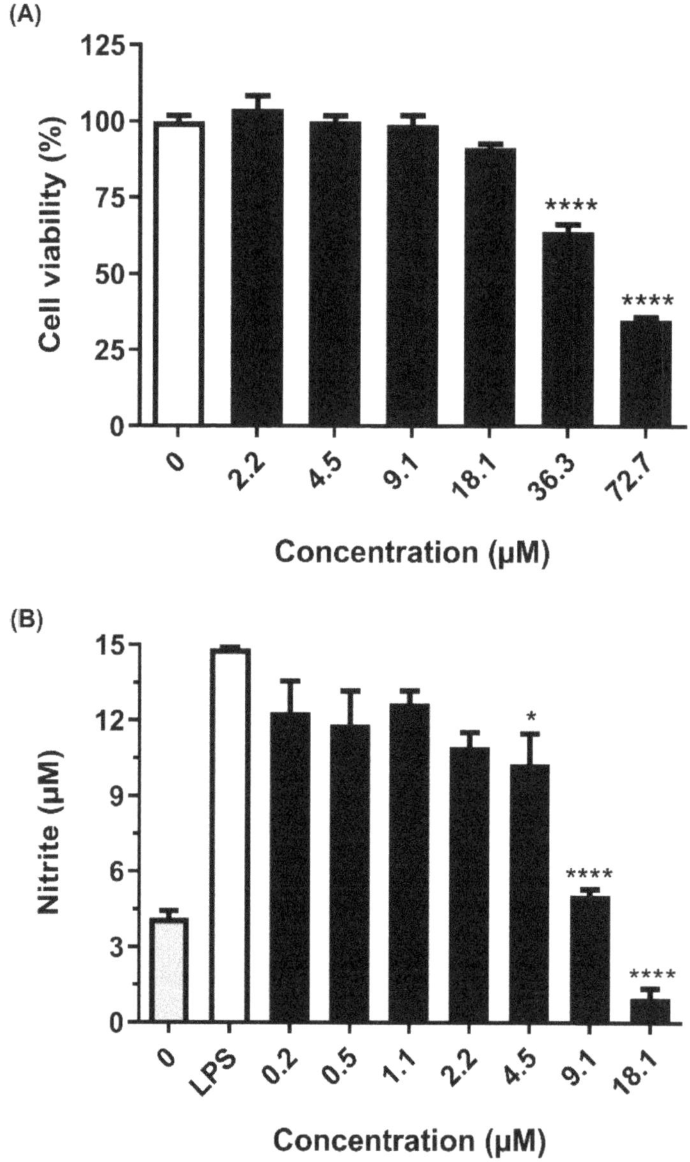

Figure 4. Effect of Ca-MAP1 peptide on (**A**) cellular viability of murine microglia BV-2 cell line and (**B**) NO production on BV-2 cell line stimulated with *E. coli* LPS. Three independent experiments were performed, in triplicate. Values are mean ± P.D.M. of three repetitions. Values of $p \leq 0.05$ and $p \leq 0.0001$ represent * and ****, respectively. The p values were compared with cells treated with control.

The Ca-MAP1 peptide showed no significant cytotoxic effects for BV-2 cells in a concentration equal to and below 18.1 μM. The anti-inflammatory activity by inhibiting nitrite production in BV-2 cells that were stimulated by LPS was evaluated in non-cytotoxic concentrations for the BV-2 cell line. The concentrations that possessed the best anti-inflammatory activities in the proposed model were the concentrations of 9.1 and 18.1 μM, which inhibited 68.08 and 93.8% of nitrite production, respectively.

2.5. Anticancer Activity

Figure 5 shows the inhibition results for the cell viability of Ca-MAP1 peptide against RD, HeLa, and NCI-H292 tumor cell lines and MRC-5 fibroblast cell line after 24 h of treatment. Ca-MAP1 inhibited 46.33% of cell viability in the NCI-H292 cell at a concentration of 36.3 μM. There was also 30.04% inhibition of cell viability in the RD cell line, 9.89% inhibition of HeLa cellular viability and 21.27% inhibition of the MRC-5 cell line at the same concentration. The IC_{50} of the Ca-MAP1 peptide is 38.4 μM to the NCI-H292 cell line.

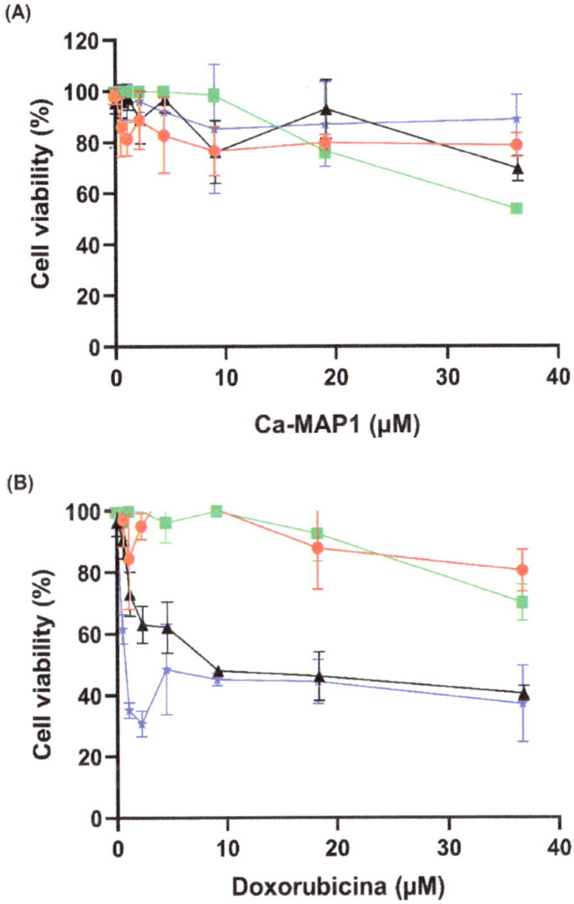

Figure 5. The cytotoxic effect of (**A**) Ca-MAP1 and (**B**) doxorubicin on the cell viability of cancer cell lines RD, HeLa, and NCI-H292, and the normal healthy MRC-5 cell line was evaluated by the MTT method. All experiments were performed in triplicate. Values are means ± P.D.M. of three repetitions. $p > 0.0001$ compared with cells treated with the control vehicle. Legend: Red circle = MRC-5, Green square = NCI-H292, Black triangle = RD, Blue star = HeLa.

The commercial antineoplastic doxorubicin in the higher concentration (20 mg.mL^{-1} or 36.7 μM) showed inhibition of 27.61, 57.30, and 68.06% to the cellular viability of NCI-H292, RD, and HeLa, respectively. The normal cellular line MRC-5 presented an inhibition of 23.98% in the concentration of 36.79 μM. The values of IC_{50} of doxorubicin to the cellular line NCI-H292 is 28.1 mg.mL^{-1} or 51.8 μM with a selectivity index of 1.44.

2.6. Circular Dichroism

The Ca-MAP1 peptide was evaluated in the presence of water, 50% trifluoroethanol (TFE), and sodium dodecyl sulphate (SDS), which are hydrophilic, hydrophobic, and anionic environments, respectively for circular dichroism analyses. The band spectrum format (Figure 6) was characterized by the occurrence of two negative band values approximately between 208 and 222 nm and one positive band at 190 nm, which are characteristic of peptides with a flexible α-helix secondary structure in the presence of SDS and TFE [22,23]. In the presence of water, it is possible to observe a negative peak in the band at 199 nm, which is close to the characteristic peak at 190 nm for random coil [24,25].

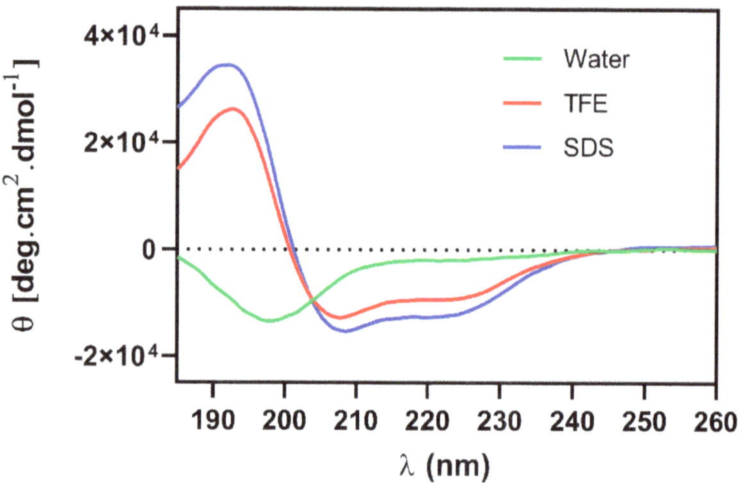

Figure 6. Circular dichroism analysis spectrum of the secondary structure of Ca-MAP1 peptide in the presence of water, TFE (50%), and SDS (30 mM), with a spectrum ranging from 185 nm to 260 nm.

2.7. Molecular Modeling

The three-dimensional structures of the Ca-MAP1 peptide were made by the I-TASSER server by comparison to analogous structures in the PDB database, where the server forms several possible models that each present a C-score, RMSD, and TM-score that indicates the reliability of the model based on the structural homology of existing sequences. The model with the best reliability was then modeled for visualization in the PyMol application (Figure 1). The predicted Ca-MAP1 three-dimensional structure validation scores obtain from I-TASSER [26], Prosa-web [27], and MolProbity [28,29] are displayed in Table 4.

Table 4. Summary of all model validation scores: C-score, RMSD, TM-score, Z-score, and Rama-Z from the predicted three-dimensional model on I-TASSER, ProSA-web, and MolProbity software version 4.5.1 (Duke University: Durham, NC, USA).

Name	C-Score	RMSD	TM-Score	Z-Score	Rama-Z
Ca-MAP1	0.05	0.6 ± 0.6Å	0.72 ± 0.11	−0.35	−1.85 ± 1.76

2.8. Molecular Dynamics

The three-dimensional model of the candidalysin derivate peptide Ca-MAP1, generated by homology modeling on the I-TASSER server, was subjected to molecular dynamics simulations to analyze its behavior in an aqueous environment; the trajectory and topology of the Ca-MAP1 atoms indicate that there is a tendency to assume a random structure. The variation of the peptide backbone indicates an increase in RMSD, which evaluates the root mean square deviation of the position of the alpha carbons of the overlapping amino acid residues over simulation time, with a range of 0–0.9 nm (Figure 7A).

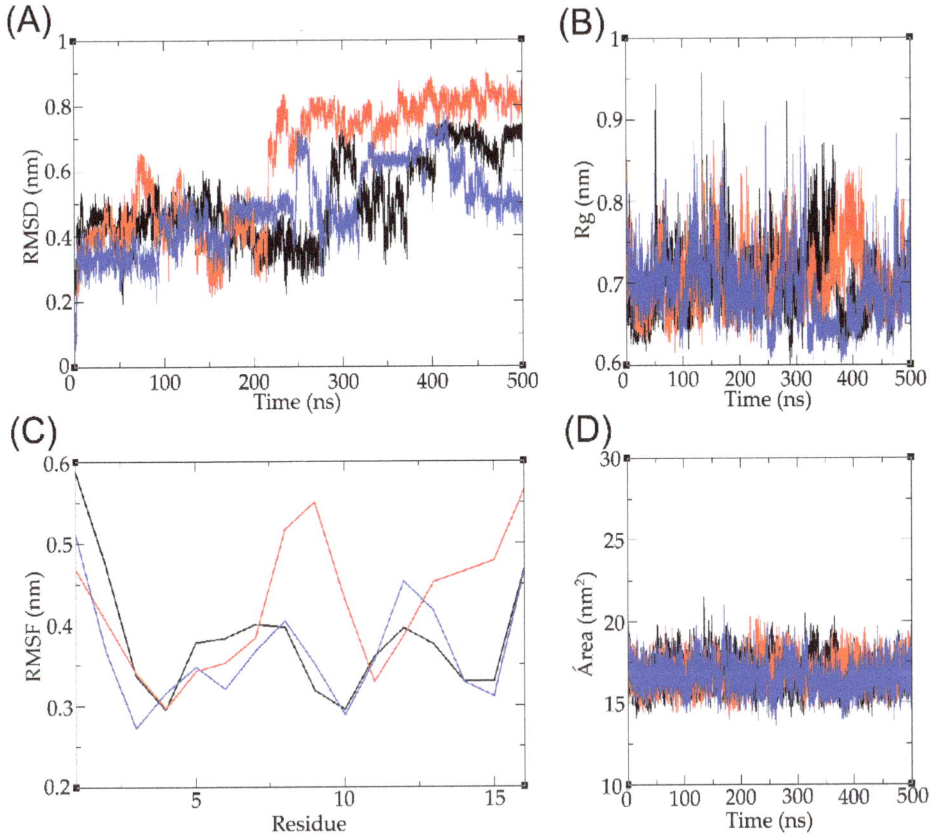

Figure 7. Molecular dynamics simulation result of Ca-MAP1 peptide in water, highlighting (**A**) root mean square deviation, (**B**) spin radius, (**C**) root mean square fluctuation, and (**D**) solvent accessible surface. The colors represent the triplicate of each parameter.

During the simulation, it is possible to observe moments in which there is an increase in the degree of decompression of the peptide. This may be due to the presence of hydrophilic residues that interact with water molecules, corroborating the increase in the diameter of the spin radius of the peptide as a whole, ranging from 0 to 0.95 nm; however, during much of the simulation, its diameter remained between approximately 0.65 and 0.8 nm (Figure 7B).

The root mean square fluctuation (RMSF) evaluates the side chain variation of each component amino acid residue of the peptide, demonstrating greater variation in the N-terminal and C-terminal regions, reaching a fluctuation of up to 0.6 nm for Ile^1 and 0.55 nm for Arg^{16}; terminal regions may vary more compared to amino acids within the sequence due to electrostatic interactions with the terminal amine and carboxyl, while more hydrophobic residues may remain more clustered and have less fluctuation (Figure 7C). Finally, the solvent-accessible surface area was calculated, indicating stability throughout the simulation, remaining approximately between 15 and 20 nm^2 (Figure 7D).

3. Discussion

Studies have shown that the physicochemical properties of peptides are correlated so that modifications in the parameters can reflect significant changes to one or more of the others. Understanding and controlling these interrelationships may be the key to designing new peptides with greater potency and specificity [9,16].

The rational peptide design proposed comprehends the reduction of length of the candidalysin, which is linked to similar maintenance antimicrobial activity parameters, and decreases toxicity [30–32].

Analyzing the physical-chemical parameters of the deposited AMPs on APD demonstrates that the main frequency of charge is +3, and there is a hydrophobicity of 45 to 55% [17]. The exchange of Gln with Lys was to keep the charge in the highest frequency on the database [17]. The replacement of Iso with Gly was to facilitate the helix-helix packing and increase electrostatic and van der Waals interaction [33]. The substitution of Lys with Leu is connected to the increase of hydrophobic moment and helical structure stabilization [34,35].

The amino acid changes caused by the definition of amphipathicity provoke the rearrangement of the hydrophobic moment, increasing when compared to candidalysin. The increase of hydrophobic moment tends to be correlated to higher efficacy of cell binding and penetration [36–38].

The low percentage of hemolysis in antibacterial, antitumor, and anti-inflammatory peptides is a remarkable point in the development of new bioactive molecules due to the low amount or absence of undesirable side effects. The Ca-MAP1 peptide, when used at low concentrations (1 to 20 μM) presents no hemolytic activity. In contrast, in the literature, the cytolytic toxin candidalysin causes complete hemolysis of erythrocytes at a concentration of 10 μM [39,40].

The Ca-MAP1 peptide inhibited the growth of *K. pneumoniae* ATCC bacteria at concentrations of 18.1 μM and acted as a bactericide at the same concentrations. The *E. coli* KPC bacteria had growth inhibited at a concentration of 9.1 μM, but all bacterial cells were eliminated at concentrations of 36.3 μM. Ciprofloxacin inhibited growth at the concentration of 32 $\mu g.mL^{-1}$ (96.5 μM) and was not shown to be bactericidal at the same concentration.

The antibiotic ciprofloxacin is not able to emit fluorescence caused by permeabilization of the *E. coli* bacteria membrane within twenty minutes, indicating that the time to cause its antibacterial activity is slow compared to the Ca-MAP1 peptide [41,42].

The Ca-MAP1 peptide was efficient in inhibiting Gram-negative bacterial growth at the lowest MIC of 9.1 μM, as in the case of *E. coli*, and at the same concentration there was relatively low hemolysis in blood cells. However, the release of lipopolysaccharide (LPS) by the lysis of Gram-negative bacteria causes sepsis and induces inflammation at picomolar concentrations [43,44].

Activated microglial cells play an important role in immune and inflammatory responses in the central nervous system and neurodegenerative diseases. Many pro-apoptotic pathways are mediated by signaling molecules that are produced during neuroinflammation. In glial cells, NF-κB, a transcription factor, initiates and regulates the expression of various inflammatory processes during inflammation, which is attributed to the pathology of various neurodegenerative diseases [45].

It is already known that LPS, which is an endotoxin in the outer membrane of Gram-negative bacteria, induces the systemic inflammatory response syndrome through toll-like receptor (TLR) signaling [4,46]. The binding of LPS to TLR4 on the surface of microglia activates several signal transduction pathways, including PI3K/AKT, MAPK, and mTOR, which ultimately lead to NF-κB activation. Activation of NF-κB then mediates the production of pro-inflammatory cytokines, chemokines, inducible nitric oxide synthase (iNOS), and COX-2, which together result in neuroinflammation [43,45]. Inhibition of NO production in LPS-induced BV-2 cells can be considered an effective treatment for CNS inflammation [47]. The NO inhibitory activities of the Ca-MAP1 peptide were evaluated in LPS-stimulated BV-2 cells, and the cytotoxic effects of these peptides were measured by MTT assay. Ca-MAP1 showed no cytotoxicity and inhibited the by-product NO (nitrite).

Chemotherapy is one of the most effective and widely used conventional treatments for cancer patients. Additionally, it is observed that 70% of cancer patients undergoing chemotherapy may develop cognitive problems during or after treatment, which affects their quality of life [48,49]. Chemotherapy is known to have adverse effects on brain function, causing dysfunctions in learning, memory, attention, motor activity, and executive function [50,51]. In addition, several studies show that tyrosine kinases, antimetabolites, microtubule inhibitors, and alkylating agents can induce neurotoxicity [52–54].

Doxorubicin belongs to the anthracycline class and is commonly used as a chemotherapy treatment for breast, hematologic, and lung cancer [55–57]. Doxorubicin exerts its antitumor effects through DNA insertion and topoisomerase II inhibition [58]. In this context, the anticancer activity of Ca-MAP1 was evaluated against tumor cell lines, RD, HeLa, and NCI-H292, and normal cell MRC-5 and BV-2 using doxorubicin as an antineoplastic standard. As we observed, the NCI-H292 strain, human lung cancer of the NCI family, the H292 type, was the one that showed the greatest capacity for resistance to doxorubicin [59,60] compared to the results of Ca-MAP1.

When comparing the IC_{50} values of doxorubicin and Ca-MAP1 in NCI-H292 cells, it showed that doxorubicin presented IC_{50} higher than that of Ca-MAP1. Moreover, Ca-MAP1 showed an IC_{50} 1.35 times less than DOX, demonstrating a higher cytotoxic activity forward NCI-H292. In addition, the derivate caused low cytotoxicity in comparison to doxorubicin to erythrocytes and human fibroblast lineage MRC-5 and neuronal microglia lineage BV-2 in ~18.1 μM.

The IC_{50} for MRC-5 was not calculated by the software because the Ca-MAP1 peptide did not demonstrate an effect on cellular viability close to fifty percent. In the case of the BV-2 cell line, the concentration of Ca-MAP1 for the NO assay was not cytotoxic. Therefore, it shows better activity than the positive control without *in vitro* deleterious consequences to the BV-2 cell line, MRC5 cell line, and erythrocytes.

After the activity assay, the next step was conformational analysis *in silico* and *in vitro*. The *in silico* studies on the creation of a structural homology-based model showed the possible formation of an α-helix, which can be validated via the Ramachandran plot [29]. The validation scores of the predicted three-dimensional structure demonstrate correct prediction based on structural homology [26,27]

Ca-MAP1 presented 92.86% of the amino acid residues in favored regions of the right-handed α-helix quadrant, 7.14% amino acid residue in the allowed region, and none in the disallowed region. In addition, the peptide Ca-MAP1 presents a Rama-Z of -1.85 ± 1.76, showing appropriated backbone geometry.

In the study of proteins by circular dichroism, it was possible to observe the characteristic spectra of peptide bonds at 190 nm, and in particular, in the bands around 208 and 222 nm, it is possible to estimate the formation of a helical structure, corroborated with the *in silico* studies [24,61].

In the circular dichroism, Ca-MAP1 demonstrated the occurrence of a positive band near 190 nm and two negative bands at approximately 208 and 222 nm. It is possible to assume that the peptide presents a secondary α-helical or 3_{10}-helical structure that has close bands in the above values in the presence of TFE and SDS [24,25,62]. The formation of a helical structure is observed in the circular dichroism of the cytolytic peptide candidalysin in the HEPES buffer [15].

The broad negative band around 199 nm, and the fact that there is no other negative band in the analysis of circular dichroism in the presence of water, show that the peptide Ca-MAP1 can form a random coil [24,25], a result that is corroborated with the molecular dynamics of the peptide RMSD, RMSF, SASA, and Rg fluctuations in the presence of water [63]. More environments, such as SDS 30 mM and TFE 50%, need to be studied *in silico* with molecular dynamics to visualize if there is a corroboration between *in silico* and *in vitro* analysis.

Several studies have demonstrated the conformational changes in hydrophobic or hydrophilic environments for antimicrobial peptides through circular dichroism [64–66]. The CD analyses show that, in a hydrophobic environment, the conformational preference for Ca-MAP1 was α-helix, indicating that its structure favors the mechanism of action of the membrane permeabilization, as demonstrated in the Sytox Green assay. This result agrees with Migliolo et al., who demonstrated that an alanine-rich peptide presents helicoidal conformation and membrane disruption [67].

Other peptides in the literature have displayed similar activity and conformational behavior; antimicrobial activity, antibacterial in particular, is well described for cationic α-helix amphipathic peptides [68–70]. Mycotoxin peptide derivates have not been widely studied, but other peptide toxin derivates have demonstrated similar characterization results; animal venoms peptide is more widely used for the design of new drugs [69,70].

The Brazilian yellow scorpion (*Tityus serrulatus*) venom peptides TsAP-1 and TsAP2 possess different activities, the first being low hemolytic and bactericide at 120 to 160 μM. The second is more highly hemolytic and bactericidal at 5 to 10 μM. The increase of net charge by adding lysin to TsAP peptides increases hemolytic activity but dramatically increases the potency of antibacterial and anticancer effects, lowering the IC_{50} from 320 μM to 5 μM [64].

The exchange of amino acids to leucine and lysine has been effective in the creation of new peptide derivates in the venom peptide Hp1404 from the scorpion *Heterometrus petersii*; the derivates showed less hemolytic effect and antibacterial effect on multidrug-resistant *Pseudomonas aeruginosa* in the concentration of 0.78 to 25 μM, and all derivates had an amphipathic cationic α-helix in membrane-mimicking environments [69].

The cationic synthetic peptides Hp-MAP1 and Hp-MAP2, derived from the amphibian (*Hylarana picturata*) peptide toxin temporin-PTa, present antibacterial activity in concentrations ranging from 2.8 to 92 μM, without a hemolytic effect on erythrocytes, and in molecular dynamics present an α-helix in the presence of hydrophobic and anionic environments and can form interactions with saline and hydrogen bounds [70]. Molecular dynamics in the presence of membranes mimetic with 1,2-dipalmitoylsn-glycerol-3-phosphatidylglycerol (DPPG-anionic) and 1,2-dipalmitoyl-sn-lyco-3 phosphatidylethanolamine (DPPE-neutral) are needed to observe the interactions with the membrane phospholipids and the Ca-MAP1 peptide.

Other compounds from animal venom have equal biotechnological importance; the Brazilian snake *Bothrops moojeni* produces an important phospholipase, A2, which can be collected due to its anticancer effects, at a concentration of 9.2 μM, in many cancers cell lines, including lung mucoepidermoid carcinoma NCI-H292 [65]. Another example is the wasp and bee venom of the species *Vespa velutina*, which possess antibacterial activity in Gram-negative and -positive bacteria, and anti-inflammatory activity in LPS-induced BV-2 inflammation at concentrations of 0.5 to 20 $\mu g.mL^{-1}$ [66].

4. Conclusions

In summary, the use of a rational design approach for the creation of Ca-MAP1, bioinspired on a fungal toxin, present here for the first time an initial characterization of viable antibacterial, anti-neuroinflammatory, and anticancer activity. The peptide Ca-MAP1 has the potential to be used for the control of exacerbated immune responses caused by bacterial infections that release toxins and by-products. It is also noteworthy that there is activity against a rare metastatic human lung cancer, NCI-H292, with a more discriminatory activity than the antineoplastic drug doxorubicin. More studies for the development of its characterization and comprehension of the underlying mechanisms of action are needed.

5. Materials and Methods

5.1. Rational Design

Candidalysin, a cytolytic mycotoxin peptide secreted from *C. albicans*, presents 32 amino acid residues, a net charge of +4, apolar residues of 56.5%, hydrophobicity of 0.679 in the Eisenberg scale, and hydrophobic moment of 0.408. Faced with the challenge of searching for information regarding candidates that are multifunctional with the potential to combat and control diseases, candidalysin presented activity on immunomodulatory and cytolytic human epithelial cells and is thus a peptide that is little characterized in the literature.

The strategy to design the Ca-MAP1 peptide (Supplementary Material Figure S1) was guided by two steps and considered three requisites: (1) a shorter length comprising charge between +3 and +4; (2) apolar amino acid residue percentage between 40 and 60% with hydrophobicity in agreement with an Eisenberg scale above 0.400; (3) hydrophobic moment above 0.400 (Supplementary Material Table S1). The second step to construct Ca-MAP1 was to correlate the amino acid residue modification with amphipathicity organization, guided by helix diagrams and C-terminal structure stability with a helical conformation preference (Supplementary Material Figure S1).

In agreement with the requisites described in the methodology, two sequences were found. One primary sequence with 14 and another with 16 amino acid residues, both located on the C-terminal functional hot spot region of the candidalysin (Supplementary Material Table S1). The peptide with a length of 14 was discarded due to a decrease in the net charge below the requirements.

Based on these two regions, amino acid glycine, leucine, and lysine frequencies, observed in the profiles of bacterial lantibiotics, plants cyclotides, and amphibian temporin antimicrobial peptide families, were used to create the derived peptide Ca-MAP1 [71]. The results demonstrated that the addition of amino acid preferences in positions 2, 3, 12, and 15, where the 2 and 3 positions are associated with amphipathicity and positions 12 and 15 are associated with helices stability, is in agreement with helices diagrams and theoretical models (Figure 1).

5.2. Peptide Synthesis

The peptides were synthesized by the solid-phase method using 9-fluorenyl-methoxycarbonyl chemistry [72], purified by reverse-phase high-performance liquid chromatography (RP-HPLC) to >98% purity on an acetonitrile/H_2O-TFA gradient, and confirmed by electrospray ionization mass spectrometry by Aminotech Company (Sorocaba, Brazil). The Ca-MAP1 peptide was solubilized in Milli-Q ultrapure water to create a stock solution, which was stored in a −20 °C freezer and used for all assays.

5.3. Hemolytic Assay

The hemolytic assay was developed according to Kim et al. (2005) [73], with modifications. Red cells were washed three times with 50 mM phosphate buffer (PBS); pH 7.4. peptide solutions were added to the erythrocyte suspension (1% by volume) at a final concentration ranging from 4 to 128 $\mu g.mL^{-1}$, which can be expressed at molar concentrations of 2.2 to 72.7 µM, in a final volume of 100 µL. The samples were incubated at room temperature for 60 min. Hemoglobin release was monitored by measuring the absorbance of the supernatant at 415 nm. Zero hemolysis (blank) was determined with red cells suspended in the presence of 50 mM PBS, pH 7.4, while a 1% (by volume) aqueous solution of Triton X-100 was used as a positive control (100% red cell lysis). This experiment was approved by CEUA under number 014/2018.

5.4. Minimum Inhibitory Concentration (MIC) and Minimum Bactericidal Concentration (MBC) Assays

The MIC assays were performed against strains of *Acinetobacter baumannii, Pseudomonas aeruginosa, Escherichia coli* (ATCC 25922 and KPC 001812446), *Klebsiella pneumoniae* (ATCC 13883), and *Staphylococcus aureus*. The bacteria were plated on Mueller Hinton agar (MHA) plates and incubated at 37 °C overnight. After this period, three isolated colonies of each bacterium were inoculated into 5 mL of Mueller Hinton broth (MHB) and incubated at 200 rpm at 37 °C overnight. Bacterial growth was monitored by a spectrophotometer at 600 nm.

MIC tests were performed by the 96-well microplate dilution method at a final bacterial concentration of 2–5 × 10^5 CFU.mL^{-1}. The peptides were tested at concentrations ranging from 4 to 128 μg.mL^{-1}, which form the molar concentrations of 2.2 to 72.7 μM for Ca-MAP1. Ciprofloxacin was used as a positive control, due to its capability to be active against all of the clinically isolated bacteria used in this study; other tested antibiotics are displayed in Supplementary Material Table S2. The positive control was used in the same concentrations as the peptides, while the bacterial suspension in MHB was used as a negative control. The microplates were incubated at 37 °C for 18 h, and readings were taken on a Multiskan Go microplate reader (Thermo Scientific: Waltham, MA, USA) at 600 nm after the incubation time. MIC was determined as the lowest peptide concentration at which there was no significant bacterial growth.

Evaluation of the minimum bactericidal concentration (MBC) was dependent on the MIC results. Three replicates of 10 μL were taken from the microplate wells, plated in MHA, and incubated at 37 °C for 24 h. The MBC was determined as the lowest peptide concentration at which no bacterial growth was detected. All experiments were performed in triplicate.

5.5. *E. coli* Membrane Permeabilization

Membrane permeability was investigated as described by Mohanram and Bhattacharjya (2016) [20] and modifications proposed by Almeida et al. (2021) [19]. In the assay, a suspension of *E. coli* KPC + 001812446, grown in MH broth for 18 h at 37 °C, was prepared with $OD_{600\ nm}$ at 0.5 in 10 mM sodium phosphate buffer, pH 7.0. Then, 280 μL of the bacterial suspension was transferred to 96-well black microplates, where 10 μL of Sytox Green at 30 μM was added and incubated for 10 min at 37 °C. Subsequently, 10 μL of the peptide Ca-MAP1, at a concentration 30 times the MIC, was added to each well, and the kinetic assay was performed for 50 min, with readings every 5 min. The assay was performed with fluorescence readout, excitation at 485 nm, and emission at 520 nm in a Varioskan Lux microplate reader (Thermo Scientific: Waltham, MA, USA). The negative control of membrane damage was performed with *E. coli* KPC + 001812446, incubated with 10 μL of 10 mM sodium phosphate buffer, pH 7.0. Three independent experiments were performed, in triplicate.

5.6. Cell Cultures

In this study, we used diploid human fetal lung cell lineage (MRC-5), murine microglia cells (BV-2), human rhabdomyosarcoma cells (RD), cervical cancer cells (HeLa), and mucoepidermoid lung carcinoma (NCI-H292), of which MRC-5, RD, HeLa, and NCI-H292 cells were acquired from the Cell Culture Center of the Adolf Lutz Institute (São Paulo-SP), and BV-2 cells were acquired from the Rio de Janeiro Cell Bank. All cell lines were stored in liquid nitrogen cryopreservation at a temperature of approximately −196 °C at the Universidade Católica Dom Bosco (UCDB: Campo Grande, MS, Brazil). MRC-5, BV-2, RD, NCI-H292, and HeLa cells were cultured in the Immunology Laboratory (UCDB) in DMEM high glucose medium (for MRC-5 and RD) and RPMI-1640 (for BV-2, NCI-H292, and HeLa), respectively, supplemented with 10% fetal bovine serum (FBS), 100 U.mL^{-1} penicillin, and 100 μg.mL^{-1} streptomycin (Gibco: Waltham, MA, USA) at 37 °C in an incubator at 5% CO_2.

5.7. Cell Viability Test Using MTT Methodology

To verify whether the peptide derivate inhibits anticancer activity, the viability of NCI-H292, HeLa, and RD cancer cells and normal cells MRC-5 and BV-2 were evaluated according to a method adapted from Mosmann (1983) [74], based on the enzymatic reduction of 3-(4,5-demethylthiazol-2-yl)-2,5-diphenyltetrazolium bromide (MTT: Sigma-Aldrich;

St Louis, MO, USA) to formazan crystals. NCI-H292, HeLa, RD, MRC-5, and BV-2 cells were plated at 1×10^4 cells. $well^{-1}$ in 96-well microplates and treated with 100 µL of different molar concentrations of Ca-MAP1 (0.5, 1.1, 2.2, 4.5, 9.1, 18.1, 36.3 µM) for 24 h. Culture medium was used as a negative control. After the incubation period, the supernatant was removed and 100 µL of MTT solution (1 $mg.mL^{-1}$ diluted in culture medium) was added to the cells. After 4 h of incubation, the formazan crystals were resuspended with 100 µL of dimethyl sulfoxide (DMSO) and read at 570 nm on a Thermo scientific reader (MultiSkan Go Model) [74]. The commercial antineoplastic doxorubicin was used as a positive control, with concentrations in serial dilution ranging from 0.3 to 20 $mg.mL^{-1}$, which can be converted to micromolar concentrations ranging from 0.5 to 36.7 µM. Three independent experiments were performed in triplicate. Cell viability was calculated from the following formula:

$$\text{Cell viability } (\%) = (\text{AbsSample} + \text{AbsNegativeControl}) \times 100$$

5.8. Inhibition of Microglial Activation by LPS

BV-2 microglia strains were plated at a density of 5×10^5 $cells.mL^{-1}$ in 96-well plates, followed by adhesion for 24 h at 37 °C in a 5% CO_2 atmosphere. After adhesion, the medium was removed and cells were stimulated with lipopolysaccharide-LPS (Final concentration of 1 $\mu g.mL^{-1}$), together with Ca-MAP1 peptide treated at various concentrations (0.2, 0.5, 1.1, 2.2, 4.5, 9.1, 18.1 µM) in a final volume of 100 $\mu L.well^{-1}$ of RPMI culture medium supplemented with 1% SFB. For the control experiment, cells were cultured with a culture medium and medium with LPS. The cells were then incubated for another 24 h at 37 °C, 5% CO_2, and the cell supernatant was collected for NO analysis with adhered cells, and cell viability assay by the MTT method was performed. The production of nitric oxide was measured by the dosage of its most stable degradation product, nitrite, using the Griess reagent [21]. For the determination of NO production, 100 µL of cell supernatant was subjected to the reaction with an equal volume of Griess reagent. For the preparation of this reagent, solutions of naphthylethylenediamine chloride (0.1%) dissolved in water and 1% sulfanilamide dissolved in H_3PO_4 (5%) were used. Just before use, the solutions were added in a 1:1 ratio, forming the Griess reagent properly. After the 10-min incubation period, the samples were read in a microplate reader at 540 nm. The calculation of nitrite concentrations was performed based on standard curves using different concentrations of Nitrite (3.12 up to 200 µM). Three independent experiments were performed in triplicate.

5.9. Circular Dichroism (CD)

The analyses were performed on a Jasco J-1100 spectropolarimeter (Jasco Inc.: Hachioji-shi, TYO, Japan) using a quartz cuvette with a 1 mm optical path. The spectrum from 260 to 185 nm was collected with steps at a resolution of 0.1 nm to 100 $nm.s^{-1}$, at 25 °C, with an average of 5 accumulated scans for each spectrum. The Ca-MAP1 peptide was prepared in a stock solution of 120 µM and was incubated under different conditions at a concentration of 30 µM. The Ca-MAP1 peptide had its secondary structure analyzed in the presence of water, 50% trifluoroethanol (TFE) to mimic a hydrophobic environment, or 30 mM SDS, an amphipathic compound that forms micelles mimicking an anionic environment. The data were converted to molar ellipticity (θ), according to the equation:

$$[\theta] = \frac{\theta}{10 * C * l * nr}$$

where θ is the ellipticity measured in milliseconds, C is the peptide concentration (M), l is the path length of the cuvette, and n_r is the number of amino acid residues. The fractional alpha-helix content, *fH*, was estimated using the equation:

$$fH = \frac{\theta 222 - \theta C}{\theta H - \theta C}$$

where θC = 2220 − 53, θH = (250 T − 44,000) (1 − 3/n), where T is the temperature in celsius and n is the number of amino acid residues in the peptide. The θC and θH values represent, respectively, the threshold values of average ellipticity at 222 nm (θ222) for a disordered and alpha-helix conformation.

5.10. Comparative Modeling and Validation

The theoretical three-dimensional models were built by fold recognition using template X-ray crystallography and the nuclear magnetic resonance of a similar primary sequence deposited in the PDB database through the I-TASSER server. Model validation was performed based on the statistical data of C-score, Z-score, and root mean square deviation or root mean square deviation of alpha-carbon position generated by the I-TASSER server. ProSA-web was used to calculate the overall quality score for the models based on score results within the ranges for native proteins [26,27]. The visualization of the predicted model was visualized by the PyMOL version 2.0 program [26,75]. The possible dihedral angles psi (Ψ) and phi (Φ) of each amino acid were calculated by MolProbity and observed in the Ramachandran plot [29,76].

5.11. Molecular Dynamics in Water

To perform the molecular dynamics simulations, the .pdb file model was generated from the modeling analyses of the I-TASSER server, and the calculations were performed by the program Gromacs version 5.0.4 (Groningen, The Netherlands) [77,78]. Molecular dynamics consists of three main stages; initially, assembly and parameterization of the system are performed, where a cubic box is generated in a vacuum, and it is filled with the solvent, composed of water molecules in the simple point charge (SPC) model and ions of Na^+ and Cl^- to neutralize the total charge of the system. After the solvation is performed, the energy minimization of the system promotes the lowest free energy of the molecules; after this process, heating at a temperature of 310 Kelvin and pressurization of the system is carried out. After the parameterization of the system, the simulation calculations of the atomic movement and topology of the atoms of the system are performed based on molecular interactions according to the Newtonian mechanics of classical physics. Finally, after performing the calculations, the data analysis is carried out, which generates the values of RMSD, RMSF, spin radius, and solvent access area. All experiments were performed in triplicate.

5.12. Statistical Analysis

The statistical significance of the experimental results was determined by one-way Student's *t*-test or one-way analysis of variance (ANOVA), followed by the Dunnett test. Values of $p < 0.05$ were considered statistically significant. GraphPad Prism version 8.0 was used for all statistical analyses.

Supplementary Materials: The following supporting information can be downloaded at: https://www.mdpi.com/article/10.3390/toxins14100696/s1, Figure S1: Database filtering technology schematic: (A) Candidalysin activity, (B) Determination of peptide length, (C) Amino acids frequency, (D) Net charge, (E) Hydrophobic content, (F) Structure type (F) Functional hot spot. The designed peptide is called Ca-MAP1. Table S1: Strategy for the selection of peptide based on physical-chemical parameters of candidalysin. Legend: <H> = Hydrophobicity in the Eisenberg scale; <μH> = Hydrophobic moment. Table S2: Antibiogram assay against Gram-negative and -positive bacteria. (+) resistant at the concentrations tested.

Author Contributions: Study conception and design, P.H.d.O.C. and A.P.d.A.B.; data collection, L.T.H.M., P.S.e.S., A.S.G., L.F.R.N.d.M. and C.F.R.d.O.; analysis and interpretation of results, M.L.R.M., C.M.E.C. and A.P.d.C.; draft manuscript preparation, P.H.d.O.C., A.P.d.A.B. and L.M. All authors have read and agreed to the published version of the manuscript.

Funding: This work was supported by the Brazilian funding agencies CNPq, CAPES number 001, and FUNDECT.

Institutional Review Board Statement: Not applicable.

Informed Consent Statement: Not applicable.

Data Availability Statement: Not applicable.

Acknowledgments: This work was supported by CAPES (Coordenação de Aperfeiçoamento de Pessoal de Nível Superior), FUNDECT (Fundação de Apoio ao Desenvolvimento da Educação, Ciência e Tecnologia do Estado de Mato Grosso do Sul), and CNPq (Conselho Nacional de Desenvolvimento Científico e Tecnológico).

Conflicts of Interest: The authors declare no conflict of interest.

References

1. Boleti, A.P.D.A.; Flores, T.M.D.O.; Moreno, S.E.; dos Anjos, L.; Mortari, M.R.; Migliolo, L. Neuroinflammation: An Overview of Neurodegenerative and Metabolic Diseases and of Biotechnological Studies. *Neurochem. Int.* **2020**, *136*, 104714. [CrossRef] [PubMed]
2. Yong, H.Y.F.; Rawji, K.S.; Ghorbani, S.; Xue, M.; Yong, V.W. The Benefits of Neuroinflammation for the Repair of the Injured Central Nervous System. *Cell. Mol. Immunol.* **2019**, *16*, 540–546. [CrossRef] [PubMed]
3. Liao, D.; Xiang, D.; Dang, R.; Xu, P.; Wang, J.; Han, W.; Fu, Y.; Yao, D.; Cao, L.; Jiang, P. Neuroprotective Effects of Dl-3-n-Butylphthalide against Doxorubicin-Induced Neuroinflammation, Oxidative Stress, Endoplasmic Reticulum Stress, and Behavioral Changes. *Oxid. Med. Cell. Longev.* **2018**, *2018*, 9125601. [CrossRef] [PubMed]
4. Boleti, A.P.D.A.; Frihling, B.E.F.; Silva, P.S.E.; Cardoso, P.H.D.O.; de Moraes, L.F.R.N.; Rodrigues, T.A.A.; Biembengute, M.E.F.; Koolen, H.H.F.; Migliolo, L. Biochemical Aspects and Therapeutic Mechanisms of Cannabidiol in Epilepsy. *Neurosci. Biobehav. Rev.* **2022**, *132*, 1214–1228. [CrossRef]
5. Batista, C.R.; Gomes, G.F.; Candelario-Jalil, E.; Fiebich, B.L.; de Oliveira, A.C. Lipopolysaccharide-Induced Neuroinflammation as a Bridge to Understand Neurodegeneration. *Int. J. Mol. Sci.* **2019**, *20*, 2293. [CrossRef] [PubMed]
6. Giuliani, A.; Pirri, G.; Rinaldi, A.C. Antimicrobial Peptides: The LPS Connection. In *Antimicrobial Peptides*; Springer: Berlin/Heidelberg, Germany, 2010; pp. 137–154.
7. Di Lorenzo, F.; De Castro, C.; Silipo, A.; Molinaro, A. Lipopolysaccharide Structures of Gram-Negative Populations in the Gut Microbiota and Effects on Host Interactions. *FEMS Microbiol. Rev.* **2019**, *43*, 257–272. [CrossRef] [PubMed]
8. Haney, E.F.; Mansour, S.C.; Hancock, R.E.W. Antimicrobial Peptides: An Introduction. *Methods Mol. Biol.* **2017**, *1548*, 3–22. [CrossRef] [PubMed]
9. Fosgerau, K.; Hoffmann, T. Peptide Therapeutics: Current Status and Future Directions. *Drug Discov. Today* **2015**, *20*, 122–128. [CrossRef]
10. Kombrink, A.; Tayyrov, A.; Essig, A.; Stöckli, M.; Micheller, S.; Hintze, J.; van Heuvel, Y.; Dürig, N.; Lin, C.; Kallio, P.T.; et al. Induction of Antibacterial Proteins and Peptides in the Coprophilous Mushroom Coprinopsis Cinerea in Response to Bacteria. *ISME J.* **2019**, *13*, 588–602. [CrossRef] [PubMed]
11. Ageitos, J.M.; Sánchez-Pérez, A.; Calo-Mata, P.; Villa, T.G. Antimicrobial Peptides (AMPs): Ancient Compounds That Represent Novel Weapons in the Fight against Bacteria. *Biochem. Pharmacol.* **2017**, *133*, 117–138. [CrossRef]
12. Künzler, M. How Fungi Defend Themselves against Microbial Competitors and Animal Predators. *PLoS Pathog.* **2018**, *14*, e1007184. [CrossRef] [PubMed]
13. Blagojevic, M.; Camilli, G.; Maxson, M.; Hube, B.; Moyes, D.L.; Richardson, J.P.; Naglik, J.R. Candidalysin Triggers Epithelial Cellular Stresses That Induce Necrotic Death. *Cell. Microbiol.* **2021**, *23*, e13371. [CrossRef] [PubMed]
14. Ho, J.; Wickramasinghe, D.N.; Nikou, S.-A.; Hube, B.; Richardson, J.P.; Naglik, J.R. Candidalysin Is a Potent Trigger of Alarmin and Antimicrobial Peptide Release in Epithelial Cells. *Cells* **2020**, *9*, 699. [CrossRef] [PubMed]
15. Moyes, D.L.; Wilson, D.; Richardson, J.P.; Mogavero, S.; Tang, S.X.; Wernecke, J.; Höfs, S.; Gratacap, R.L.; Robbins, J.; Runglall, M.; et al. Candidalysin Is a Fungal Peptide Toxin Critical for Mucosal Infection. *Nature* **2016**, *532*, 64–68. [CrossRef]
16. Torres, M.D.T.; Sothiselvam, S.; Lu, T.K.; de la Fuente-Nunez, C. Peptide Design Principles for Antimicrobial Applications. *J. Mol. Biol.* **2019**, *431*, 3547–3567. [CrossRef]
17. Wang, G.; Li, X.; Wang, Z. APD3: The Antimicrobial Peptide Database as a Tool for Research and Education. *Nucleic Acids Res.* **2016**, *44*, D1087–D1093. [CrossRef]
18. Gautier, R.; Douguet, D.; Antonny, B.; Drin, G. HELIQUEST: A Web Server to Screen Sequences with Specific Alpha-Helical Properties. *Bioinformatics* **2008**, *24*, 2101–2102. [CrossRef]
19. Almeida, C.V.; de Oliveira, C.F.R.; dos Santos, E.L.; dos Santos, H.F.; Júnior, E.C.; Marchetto, R.; da Cruz, L.A.; Ferreira, A.M.T.; Gomes, V.M.; Taveira, G.B.; et al. Differential Interactions of the Antimicrobial Peptide, RQ18, with Phospholipids and Cholesterol Modulate Its Selectivity for Microorganism Membranes. *Biochim. Biophys. Acta-Gen. Subj.* **2021**, *1865*, 129937. [CrossRef]
20. Mohanram, H.; Bhattacharjya, S. Salt-Resistant Short Antimicrobial Peptides. *Biopolymers* **2016**, *106*, 345–356. [CrossRef]
21. Green, L.C.; Wagner, D.A.; Glogowski, J.; Skipper, P.L.; Wishnok, J.S.; Tannenbaum, S.R. Analysis of Nitrate, Nitrite, and [15N]Nitrate in Biological Fluids. *Anal. Biochem.* **1982**, *126*, 131–138. [CrossRef]

22. Silva, O.N.; Porto, W.F.; Migliolo, L.; Mandal, S.M.; Gomes, D.G.; Holanda, H.H.S.; Silva, R.S.P.; Dias, S.C.; Costa, M.P.; Costa, C.R.; et al. Cn-AMP1: A New Promiscuous Peptide with Potential for Microbial Infections Treatment. *Biopolymers* **2012**, *98*, 322–331. [CrossRef] [PubMed]
23. Migliolo, L.; Felício, M.R.; Cardoso, M.H.; Silva, O.N.; Xavier, M.-A.E.; Nolasco, D.O.; de Oliveira, A.S.; Roca-Subira, I.; Vila Estape, J.; Teixeira, L.D.; et al. Structural and Functional Evaluation of the Palindromic Alanine-Rich Antimicrobial Peptide Pa-MAP2. *Biochim. Biophys. Acta* **2016**, *1858 Pt A*, 1488–1498. [CrossRef]
24. Greenfield, N.J. Using Circular Dichroism Spectra to Estimate Protein Secondary Structure. *Nat. Protoc.* **2006**, *1*, 2876–2890. [CrossRef] [PubMed]
25. Lopes, J.L.S.; Miles, A.J.; Whitmore, L.; Wallace, B.A. Distinct Circular Dichroism Spectroscopic Signatures of Polyproline II and Unordered Secondary Structures: Applications in Secondary Structure Analyses. *Protein Sci.* **2014**, *23*, 1765–1772. [CrossRef] [PubMed]
26. Yang, J.; Yan, R.; Roy, A.; Xu, D.; Poisson, J.; Zhang, Y. The I-TASSER Suite: Protein Structure and Function Prediction. *Nat. Methods* **2015**, *12*, 7–8. [CrossRef]
27. Wiederstein, M.; Sippl, M.J. ProSA-Web: Interactive Web Service for the Recognition of Errors in Three-Dimensional Structures of Proteins. *Nucleic Acids Res.* **2007**, *35* (Suppl. 2), 407–410. [CrossRef]
28. Davis, I.W.; Leaver-Fay, A.; Chen, V.B.; Block, J.N.; Kapral, G.J.; Wang, X.; Murray, L.W.; Arendall III, W.B.; Snoeyink, J.; Richardson, J.S. MolProbity: All-Atom Contacts and Structure Validation for Proteins and Nucleic Acids. *Nucleic Acids Res.* **2007**, *35* (Suppl. 2), W375–W383. [CrossRef]
29. Williams, C.J.; Headd, J.J.; Moriarty, N.W.; Prisant, M.G.; Videau, L.L.; Deis, L.N.; Verma, V.; Keedy, D.A.; Hintze, B.J.; Chen, V.B.; et al. MolProbity: More and Better Reference Data for Improved All-atom Structure Validation. *Protein Sci.* **2018**, *27*, 293–315. [CrossRef]
30. Yan, J.; Bhadra, P.; Li, A.; Sethiya, P.; Qin, L.; Tai, H.K.; Wong, K.H.; Siu, S.W.I. Deep-AmPEP30: Improve Short Antimicrobial Peptides Prediction with Deep Learning. *Mol. Ther. Nucleic Acids* **2020**, *20*, 882–894. [CrossRef]
31. Kim, H.; Jang, J.H.; Kim, S.C.; Cho, J.H. De Novo Generation of Short Antimicrobial Peptides with Enhanced Stability and Cell Specificity. *J. Antimicrob. Chemother.* **2014**, *69*, 121–132. [CrossRef]
32. Ramesh, S.; Govender, T.; Kruger, H.G.; de la Torre, B.G.; Albericio, F. Short AntiMicrobial Peptides (SAMPs) as a Class of Extraordinary Promising Therapeutic Agents. *J. Pept. Sci.* **2016**, *22*, 438–451. [CrossRef] [PubMed]
33. Jiang, Z.; Mant, C.T.; Vasil, M.; Hodges, R.S. Role of Positively Charged Residues on the Polar and Non-Polar Faces of Amphipathic α-Helical Antimicrobial Peptides on Specificity and Selectivity for Gram-Negative Pathogens. *Chem. Biol. Drug Des.* **2018**, *91*, 75–92. [CrossRef] [PubMed]
34. Béven, L.; Castano, S.; Dufourcq, J.; Wieslander, A.; Wróblewski, H. The Antibiotic Activity of Cationic Linear Amphipathic Peptides: Lessons from the Action of Leucine/Lysine Copolymers on Bacteria of the Class Mollicutes. *Eur. J. Biochem.* **2003**, *270*, 2207–2217. [CrossRef]
35. Chen, Y.; Guarnieri, M.T.; Vasil, A.I.; Vasil, M.L.; Mant, C.T.; Hodges, R.S. Role of Peptide Hydrophobicity in the Mechanism of Action of Alpha-Helical Antimicrobial Peptides. *Antimicrob. Agents Chemother.* **2007**, *51*, 1398–1406. [CrossRef] [PubMed]
36. Torres, M.D.T.; Pedron, C.N.; Higashikuni, Y.; Kramer, R.M.; Cardoso, M.H.; Oshiro, K.G.N.; Franco, O.L.; Silva Junior, P.I.; Silva, F.D.; Oliveira Junior, V.X.; et al. Structure-Function-Guided Exploration of the Antimicrobial Peptide Polybia-CP Identifies Activity Determinants and Generates Synthetic Therapeutic Candidates. *Commun. Biol.* **2018**, *1*, 221. [CrossRef]
37. Takechi-Haraya, Y.; Ohgita, T.; Kotani, M.; Kono, H.; Saito, C.; Tamagaki-Asahina, H.; Nishitsuji, K.; Uchimura, K.; Sato, T.; Kawano, R.; et al. Effect of Hydrophobic Moment on Membrane Interaction and Cell Penetration of Apolipoprotein E-Derived Arginine-Rich Amphipathic α-Helical Peptides. *Sci. Rep.* **2022**, *12*, 4959. [CrossRef]
38. Porto, W.F.; Ferreira, K.C.V.; Ribeiro, S.M.; Franco, O.L. Sense the Moment: A Highly Sensitive Antimicrobial Activity Predictor Based on Hydrophobic Moment. *Biochim. Biophys. Acta-Gen. Subj.* **2022**, *1866*, 130070. [CrossRef]
39. Höfs, S. *Identification of Candidalysin: A Candida Albicans Peptide Toxin Involved in Epithelial Damage*; Friedrich-Schiller-Universität: Jena, Germany, 2016.
40. Oddo, A.; Hansen, P.R. Hemolytic Activity of Antimicrobial Peptides. In *Antimicrobial Peptides*; Springer: Berlin/Heidelberg, Germany, 2017; pp. 427–435.
41. Shakil, M.M.; Hossain, M.A.J.; Karal, M.A.S.; Islam, M.Z.; Chowdhury, M.S.; Hasan; Haque, M.A.; Alam, M.M. Comparative Time-Kill Assessment of Cationic Antimicrobial Peptide and Fluoroquinolone against Gram-Negative Bacteria. *EC Pharmacol. Toxicol.* **2020**, *8*, 117–124.
42. Mortimer, F.C.; Mason, D.J.; Gant, V.A. Flow Cytometric Monitoring of Antibiotic-Induced Injury in Escherichia Coli Using Cell-Impermeant Fluorescent Probes. *Antimicrob. Agents Chemother.* **2000**, *44*, 676–681. [CrossRef]
43. Wang, W.Y.; Tan, M.S.; Yu, J.T.; Tan, L. Role of Pro-Inflammatory Cytokines Released from Microglia in Alzheimer's Disease. *Ann. Transl. Med.* **2015**, *3*, 136. [CrossRef]
44. Wang, X.; Rousset, C.I.; Hagberg, H.; Mallard, C. Lipopolysaccharide-Induced Inflammation and Perinatal Brain Injury. *Semin. Fetal Neonatal Med.* **2006**, *11*, 343–353. [CrossRef] [PubMed]
45. Shabab, T.; Khanabdali, R.; Moghadamtousi, S.Z.; Kadir, H.A.; Mohan, G. Neuroinflammation Pathways: A General Review. *Int. J. Neurosci.* **2017**, *127*, 624–633. [CrossRef] [PubMed]

46. Xie, L.; Huang, Z.; Meng, H.; Shi, X.; Xie, J. Immunomodulation Effect of Polysaccharides from Liquid Fermentation of Monascus Purpureus 40269 via Membrane TLR-4 to Activate the MAPK and NF-KB Signaling Pathways. *Int. J. Biol. Macromol.* **2022**, *201*, 480–491. [CrossRef] [PubMed]
47. Oh, Y.-C.; Jeong, Y.H.; Li, W.; Go, Y. Angelicae Gigantis Radix Regulates LPS-Induced Neuroinflammation in BV2 Microglia by Inhibiting NF-KB and MAPK Activity and Inducing Nrf-2 Activity. *Molecules* **2019**, *24*, 3755. [CrossRef] [PubMed]
48. Zandbergen, N.; de Rooij, B.H.; Vos, M.C.; Pijnenborg, J.M.A.; Boll, D.; Kruitwagen, R.F.P.M.; van de Poll-Franse, L.V.; Ezendam, N.P.M. Changes in Health-Related Quality of Life among Gynecologic Cancer Survivors during the Two Years after Initial Treatment: A Longitudinal Analysis. *Acta Oncol.* **2019**, *58*, 790–800. [CrossRef]
49. Xu, Z.; Luo, F.; Wang, Y.; Zou, B.; Man, Y.; Liu, J.; Li, H.; Arshad, B.; Li, H.; Li, S. Cognitive Impairments in Breast Cancer Survivors Treated with Chemotherapy: A Study Based on Event-Related Potentials. *Cancer Chemother. Pharmacol.* **2020**, *85*, 61–67. [CrossRef]
50. Yamada, T.H.; Denburg, N.L.; Beglinger, L.J.; Schultz, S.K. Neuropsychological Outcomes of Older Breast Cancer Survivors: Cognitive Features Ten or More Years after Chemotherapy. *J. Neuropsychiatry Clin. Neurosci.* **2010**, *22*, 48–54. [CrossRef]
51. Wang, L.; Apple, A.C.; Schroeder, M.P.; Ryals, A.J.; Voss, J.L.; Gitelman, D.; Sweet, J.J.; Butt, Z.A.; Cella, D.; Wagner, L.I. Reduced Prefrontal Activation during Working and Long-term Memory Tasks and Impaired Patient-reported Cognition among Cancer Survivors Postchemotherapy Compared with Healthy Controls. *Cancer* **2016**, *122*, 258–268. [CrossRef]
52. Groves, T.R.; Farris, R.; Anderson, J.E.; Alexander, T.C.; Kiffer, F.; Carter, G.; Wang, J.; Boerma, M.; Allen, A.R. 5-Fluorouracil Chemotherapy Upregulates Cytokines and Alters Hippocampal Dendritic Complexity in Aged Mice. *Behav. Brain Res.* **2017**, *316*, 215–224. [CrossRef]
53. Lee, B.E.; Choi, B.Y.; Hong, D.K.; Kim, J.H.; Lee, S.H.; Kho, A.R.; Kim, H.; Choi, H.C.; Suh, S.W. The Cancer Chemotherapeutic Agent Paclitaxel (Taxol) Reduces Hippocampal Neurogenesis via down-Regulation of Vesicular Zinc. *Sci. Rep.* **2017**, *7*, 11667. [CrossRef]
54. Jiang, Z.-G.; Winocur, G.; Wojtowicz, J.M.; Shevtsova, O.; Fuller, S.; Ghanbari, H.A. PAN-811 Prevents Chemotherapy-Induced Cognitive Impairment and Preserves Neurogenesis in the Hippocampus of Adult Rats. *PLoS ONE* **2018**, *13*, e0191866. [CrossRef] [PubMed]
55. Hernandez-Aya, L.F.; Gonzalez-Angulo, A.M. Adjuvant Systemic Therapies in Breast Cancer. *Surg. Clin.* **2013**, *93*, 473–491. [CrossRef] [PubMed]
56. Dos Santos, J.M.; Alfredo, T.M.; Antunes, K.Á.; Cunha, J.D.S.M.D.; Costa, E.M.A.; Lima, E.S.; Silva, D.B.; Carollo, C.A.; Schmitz, W.O.; Boleti, A.P.D.A. Guazuma Ulmifolia Lam. Decreases Oxidative Stress in Blood Cells and Prevents Doxorubicin-Induced Cardiotoxicity. *Oxid. Med. Cell. Longev.* **2018**, *2018*, 2935051. [CrossRef]
57. Olmedo, M.E.; Forster, M.; Moreno, V.; López-Criado, M.P.; Braña, I.; Flynn, M.; Doger, B.; de Miguel, M.; López-Vilariño, J.A.; Núñez, R. Efficacy and Safety of Lurbinectedin and Doxorubicin in Relapsed Small Cell Lung Cancer. Results from an Expansion Cohort of a Phase I Study. *Investig. New Drugs* **2021**, *39*, 1275–1283. [CrossRef]
58. McGowan, J.V.; Chung, R.; Maulik, A.; Piotrowska, I.; Walker, J.M.; Yellon, D.M. Anthracycline Chemotherapy and Cardiotoxicity. *Cardiovasc. Drugs Ther.* **2017**, *31*, 63–75. [CrossRef] [PubMed]
59. Guimarães, C.I.F.D.S.; André, S.A.; Correia, J.M.F.; Matos, C.L.A.M.D.; Nogueira, F.J.D. Carcinoma Mucoepidermóide Do Pulmão—Estado Da Arte a Propósito de Um Caso Clínico Mucoepidermoid Carcinoma of the Lung—State of Art with Regard to a Case Report. *Rev. Soc. Bras. Clin. Med.* **2016**, *14*, 41–44.
60. Kashkin, K.N.; Musatkina, E.A.; Komelkov, A.V.; Favorskaya, I.A.; Trushkin, E.V.; Shleptsova, V.A.; Sakharov, D.A.; Vinogradova, T.V.; Kopantzev, E.P.; Zinovyeva, M.V. Expression Profiling and Putative Mechanisms of Resistance to Doxorubicin of Human Lung Cancer Cells. In *Doklady. Biochemistry and Biophysics*; Springer Nature BV: London, UK, 2010; Volume 430, p. 20.
61. Kelly, S.M.; Jess, T.J.; Price, N.C. How to Study Proteins by Circular Dichroism. *Biochim. Biophys. Acta-Proteins Proteom.* **2005**, *1751*, 119–139. [CrossRef]
62. Kaminský, J.; Kubelka, J.; Bour, P. Theoretical Modeling of Peptide α-Helical Circular Dichroism in Aqueous Solution. *J. Phys. Chem. A* **2011**, *115*, 1734–1742. [CrossRef]
63. Khezri, A.; Karimi, A.; Yazdian, F.; Jokar, M.; Mofradnia, S.R.; Rashedi, H.; Tavakoli, Z. Molecular Dynamic of Curcumin/Chitosan Interaction Using a Computational Molecular Approach: Emphasis on Biofilm Reduction. *Int. J. Biol. Macromol.* **2018**, *114*, 972–978. [CrossRef]
64. Guo, X.; Ma, C.; Du, Q.; Wei, R.; Wang, L.; Zhou, M.; Chen, T.; Shaw, C. Two Peptides, TsAP-1 and TsAP-2, from the Venom of the Brazilian Yellow Scorpion, Tityus Serrulatus: Evaluation of Their Antimicrobial and Anticancer Activities. *Biochimie* **2013**, *95*, 1784–1794. [CrossRef]
65. Frihling, B.E.F.; Boleti, A.P.D.A.; de Oliveira, C.F.R.; Sanches, S.C.; Cardoso, P.H.d.O.; Verbisck, N.; Macedo, M.L.R.; Rita, P.H.S.; Carvalho, C.M.E.; Migliolo, L. Purification, Characterization and Evaluation of the Antitumoral Activity of a Phospholipase A2 from the Snake Bothrops Moojeni. *Pharmaceuticals* **2022**, *15*, 724. [CrossRef] [PubMed]
66. Yun, H.S.; Oh, J.; Lim, J.S.; Kim, H.J.; Kim, J.-S. Anti-Inflammatory Effect of Wasp Venom in BV-2 Microglial Cells in Comparison with Bee Venom. *Insects* **2021**, *12*, 297. [CrossRef] [PubMed]
67. Migliolo, L.; Silva, O.N.; Silva, P.A.; Costa, M.P.; Costa, C.R.; Nolasco, D.O.; Barbosa, J.A.R.G.; Silva, M.R.R.; Bemquerer, M.P.; Lima, L.M.P.; et al. Structural and Functional Characterization of a Multifunctional Alanine-Rich Peptide Analogue from Pleuronectes Americanus. *PLoS ONE* **2012**, *7*, e47047. [CrossRef] [PubMed]

68. Hong, J.; Lu, X.; Deng, Z.; Xiao, S.; Yuan, B.; Yang, K. How Melittin Inserts into Cell Membrane: Conformational Changes, Inter-Peptide Cooperation, and Disturbance on the Membrane. *Molecules* **2019**, *24*, 1775. [CrossRef]
69. Kim, M.K.; Kang, H.K.; Ko, S.J.; Hong, M.J.; Bang, J.K.; Seo, C.H.; Park, Y. Mechanisms Driving the Antibacterial and Antibiofilm Properties of Hp1404 and Its Analogue Peptides against Multidrug-Resistant Pseudomonas Aeruginosa. *Sci. Rep.* **2018**, *8*, 1763. [CrossRef]
70. Souza E Silva, P.; Ferreira, M.A.; de Moraes, L.F.R.; de Barros, E.; Preza, S.L.E.; Cardoso, M.H.; Franco, O.L.; Migliolo, L. Synthetic Peptides Bioinspired in Temporin-PTa with Antibacterial and Antibiofilm Activity. *Chem. Biol. Drug Des.* **2022**, *100*, 51–63. [CrossRef]
71. Mishra, B.; Wang, G. Ab Initio Design of Potent Anti-MRSA Peptides Based on Database Filtering Technology. *J. Am. Chem. Soc.* **2012**, *134*, 12426–12429. [CrossRef]
72. Merrifield, R.B. Solid Phase Synthesis. *Science* **1986**, *232*, 341–348. [CrossRef]
73. Kim, J.Y.; Park, S.C.; Kim, M.H.; Lim, H.T.; Park, Y.; Hahm, K.S. Antimicrobial Activity Studies on a Trypsin–Chymotrypsin Protease Inhibitor Obtained from Potato. *Biochem. Biophys. Res. Commun.* **2005**, *330*, 921–927. [CrossRef]
74. Mosmann, T. Rapid Colorimetric Assay for Cellular Growth and Survival: Application to Proliferation and Cytotoxicity Assays. *J. Immunol. Methods* **1983**, *65*, 55–63. [CrossRef]
75. DeLano, W.L. Pymol: An Open-Source Molecular Graphics Tool. *CCP4 Newsl. Protein Crystallogr.* **2002**, *40*, 82–92.
76. Lovell, S.C.; Davis, I.W.; Arendall, W.B., III.; De Bakker, P.I.W.; Word, J.M.; Prisant, M.G.; Richardson, J.S.; Richardson, D.C. Structure Validation by Cα Geometry: ϕ, ψ and Cβ Deviation. *Proteins Struct. Funct. Bioinform.* **2003**, *50*, 437–450. [CrossRef] [PubMed]
77. Lindahl, E.; Hess, B.; Van Der Spoel, D. GROMACS 3.0: A Package for Molecular Simulation and Trajectory Analysis. *J. Mol. Model.* **2001**, *7*, 306–317. [CrossRef]
78. Abraham, M.J.; van der Spoel, D.; Lindahl, E.; Hess, B. *GROMACS User Manual Version 5.0.4*; Royal Institute of Technology and Uppsala University: Uppsala, Sweden, 2014.

toxins

MDPI

Article

The Inhibitory Effect of *Pseudomonas stutzeri* YM6 on *Aspergillus flavus* Growth and Aflatoxins Production by the Production of Volatile Dimethyl Trisulfide

An-Dong Gong [1,2], Yin-Yu Lei [2], Wei-Jie He [1], Yu-Cai Liao [1,3], Ling Ma [1,3,4], Tian-Tian Zhang [1,3,4] and Jing-Bo Zhang [1,3,4,*]

1 Molecular Biotechnology Laboratory of Triticeae Crops, Huazhong Agricultural University, Wuhan 430070, China
2 College of Life Science, Xinyang Normal University, Xinyang 464000, China
3 College of Plant Science and Technology, Huazhong Agricultural University, Wuhan 430070, China
4 Hubei Key Laboratory of Plant Pathology, Wuhan 430070, China
* Correspondence: jingbozhang@mail.hzau.edu.cn; Tel.: +86-27-87283008

Abstract: *Aspergillus flavus* and the produced aflatoxins cause great hazards to food security and human health across all countries. The control of *A. flavus* and aflatoxins in grains during storage is of great significance to humans. In the current study, bacteria strain YM6 isolated from sea sediment was demonstrated effective in controlling *A. flavus* by the production of anti-fungal volatiles. According to morphological characteristics and phylogenetic analysis, strain YM6 was identified as *Pseudomonas stutzeri*. YM6 can produce abundant volatile compounds which could inhibit mycelial growth and conidial germination of *A. flavus*. Moreover, it greatly prevented fungal infection and aflatoxin production on maize and peanuts during storage. The inhibition rate was 100%. Scanning electron microscopy further supported that the volatiles could destroy the cell structure of *A. flavus* and prevent conidia germination on the grain surface. Gas chromatography/mass spectrometry revealed that dimethyl trisulfide (DMTS) with a relative abundance of 13% is the most abundant fraction in the volatiles from strain YM6. The minimal inhibitory concentration of DMTS to *A. flavus* conidia is 200 μL/L (compound volume/airspace volume). Thus, we concluded that *Pseudomonas stutzeri* YM6 and the produced DMTS showed great inhibition to *A. flavus*, which could be considered as effective biocontrol agents in further application.

Keywords: *Aspergillus flavus*; aflatoxins; *Pseudomonas stutzeri*; dimethyl trisulfide; post-harvest control

Key Contribution: *Pseudomonas stutzeri* YM6 and the produced dimethyl trisulfide showed great inhibition to *A. flavus* and aflatoxins in grains, which could be considered as effective biocontrol agents in further application.

1. Introduction

A. flavus, as a harmful phytopathogen, can infect peanuts, maize, and soybeans as well as their products, and produce highly toxic aflatoxins in the field and storage conditions [1]. Aflatoxins are considered unavoidable contaminants of human food and animal feed by the US Food and Drug Administration (FDA), which contaminate over 25% of the world's crops annually [1] and cause great hazards to food security and human health [2]. Aflatoxin B1 (AFB1) is the most toxic member of the mycotoxin group, and its toxicity is 10 times that of potassium cyanide and 68 times that of arsenic [3]. The AFB1 is classified as a group I carcinogen by the International Agency for Research on Cancer (IARC) for its high degree of toxicity, and it may cause liver and lung carcinogen, and even acute death in humans [4].

It is estimated that more than 5 billion people are exposed to the harm of aflatoxins annually [5]. In the 1960s, hundreds of thousands of turkeys died from the aflatoxin-contaminated grain diet, which was known as the proverbial "Turkey X disease" [6].

Aflatoxins also caused serious economic losses and health hazards. For example, in the USA, financial losses due to aflatoxin contamination were estimated to be hundreds of million dollars annually, with maize and peanut being the most seriously affected food crops [7]. Of the 550,000–600,000 de novo liver cancer cases worldwide each year, it is estimated that 25,200–155,000 may be associated with aflatoxin exposure [8]. Countries considered endemic for hepatitis B virus infection (such as China, India, etc.) experienced more hepatocellular carcinoma and high death rates caused by aflatoxin [9]. For example, China has one of the highest populations of HBV carriers in the world, i.e., nearly 10% of the whole population. It was reported that over 370,000 people die of liver cancer every year, which accounts for more than 50% of the world's liver cancer deaths [9].

Because of the great risks of aflatoxins to humans, more than 100 countries have legislatively defined the maximum levels in grains. The maximum tolerable limits for aflatoxins in maize and groundnut in the EU, the East African Community (EAC), and the USA are 4, 10, and 20 μg/kg, respectively [10,11]. In China, the maximum level for AFB1 is 20 ppb, with no regulations for total aflatoxins until now [5].

Legislations aimed to reduce the potential damage of aflatoxins are necessary. However, the more effective methods to avoid risks are preventing *A. flavus* growth and aflatoxins production in the field and storage. The methods used in controlling *A. flavus* and aflatoxins were the breeding of resistant plant varieties, chemical fungicides, and biological control agents. However, owing to the limited resistant breeds in nature and the lengthy breeding process, the effort to control the disease through plant breeding is less effective. While the approach of chemical fungicides is an efficient and quick controlling method against *A. flavus* and aflatoxin compared to plant breeding, it can cause pathogen resistance and pose potential health risks with pesticide residue [12]. It is worth noting that biological agents were considered as candidate alternatives in controlling *A. flavus* and aflatoxins in the field and storage. Some research has shown that certain microbes, such as non-aflatoxigenic *A. flavus* strain and bacteria, can prevent *A. flavus* infection and aflatoxins production in the field [13,14]. These microbes have been gradually utilized in controlling *A. flavus* and aflatoxins in practice. Additionally, the microbial metabolites produced by microorganisms such as peptides, iturins, fengycin, and surfactants were also effective against *A. flavus* [15,16]. Among the identified metabolites, low-molecular-weight volatile compounds can quickly be evaporated and evenly distributed throughout the grain storage space [17]. In our previous work, the microbes including *Staphylococcus saprophyticus* L-38, *Serratia marcescens* Pt-3, *Enterobacter asburiae* Vt-7, and *Alcaligenes faecalis* N1-4 were demonstrated useful in controlling *A. flavus* and aflatoxins during storage by the production of antifungal volatiles [5,18–20]. In an effort to identify more novel strains for biocontrol, we isolated and identified one bacterial strain of *Pseudomonas stutzeri* YM6 from the collected marine sediment. The abundant anti-fungal volatile dimethyl trisulfide (DMTS) produced by YM6 showed strong anti-fungal activity against *A. flavus* and seven other important fungal pathogens in face-to-face dual cultural tests. It also greatly inhibited *A. flavus* growth and aflatoxins production in maize and peanut kernels during storage. The minimal inhibitory concentration of DMTS to *A. flavus* was 200 μL/L. The novel agents identified in the previous study can control *A. flavus* growth and aflatoxin production in grains during storage.

2. Results

2.1. Identification of Strain YM6

The colony of strain YM6 was opaque and pale yellow with a smooth surface and entire edge on the NA medium. YM6 was identified as a gram-negative, short-rod bacterium under the microscope. The 16S rDNA sequence of YM6 was homologous to three species (*P. stutzeri*, *P. putida*, and *P. xanthomarina*) of *Pseudomonas* spp. Eight strains of three species with great similarity to YM6 were selected to construct the phylogenetic tree. Strain YM6 and *P. stutzeri* 28a42 (AJ312165.1) showed the highest level of homology in terms of clade-level. Thus, the YM6 was identified as a strain of *P. stutzeri* (Figure 1). Its sequence was submitted to the NCBI database (accession number KF135442).

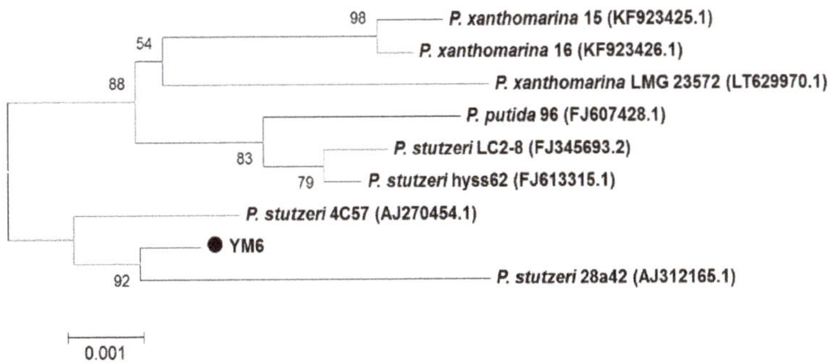

Figure 1. Neighbor-joining phylogenetic tree based on 16S rDNA sequence of strain YM6 and other homologous strains retrieved from the NCBI database. The scale bar indicates the number of substitutions per base position.

2.2. Inhibition of *Pseudomonas stutzeri* YM6 against *A. flavus*

Strain YM6 can greatly inhibit the growth of *A. flavus* in face-to-face dual culture without contact. The growth diameter of *A. flavus* inoculated on PDA medium was 4.5 cm at 5 dpi (days post inoculation). In the treatment group, YM6 inoculated on the NA medium produced abundant volatiles and greatly inhibited the mycelia growth of *A. flavus* in the PDA medium. The mycelia showed no signs of growth with YM6 treatment (Figure 2A). The inhibitory rate of volatiles from YM6 on mycelia growth of *A. flavus* was 100% at 5 dpi.

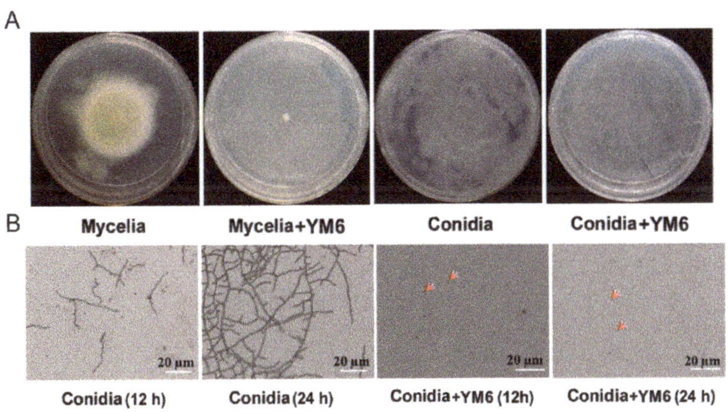

Figure 2. The inhibitory effect of strain YM6 on the growth of *A. flavus* conidia and mycelia in sealed Petri dishes. (**A**). The growth of *A. flavus* mycelia/conidia cultured on the PDA plate was completely inhibited by volatiles produced by YM6 on NA plate in face-to-face dual culture. The two dishes were placed face-to-face without physical contact and cultured at 28 °C for 5 days. (**B**). The germination of *A. flavus* conidia on PDA plate was completely inhibited by volatiles from YM6 under microscopy in 24 h. The red arrows point to the position of *A. flavus* conidia under microscopy.

Similar results were also observed in the inhibitory tests of YM6 on conidia germination of *A. flavus*. In the control group, the *A. flavus* conidia inoculated on PDA medium germinated quickly, producing germ tubes that grow to form hyphae in 12 h (Figure 2B). In the subsequent 12 h, the hyphae branched outward, forming an extensive network that resembles the branches of a tree, and covered the surface of the PDA medium (Figure 2B).

In the treatment group, the volatiles from YM6 inhibited the germination of *A. flavus* conidia with no formation of germ tubes during 24 h (Figure 2B). Thus, the inhibitory rate of volatiles from YM6 on conidia germination of *A. flavus* was 100% in dual culture.

2.3. Biocontrol Activity of YM6 against *A. flavus* and Aflatoxins in Peanut and Maize during Storage

In the control groups, the peanut and maize kernels were severely infected by *A. flavus*. There were a large number of green mycelia of *A. flavus* on the surface of peanut and maize seeds. Additionally, the number of infected kernels at a higher level of water activity (a_w) outnumbered that at a lower level of a_w. At a_w of 0.9, the disease incidence was 100% in the control groups of maize and peanuts. At a_w of 0.8 and 0.7, the disease incidence of the peanuts group was 82%, and of the maize control group was 32% (Figure 3A,B). In comparison, the infection of *A. flavus* was greatly inhibited in the presence of YM6. At higher a_w (0.9), peanut kernels were slightly infected by *A. flavus* with a disease incidence of 32%, while no infection was observed in the maize group. At lower a_w (0.7 and 0.8), no disease symptom was observed in maize and peanut kernels (Figure 3). The results clearly showed that the disease incidence rate in the control group was significantly higher than that in the YM6 treatment group. Based on these observations, we conclude that the volatiles from YM6 can greatly inhibit the infection of *A. flavus* conidia, and significantly prevent the disease development in peanut and maize kernels.

The aflatoxins in peanut and maize samples were also determined through quantitative analysis. In the control group, the total amounts of aflatoxins in peanut samples were 99.49, 330.17, and 1767.61 ppb at a_w of 0.740, 0.859, and 0.923, respectively. Similar phenomenon was also observed in maize samples that the content of aflatoxins was 27.09, 178.39, and 466.13 ppb at a_w of 0.785, 0.866, and 0.934, respectively (Table 1). On the one hand, these results showed that *A. flavus* produced more aflatoxins (with maximum AFB1 of over 80%) in crop seeds at higher a_w in the control group. On the other hand, no aflatoxin was detected in maize samples under three a_w, and peanut samples under a_w 0.740, 0.859 with the treatment of YM6. A little amount of aflatoxin (3.74 ppb) was detected in peanuts at a_w of 0.923 (Table 1). Therefore, the volatiles from YM6 are able to inhibit the production of aflatoxins by *A. flavus* in peanut and maize samples at higher a_w under storage conditions.

Table 1. Quantitative analysis of aflatoxins in inoculated maize and peanut kernels.

	Value of Water Activity	Treatment	Aflatoxin Concentration/ppb		
			AFB1	AFB2	Total
Peanut	0.740	Control	85.41 ± 3.21	14.08 ± 0.76	99.49
		YM6	—	—	—
	0.859	Control	299.36 ± 17.55	30.81 ± 7.83	330.17
		YM6	—	—	—
	0.923	Control	1575.19 ± 67.32	192.42 ± 9.22	1767.61
		YM6	3.74 ± 0.24 *	—	3.74
Maize	0.785	Control	24.92 ± 2.60	2.17 ± 0.13	27.09
		YM6	—	—	—
	0.866	Control	156.62 ± 4.63	21.77 ± 1.56	178.39
		YM6	—	—	—
	0.934	Control	402.31 ± 9.32	63.82 ± 4.35	466.13
		YM6	—	—	—

The concentration of aflatoxin was shown as average ± SE. Total means the sum of AFB1 and AFB2. — means that the aflatoxin was not detected with the minimum detection limit of 0.2 ppb. * means significant difference at $p < 0.05$ compared to control group.

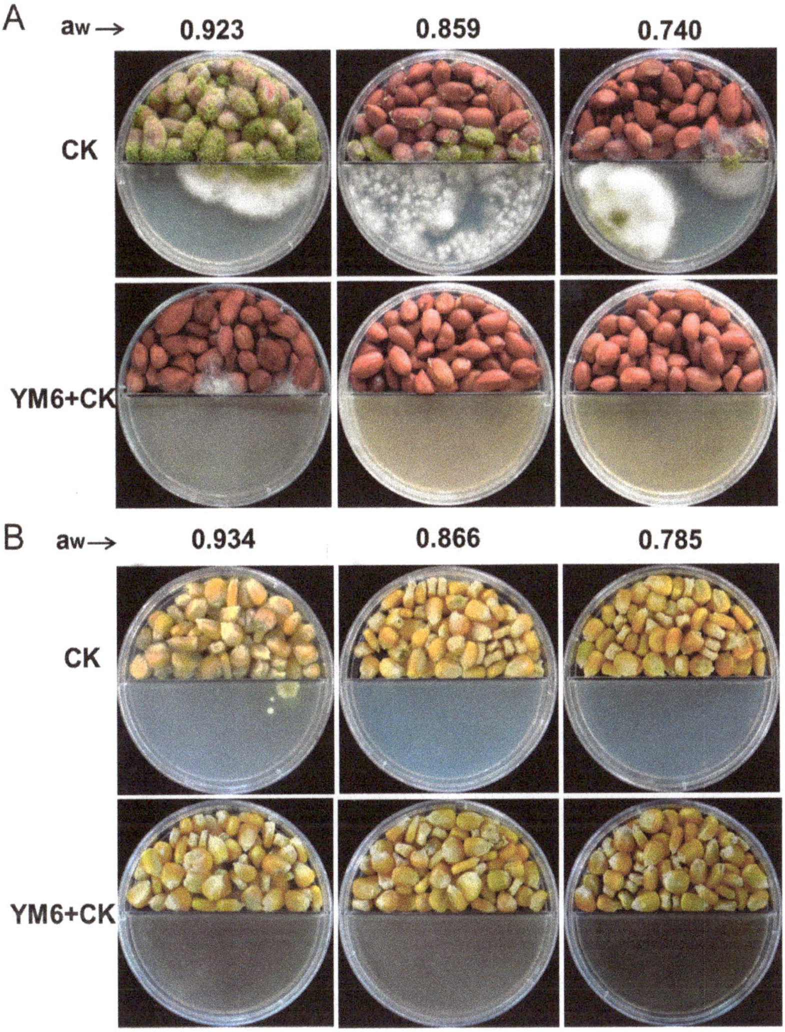

Figure 3. Biocontrol efficiency of strain YM6 against *A. flavus* infection in peanut (**A**) and maize (**B**) kernel. The grains of peanut and maize inoculated with A. flavus condia were placed on one side of separated Petri dishes, challenged with (*A. flavus* + YM6) or without (*A. flavus*) the presence of strain YM6 on the NA plate on the other side of the separated Petri dishes.

2.4. Analysis of A. flavus Cell Structure Affected by YM6

The experiment inoculated peanuts with conidia of *A. flavus* at a_W of 0.923 and cultured them at 28 °C for 5 days. The phenotype of conidia was analyzed through a scanning electron microscope and the results indicated conidial germination and transition from conidia to hyphae in the control group. The peanut seed coat was found to be covered by abundant mycelia. The hyphae also formed conidiophores and produced large amounts of conidia which induced secondary infection. The conidia were uniform in shape with fertile spherical proliferations on the surface. In contrast, only a few severely dehydrated conidia were found on the surface of peanuts in the YM6 group. The conidia were too deformed to germinate into hyphae (Figure 4). The observation suggested that under non-contact conditions, the volatiles from YM6 can effectively prevent *A. flavus* infection and destroy the cell structure of *A. flavus* conidia.

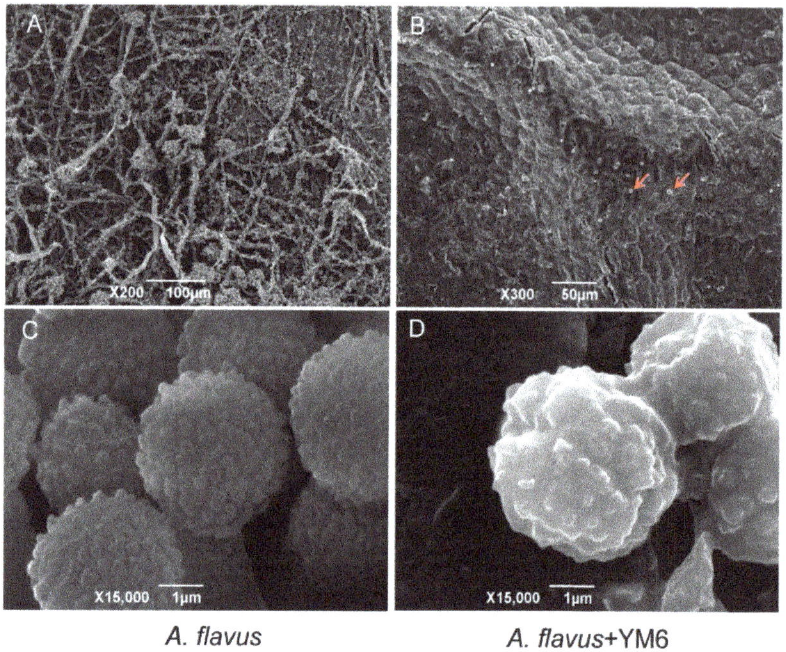

Figure 4. Ultra-structure of *A. flavus* conidia inoculated on peanut seeds affected by the volatiles from YM6. *A. flavus* conidia were inoculated on peanut seed coat and cultured for 5 days with (*A. flavus* + YM6) or without (*A. flavus*) the presence of YM6. In control group, the conidia on peanut coat could germinate to hyphae and phialids-conidiophores (**A**), and produce abundant conidia (**C**). The conidia on peanut surface cannot germinate to hyphae (**B**) and showed deformed character (**D**) in YM6 group. The red arrows pointed to the position of *A. flavus* conidia on the surface of the peanut coat.

2.5. Chemical Identification of Volatiles Produced by the YM6 Strain

GC-MS analysis showed that the YM6 strain could produce abundant volatile substances (Figure 5). These volatiles were dimethyl trisulfide, 1-(trimethylsilyl)-1-propyne oxalic acid, 1-methyl-2-pentyl-cyclohexane, isobutyl pentadactyl ester, undecane, and 3,5-dimethyl-Isoxazole (Table 2). These substances with a molecular weight ranging from 97 to 356 Dalton (D) can easily volatilize. Only one type of substance was characterized based on comparison with the library NIST 08, with a similarity higher than 90% and large relative abundance (over 1%, peak area/sum area of all peaks), which was considered as a candidate volatile from YM6 and the key to the anti-fungal activity. The volatile was further identified to be DMTS based on the comparison of retention time and fragment ions with the commercial standard.

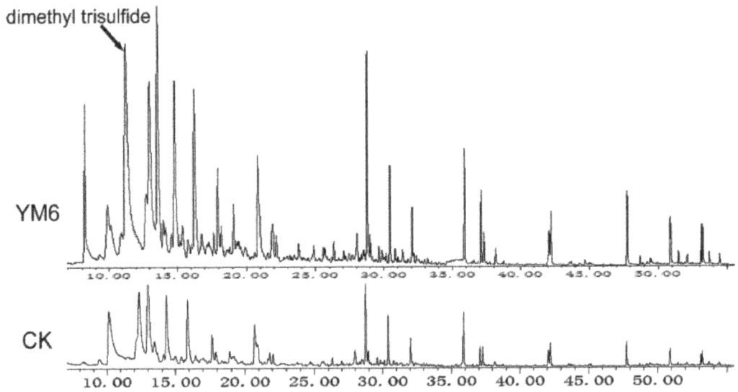

Figure 5. GC-MS analysis of volatiles emitted from strain YM6. Strain YM6 grown on NA plate for 48 h to produce volatile compounds. The compounds detected in YM6 spectra (YM6) omitting the same compounds in blank NA medium (CK) were considered the authentic compound.

Table 2. Identification of volatiles emitted from strain YM6 through GC-MS system.

No.	Compounds	Retention Time/min	Peak Area /% [a]	Similarity /% [b]	Mass /m·z^{-1}
1	Dimethyl trisulfide	11.238	13.877	90	125.963
2	1-(trimethylsilyl)-1-propyne	14.797	7.451	52	112.071
3	Oxalic acid, isobutyl pentadecyl ester	16.777	0.962	72	356.293
4	1-methyl-2-pentyl- Cyclohexane	17.131	0.122	76	168.188
5	Undecane	18.785	0.200	93	156.188
6	3,5-dimethyl- Isoxazole	30.447	0.884	72	97.053
7	2,4,4,6,6,8,8-Heptamethyl-2-nonene	30.824	0.485	60	224.252
8	Butylated hydroxytoluene	37.084	0.992	97	220.183

[a] The peak area of the identified compound to the total area of all peaks. [b] the mass spectrum of identified compound compared to the spectrum of standard substance in NIST 08 database, respectively. Mass means mass to charge ratio.

2.6. Minimal Inhibitory Concentration of DMTS against *A. flavus*

DMTS purchased from Sigma was used to test the inhibitory effect against *A. flavus*. The result revealed that DMTS could significantly inhibit the growth of *A. flavus* in confined spaces. The mycelia of *A. flavus* spread quickly to the edge of the Petri dishes and produced abundant green conidia at 5 dpi in the control group. Additionally, DMTS showed increased antagonistic effects against *A. flavus* with elevated concentrations. At a lower concentration of 50 μL/L (compound volume/airspace volume), the inhibition rate was only 7.5%. At 100 μL/L, the inhibition rate was 20.93%. DMTS completely inhibited the growth of *A. flavus* at 200 μL/L. Thus, the minimal inhibitory concentration (MIC) for DMTS against *A. flavus* was 200 μL/L (Figure 6).

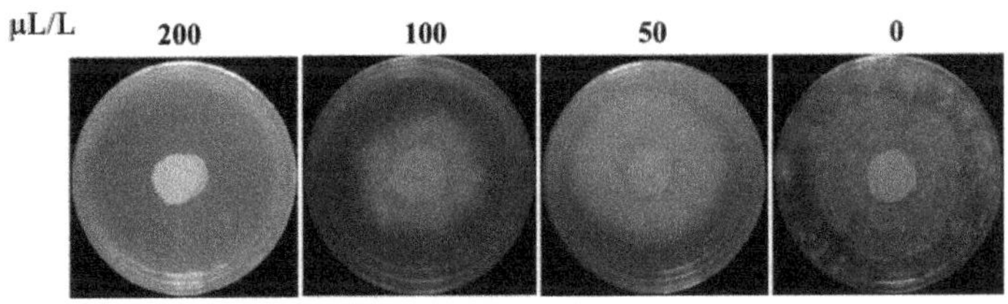

Figure 6. The minimal inhibitory concentration analysis of DMTS against *A. flavus* on PDA plates. The concentration of DMTS was 200, 100, and 50 μL/L (compound volume/airspace volume, *v*/*v*). The *A. flavus* conidia inoculated to the round paper in the center of the PDA plate was used as a control.

3. Discussion

A. flavus is a globally distributed saprophytic fungus that infects many important crops, such as maize, peanuts, and soybeans during storage [20]. Additionally, it produces highly toxic and carcinogenic aflatoxins which threaten food security and human health. Hence, identifying safe and efficient agents to control *A. flavus* and aflatoxins in crops during storage is of significance and worth further study.

Microbes as important organisms are widely distributed in the natural environment. They can produce abundant secondary metabolites which have been *extensively* used in control plant pathogens such as *Fusarium graminearum* [21], *Penicillium digitatum* [22], *Rhizoctonia solani* [23], *Penicillium digitatum* [24], and *A. flavus* [25]. It is difficult to prevent *A. flavus* infection and aflatoxins production in grains during storage as *A. flavus* possesses great spore production, discharge, and dispersal capacities. Many chemical fungicides, although effective in controlling *A. flavus* and aflatoxins in storage, exhibit detrimental effects on food security and human health due to food contamination caused by pesticide residue. Compared to stable pesticides, compounds with smaller molecular sizes evaporate faster and can be evenly distributed in the airspace of storage, which is crucial to the control of *A. flavus* and aflatoxins. In 2015, our lab first demonstrated that volatile organic compounds (VOCs) produced by *Shewanella algae* strain YM8 [26] could greatly inhibit *A. flavus* growth and aflatoxin production in grains during storage. Some other microbes also have been proved with the capacity to prevent *A. flavus* infection and aflatoxin production in grains such as *Streptomyces alboflavus* [27], *Streptomyces yanglinensis* [28], *Bacillus megaterium* [29], *Pseudomonas protegens* [29], *Staphylococcus saprophyticus* [18], and *Serratia marcescens* [19]. However, there was no report of volatiles from *Pseudomonas stutzeri* regarding their effects against *A. flavus* and aflatoxins.

In the current study, YM6, which is a strain of *Pseudomonas stutzeri*, was isolated from marine sediment. It can produce volatiles that can greatly inhibit *A. flavus* growth and aflatoxins production in storage. The volatiles from strain YM6 exhibited a significant antifungal effect as they could inhibit the mycelial growth and spore germination of *A. flavus* by 100%, and severely damage the cell structure of *A. flavus*. The key antimicrobial substance produced by YM6 was identified as DMTS, with a minimal inhibitory concentration against *A. flavus* being 200 μL/L.

We also found that DMTS could deform the conidia of *A. flavus* and prevent the cell germination, but the inhibitory mechanism of DMTS for *A. flavus* remains unknown. However, on the basis of the study by Tang et al. [30], DMTS could cause deterioration of subcellular structures of *Colletotrichum gloeosporioides*, such as cell walls, plasma membranes, Golgi bodies, and mitochondria, as well as contribute to the leakage of protoplasm and cell death. Moreover, Zuo et al. [31] found that, in *Fusarium oxysporum* cells, DMTS could affect the glycosylation and ROS accumulation, inhibit steroid biosynthesis and glycerophospholipids metabolism, disrupt the cell membrane integrity, and finally result in the cell death. Taken together, DMTS could enter into fungal cells, disrupt cell structures, and inhibit cell growth. These results provided important evidence for understanding the inhibitory mechanism of DMTS for *A. flavus*, whereas the molecular target as well as interacted proteins of DMTS have not been investigated in fungal cells, which is worth further investigation.

Many volatiles produced by microbes have been demonstrated effective in controlling *A. flavus* and aflatoxins during storage, such as DMTS, dimethyl disulfide, methyl isovalerate, 1-Pentanol, Phenylethyl Alcohol and 3, 3-Dimethyl-1,2-epoxybutane, et al. [5,18,20,26]. Among these, sulfide exhibits have been demonstrated as excellent biocontrol effects in controlling soil-borne diseases by effectively inhibiting several pathogens and nematodes in soil [32,33]. DMTS, as an active substance, is a colorless or pale yellow substance with a strong mint odor, which exists in fresh onions, Chinese chive [34], and produced by some microbes [26]. However, this paper is the first to report that DMTS was produced by *Pseudomonas stutzeri*. DMTS is commonly used as permissible food additives in seasoning, gravy, and soup in many countries. DMTS can be applied in the storage of grain and oil products for its fast evaporation and dispersal capacities, and even distribution in confinement spaces to control plant pathogens and mycotoxins.

In conclusion, we were the first to prove that strain YM6 of *Pseudomonas stutzeri* showed great inhibition on *A. flavus* and aflatoxins in grains during storage by producing volatile DMTS. As a result, we deduced that the bacterial strain of YM6 and one of its VOCs (DMTS) could be used predominantly as a bio-control agent for crop protection in post-harvest stage. In addition, the microbe-associated volatiles showed great antifungal activity to *A. flavus* in laboratory investigation, but it is still a long distance from lab to commercial application in grains during storage. Some important questions should be answered before commercial application. For example, different to the ventilated requirement in large-scale granaries and small barns nowadays, the application of volatiles developed here needed to be kept in airtight storage condition. Apart from that, some other queries should also be solved before commercial application such as the mass production and enrichment of volatiles, usage cost limitation, usage dosage and persistence, volatile residues, as well as biosecurity to livestock and humans. These limits presented here provide us with further research directions.

4. Materials and Methods

4.1. Microbes and Plants

Bacteria YM6 was isolated from sea sediment in the Yellow Sea of China. YM6 was cultured in nutrient agar (NA) medium at 37 °C in the dark for 24 h. *Aspergillus flavus* strain 535 isolated from diseased peanuts was stored in our lab and inoculated onto potato dextrose agar (PDA) medium and cultured at 28 °C in the dark for anti-fungal tests [26].

Peanuts (cultivar Silihong) were purchased from supermarkets and used for inoculation tests. Dimethyl trisulfide (DMTS, CAS: 03658-80-8) was purchased from Sigma (Sigma-Aldrich, St. Louis, MO, USA) and used as the chemical standard for volatile identification.

The experiments were carried out in Molecular Biotechnology Laboratory of Triticeae Crops, Huazhong Agricultural University, Wuhan, China.

4.2. DNA Extraction and Phylogenic Analysis

The strain YM6 was inoculated onto NB medium and cultured in a flask at 28 °C at 200 rpm in the dark for 24 h. The suspension was centrifuged at 12,000 rpm for 10 min. The collected cells were used for genomic DNA extraction [26]. Then, 16S rDNA was amplified by PCR and amplified fragments were sequenced by Shanghai Sangon Biological Technology Company with the following sequencing primers: 27F (AGAGTTTGATCCTG-GCTCAG) and 1541R (AAGGAGGTGATCCAG CCGC). The PCR was performed under the following conditions: initial denaturation at 94 °C for 5 min; followed by 30 cycles of 94 °C for 30 s, 55 °C for 30 s, and 72 °C for 40 s; and then 72 °C for 10 min. The 16S rDNA sequences of YM6 were submitted and aligned in the GenBank database. The 16S rDNA of bacteria with similarity over 95% to the sequences of YM6 were used for the construction of a phylogenic tree. The phylogenic tree of homologous strains was constructed by MEGA software with the neighbor-joining method.

4.3. Inhibitory Effect of YM6 on Mycelial Growth of *A. flavus*

A single YM6 colony was streaked over the NA medium surface and cultured at 37 °C in the dark for 24 h. The sterile water was added in NA medium to wash off the bacteria cells on the NA surface with an adjusted concentration of 10^8 CFU/mL for a further anti-fungal test.

A. flavus was cultured on PDA medium for 4 days and produced abundant conidia. The conidia were then inoculated in PDB medium and cultured at 28 °C at 200 rpm for 3 days to produce mycelial pellets. Fresh *A. flavus* conidia and mycelial pellets were used in the face-to-face dual cultural test to analyze the antagonistic activity of YM6.

The regular size of *A. flavus* pellets were inoculated in the center of PDA plates, respectively. Fresh YM6 cells (100 μL, 10^8 CFU·mL^{-1}) were spread on the surface of the

NA plate. Then, the PDA plate inoculated with *A. flavus* pellet was placed above the NA plate containing YM6. The two plates were placed face-to-face and sealed with two-layer tapes. The PDA plate inoculated with *A. flavus* pellet and challenged with NA medium was used as a control. Each treatment was repeated three times. All plates were cultured at 28 °C in the dark for 5 days. The mycelium diameters of *A. flavus* and the inhibition rate of YM6 in these treatments were calculated as follows:

Inhibition rate (%) = [(mycelium diameter in control group − mycelium diameter in YM6 group)/mycelium diameter in control group] × 100.

4.4. Inhibitory Effect of YM6 on the Conidia Germination of *A. flavus*

FTF dual cultural test was used to analyze the inhibitory effect of YM6 on conidia germination of *A. flavus*. A total of 100 μL of *A. flavus* spores (5×10^5 CFU/mL) was evenly coated onto cellophane covering the surface of the PDA plate, and the YM6 (100 μL, 10^8 CFU/mL) was coated onto the surface of the NA plate. The two plates were placed FTF, and the PDA plate was inoculated with *A. flavus* conidia on top. The PDA plates inoculated with *A. flavus* conidia challenged with NA medium were used as controls. Each pair of the plates was sealed with two-layer tapes and cultured at 28 °C in the dark for 24 h. The character characteristics of conidia on cellophane were detected on a microscope (Hitech Instruments Co. Ltd., Shanghai, China), and the germination rate of conidia in the control and YM6 group was determined. The inhibition rate was calculated as follows:

Inhibition rate (%) = [(conidia germination rate in control − conidia germination rate in YM6 treatment)/control conidia germination rate] × 100.

4.5. Biocontrol Activity of YM6 against *A. flavus* and Aflatoxins in Peanut and Maize

Maize and peanut kernels (100 g each) were grouped into 3 sets respectively and transferred into six flasks (250 mL) to be sterilized at an environment of 121 °C and 1.01 MPa for 20 min. The kernels were then cooled to room temperature. Fresh *A. flavus* conidia (1 mL, 5×10^5 CFU·mL^{-1}) was added into each flask and mixed for 10 min. Three flasks containing peanut kernels were filled with sterilized water; the water activity was determined at 0.785, 0.866, and 0.934 through an electronic dewpoint water activity meter, Aqualab Series 3 model TE (Decagon Devices, Pullman, Washington, DC, USA), respectively [26]. The values of water activity in flasks containing maize were measured at 0.740, 0.859, and 0.923, respectively. The maize and peanut kernels were used to test the biocontrol activity of YM6. The kernels in each flask were equally grouped into two sets, with one set placed in one compartment of the Petri dish (two in total). The other compartment containing NA medium was spread with YM6 strain (50 μL, 10^9 CFU/mL). The kernels in one compartment challenged with NA medium (in the other compartment) were used as controls. All Petri dishes were sealed and cultured at 28 °C in darkness. The disease incidence of peanut and maize was calculated at 7 dpi.

Aflatoxins in maize and peanuts were extracted with acetonitrile/water and analyzed through ultra-performance liquid chromatography and mass spectrometry (UPLC-MS, Thermo Scientific, New York, NY, USA) [26].

4.6. Structural analysis of *A. flavus* Treated by Volatile from YM6

A. flavus spores were inoculated to the peanut surface with a water activity of 0.934 and challenged with YM6 through FTF dual culture for 5 days. The *A. flavus* inoculated on peanuts surface without the presence of YM6 was used as a control. The peanut seeds were fumigated with osmic acid (0.1%, *v/v*) for 1 h, and then placed at room temperature for 3 h [35]. A small piece of the peanut coat (0.5 cm × 0.5 cm) was removed with a dissecting knife, clamped with stubs, and coated with a layer of gold for examination of SEM (JSM-6390, Hitachi Corporation, Tokyo, Japan) to observe the ultra-structure of *A. flavus*.

4.7. Identification of Volatiles from the Strain YM6

The strain YM6 was inoculated onto the surface of the NA medium in a 100 mL flask. The flask containing NA medium without YM6 inoculation was used as a control. All flasks were sealed with double-layer plastic film and cultured at 28 °C in the dark for 48 h. All the experiments were repeated twice. Then, the flasks were incubated in a water bath of 40 °C for 30 min. The volatile substance was extracted by solid phase micro-extraction (SPME, divinylbenzene/carboxy/polydimethylsiloxane) and analyzed by gas chromatography-mass spectrometry (GC-MS) (Agilent Technologies, Santa Clara, CA, USA). For volatile enrichment, the metal head of SPME was inserted through the plastic film into the flask. The coated fiber in the metal head was pushed out and placed in the center position above the NA medium to absorb the volatiles for 50 min. The coated fiber was retracted and transferred into the GC-MS system for volatile analysis. The detection parameter was used as below [26].

The splitless injection of GC was used in the GC-MS analysis. The carrier gas was helium; the inlet temperature was 250 °C. The oven was set for procedures as follows: started at 40 °C for 3 min; heated up to 160 °C at the rate of 3 °C/min for 2 min; and then raised to 220 °C at the rate of 8 °C/min for 3 min. The MS analysis was performed with an EI source under a temperature of 230 °C. The quadrupole temperature was set at 150 °C. The collision energy was 70 eV. The mass range was set from 50 to 500 amu. The detected volatiles were verified by aligning in the National Institution of Standards and Technology (NIST 08) database, and the retention time was compared with authentic compound purchased from the company.

4.8. Inhibition Effect of DMTS on *A. flavus*

The volatile substance produced by YM6 showed a significant antagonistic effect on *A. flavus* growth. DMTS (liquid state, purity $\geq$ 98%), the major component emitted by YM6, was purchased from Sigma and tested as the anti-fungal agent against *A. flavus*. The anti-fungal analysis was conducted through the FTF dual cultural test. A piece of circular filter paper (5 cm diameter), moistened by DTMS, was placed in a Petri dish. The final concentration of DMTS in each treatment was adjusted to 200 μL/L, 100 μL/L, and 50 μL/L, respectively. The other Petri dish containing PDA medium was inoculated with *A. flavus* mycelia. The dish containing *A. flavus* was placed FTF above the dish containing DMTS and sealed with two layers of tape. The dish of *A. flavus* co-cultured with the same volume of water was used as a control. All plates were cultured at 28 °C in darkness for 5 days. The mycelium diameter of *A. flavus* was measured and the inhibition rate was calculated.

4.9. Data Analysis

All experiments were conducted in duplicate. The amount of aflatoxin was evaluated and shown as mean ± SE. The significant differences between mean values were determined using Duncan's multiple range test ($p < 0.05$) following one-way analysis of variance (ANOVA). The statistical analysis was performed using SPSS 16.0 software (SPSS Inc., Chicago, IL, USA).

Author Contributions: Conceptualization, A.-D.G. and Y.-C.L.; Methodology, Y.-Y.L.; Formal Analysis, A.-D.G., W.-J.H. and L.M.; Investigation, A.-D.G. and T.-T.Z.; Writing—Original Draft Preparation, A.-D.G. and J.-B.Z.; Writing—Review & Editing, A.-D.G. and J.-B.Z. All authors contributed to editorial changes in the manuscript. All authors have read and agreed to the published version of the manuscript.

Funding: This work was supported by Key Program of Action Plan to Revitalize Inner Mongolia through Science and Technology [2022EEDSKJXM011, 2022EEDSKJXM011-3]; the National Natural Science Foundation of Henan [222300420519, 222301420111]; Scientific and Technological Project in Henan Province [212102110447]; and Nanhu Scholars Program for Young Scholars of XYNU.

Institutional Review Board Statement: Not applicable.

Informed Consent Statement: Not applicable.

Data Availability Statement: Not applicable.

Conflicts of Interest: The authors declare no conflict of interest.

References

1. Alshannaq, A.F.; Gibbons, J.G.; Lee, M.K.; Han, K.H.; Hong, S.B.; Yu, J.H. Controlling aflatoxin contamination and propagation of *Aspergillus flavus* by a soy-fermenting *Aspergillus oryzae* strain. *Sci. Rep.* **2018**, *8*, 16871. [CrossRef] [PubMed]
2. Waliyar, F.; Umeh, V.; Traore, A.; Osiru, M.; Ntare, B.; Diarra, B.; Kodio, O.; Kumar, K.V.K.; Sudini, H. Prevalence and distribution of aflatoxin contamination in groundnut (*Arachis hypogaea* L.) in Mali, West Africa. *Crop. Prot.* **2015**, *70*, 1–7. [CrossRef]
3. Ma, F.; Cai, X.F.; Mao, J.; Yu, L.; Li, P.W. Adsorptive removal of aflatoxin B1 from vegetable oils via novel adsorbents derived from a metal-organic framework. *J. Hazard. Mater.* **2021**, *412*, 125170. [CrossRef] [PubMed]
4. Asters, M.C.; Williams, W.P.; Perkins, A.D.; Mylroie, J.E.; Windham, G.L.; Shan, X. Relating significance and relations of differentially expressed genes in response to *Aspergillus flavus* infection in maize. *Sci. Rep.* **2014**, *4*, 4815. [CrossRef]
5. Gong, A.D.; Wu, N.N.; Kong, X.W.; Zhang, Y.M.; Hu, M.J.; Gong, S.J.; Dong, F.Y.; Wang, J.H.; Zhao, Z.Y.; Liao, Y.C. Inhibitory effect of volatiles emitted from *Alcaligenes faecalis* N1-4 on *Aspergillus flavus* and aflatoxins in storage. *Front. Microbiol.* **2019**, *10*, 1419. [CrossRef]
6. Pandey, M.K.; Kumar, R.; Pandey, A.K.; Soni, P.; Gangurde, S.S.; Sudini, H.K.; Fountain, J.C.; Liao, B.; Desmae, H.; Okori, P.; et al. Mitigating aflatoxin contamination in groundnut through a combination of genetic resistance and post-harvest management practices. *Toxins* **2019**, *11*, 315. [CrossRef]
7. Mitchell, N.J.; Bowers, E.; Hurburgh, C.; Wu, F. Potential economic losses to the USA corn industry from aflatoxin contamination. *Food Addit. Contam. Part A* **2016**, *33*, 540–550. [CrossRef]
8. Marchese, S.; Polo, A.; Ariano, A.; Velotto, S.; Costantini, S.; Severino, L. Aflatoxin B1, and M1: Biological properties and their involvement in cancer development. *Toxins* **2018**, *10*, 214. [CrossRef]
9. Chen, J.G.; Egner, P.A.; Ng, D.; Jacobson, L.P.; Muñoz, A.; Zhu, Y.R.; Qian, G.S.; Wu, F.; Yuan, J.M.; Groopman, J.D. Reduced aflatoxin exposure presages decline in liver cancer mortality in an endemic region of China. *Cancer Prev. Res.* **2013**, *6*, 1038–1045. [CrossRef]
10. East African Community (EAC). *East African Standards: Maize Grains–Specification*; EAC: Dar es Salaam, Tanzania, 2011; Available online: https://law.resource.org/pub/eac/ibr/eas.2.2011.html (accessed on 2011).
11. European Commission (EC). *Guidance Document for Competent Authorities for the Control of Compliance with EU Legislation on Aflatoxins*; Regulation No. 1152/2009; EC: Brussels, Belgium, 2010; Available online: https://food.ec.europa.eu/system/files/2021-05/cs_contaminants_catalogue_alfatoxins_guidance-2010_en.pdf (accessed on 27 November 2010).
12. Hua, H.J.; Xing, F.G.; Selvaraj, J.N.; Wang, Y.; Zhao, Y.J.; Zhou, L.; Liu, X.; Liu, Y. Inhibitory effect of essential oils on *Aspergillus ochraceus* growth and ochratoxin A production. *PLoS ONE* **2014**, *9*, e108285. [CrossRef]
13. Tran-Dinh, N.; Pitt, J.I.; Markwell, P.J. Selection of non-toxigenic strains of *Aspergillus flavus* for biocontrol of aflatoxins in maize in Thailand. *Biocontrol. Sci. Technol.* **2014**, *24*, 652–661. [CrossRef]
14. Kong, Q.; Shan, S.H.; Liu, Q.Z.; Wang, X.D.; Yu, F.T. Biocontrol of *Aspergillus flavus* on peanut kernels by use of a strain of marine *Bacillus megaterium*. *Int. J. Food Microbiol.* **2010**, *139*, 31–35. [CrossRef] [PubMed]
15. Chen, Y.J.; Kong, Q.; Liang, Y. Three newly identified peptides from *Bacillus megaterium* strongly inhibit the growth and aflatoxin B1 production of *Aspergillus flavus*. *Food Control* **2019**, *95*, 41–49. [CrossRef]
16. Farzaneh, M.; Shi, Z.Q.; Ahmadzadeh, M.; Hu, L.B.; Ghassempour, A. Inhibition of the *Aspergillus flavus* growth and aflatoxin B1 contamination on pistachio nut by fengycin and surfactin-producing *Bacillus subtilis* UTBSP1. *Plant Pathol. J.* **2016**, *32*, 209. [CrossRef]
17. Passone, M.A.; Etcheverry, M. Antifungal impact of volatile fractions of *Peumus boldus* and *Lippia turbinata* on *Aspergillus* section Flavi and residual levels of these oils in irradiated peanut. *Int. J. Food Microbiol.* **2014**, *168*, 17–23. [CrossRef]
18. Gong, A.D.; Sun, G.J.; Zhao, Z.Y.; Liao, Y.C.; Zhang, J.B. *Staphylococcus saprophyticus* L-38 produces volatile 3, 3-dimethyl-1, 2-epoxybutane with strong inhibitory activity against *Aspergillus flavus* germination and aflatoxin production. *World Mycotoxin J.* **2020**, *13*, 247–258. [CrossRef]
19. Gong, A.D.; Wang, G.Z.; Sun, Y.K.; Song, M.G.; Dimuna, C.; Gao, Z.; Wang, H.L.; Yang, P. Dual activity of *Serratia marcescens* Pt-3 in phosphate-solubilizing and production of anti-fungal volatiles. *BMC Microbiol.* **2022**, *22*, 1–11. [CrossRef]
20. Gong, A.D.; Dong, F.Y.; Hu, M.J.; Kong, X.W.; Wei, F.F.; Gong, S.J.; Zhang, Y.M.; Zhang, J.B.; Wu, A.B.; Liao, Y.C. Antifungal activity of volatile emitted from *Enterobacter asburiae* Vt-7 against *Aspergillus flavus* and aflatoxins in peanuts during storage. *Food Control* **2019**, *106*, 106718. [CrossRef]
21. Gong, A.D.; Li, H.P.; Yuan, Q.S.; Song, X.S.; Yao, W.; He, W.J.; Zhang, J.B.; Liao, Y.C. Antagonistic mechanism of iturin A and plipastatin A from *Bacillus amyloliquefaciens* S76-3 from wheat spikes against *Fusarium graminearum*. *PLoS ONE* **2015**, *10*, e0116871. [CrossRef]
22. Ahima, J.; Zhang, X.Y.; Yang, Q.Y.; Zhao, L.N.; Tibiru, A.M.; Zhang, H.Y. Biocontrol activity of *Rhodotorula mucilaginosa* combined with salicylic acid against *Penicillium digitatum* infection in oranges. *Biol. Control* **2019**, *135*, 23–32. [CrossRef]

23. Yang, J.H.; Zhang, W.W.; Zhuang, Y.Q.; Xiao, T. Biocontrol activities of bacteria from cowdung against the rice sheath blight pathogen. *J. Plant Dis. Prot.* **2017**, *124*, 131–141. [CrossRef]
24. Noreen, R.; Ali, S.A.; Hasan, K.A.; Sultana, V.; Ara, J.; Ehteshamul-Haque, S. Evaluation of biocontrol potential of epiphytic yeast against postharvest *Penicillium digitatum* rot of stored Kinnow fruit (*Citrus reticulata*) and their effect on its physiochemical properties. *Postharvest Biol. Technol.* **2019**, *148*, 38–48.
25. Afsharmanesh, H.; Ahmadzadeh, M.; Javan-Nikkhah, M.; Behboudi, K. Improvement in biocontrol activity of *Bacillus subtilis* UTB1 against *Aspergillus flavus* using gamma-irradiation. *Crop. Prot.* **2014**, *60*, 83–92. [CrossRef]
26. Gong, A.D.; Li, H.P.; Shen, L.; Zhang, J.B.; Wu, A.B.; He, W.J.; Yuan, Q.S.; He, J.D.; Liao, Y.C. The *Shewanella algae* strain YM8 produces volatiles with strong inhibition activity against *Aspergillus* pathogens and aflatoxins. *Front. Microbiol.* **2015**, *6*, 1091. [CrossRef] [PubMed]
27. Yang, M.G.; Lu, L.F.; Pang, J.; Hu, Y.L.; Guo, Q.B.; Li, Z.J.; Wu, S.F.; Liu, H.H.; Wang, C.L. Biocontrol activity of volatile organic compounds from *Streptomyces alboflavus* TD-1 against *Aspergillus flavus* growth and aflatoxin production. *J. Microbiol.* **2019**, *57*, 396–404. [CrossRef] [PubMed]
28. Lyu, A.; Yang, L.; Wu, M.D.; Zhang, J.; Li, G.Q. High efficacy of the volatile organic compounds of *Streptomyces yanglinensis* 3-10 in suppression of *Aspergillus* contamination on peanut kernels. *Front. Microbiol.* **2020**, *11*, 142. [CrossRef]
29. Mannaa, M.; Oh, J.Y.; Kim, K.D. Biocontrol activity of volatile-producing *Bacillus megaterium* and *Pseudomonas protegens* against *Aspergillus flavus* and aflatoxin production on stored rice grains. *Mycobiology* **2017**, *45*, 213–219. [CrossRef]
30. Tang, L.H.; Mo, J.Y.; Guo, T.X.; Huang, S.P.; Li, Q.L.; Ning, P.; Hsiang, T. In vitro antifungal activity of dimethyl trisulfide against *Colletotrichum gloeosporioides* from mango. *World J. Microbiol. Biotechnol.* **2020**, *36*, 4. [CrossRef]
31. Zuo, C.W.; Zhang, W.N.; Chen, Z.J.; Chen, B.H.; Huang, Y.H. RNA sequencing reveals that endoplasmic reticulum stress and disruption of membrane integrity underlie dimethyl trisulfide toxicity against *Fusarium oxysporum* f. sp. *cubense tropical race 4*. *Front. Microbiol.* **2017**, *8*, 1365. [CrossRef]
32. Myrta, A.; Sasanelli, N.; Dongiovanni, C.; Santori, A. Control of the root-knot nematode *Meloidogyne incognita* by dimethyl disulfide (DMDS) applied in drip irrigation on melon and tomato in Apulia and Basilicata (Italy). *Acta Hortic.* **2014**, *1044*, 401–404.
33. Papazlatani, C.; Rousidou, C.; Katsoula, A.; Kolyvas, M.; Genitsaris, S.; Papadopoulou, K.; Karpouzas, D.G. Assessment of the impact of the fumigant dimethyl disulfide on the dynamics of major fungal plant pathogens in greenhouse soils. *Eur. J. Plant Pathol.* **2016**, *146*, 391–400. [CrossRef]
34. Zhang, H.; Mallik, A.; Zeng, R.S. Control of Panama disease of banana by rotating and intercropping with Chinese chive (*Allium tuberosum* Rottler): Role of plant volatiles. *J. Chem. Ecol.* **2013**, *39*, 243–252. [CrossRef] [PubMed]
35. Zhang, Q.H.; Yang, L.; Zhang, J.; Wu, M.D.; Chen, W.D.; Jiang, D.H.; Li, G.Q. Production of anti-fungal volatiles by non-pathogenic *Fusarium oxysporum* and its efficacy in suppression of Verticillium wilt of cotton. *Plant Soil.* **2015**, *392*, 101–114. [CrossRef]

toxins

MDPI

Article

Origanum vulgare Essential Oil Modulates the AFB1-Induced Oxidative Damages, Nephropathy, and Altered Inflammatory Responses in Growing Rabbits

Mona A. Hassan [1], Azza M. A. Abo-Elmaaty [2], Asmaa W. Zaglool [3], Sally A. M. Mohamed [4], Shimaa M. Abou-Zeid [5], Mayada R. Farag [1,*], Mahmoud Alagawany [6,*], Alessandro Di Cerbo [7], Mahmoud M. Azzam [8], Rashed Alhotan [8] and Enas EL-Hady [9]

1 Department of Forensic Medicine and Toxicology, Faculty of Veterinary Medicine, Zagazig University, Zagazig 44519, Egypt
2 Department of Pharmacology, Faculty of Veterinary Medicine, Zagazig University, Zagazig 44519, Egypt
3 Animal Wealth Development Department, Faculty of Veterinary Medicine, Zagazig University, Zagazig 44519, Egypt
4 Department of Histology and Cytology, Faculty of Veterinary Medicine, Zagazig University, Zagazig 44519, Egypt
5 Department of Forensic Medicine and Toxicology, Faculty of Veterinary Medicine, University of Sadat City, Sadat City 6012201, Egypt
6 Poultry Department, Faculty of Agriculture, Zagazig University, Zagazig 44519, Egypt
7 School of Biosciences and Veterinary Medicine, University of Camerino, 62024 Matelica, Italy
8 Department of Animal Production, College of Food & Agriculture Sciences, King Saud University, Riyadh 11451, Saudi Arabia
9 Department of Anatomy and Embryology, Faculty of Veterinary Medicine, Zagazig University, Zagazig 44519, Egypt
* Correspondence: mragabfa@zu.edu.eg or dr.mayadarf@gmail.com (M.R.F.); mmalagwany@zu.edu.eg (M.A.)

Abstract: The current study was performed to investigate the toxic effects of aflatoxin B1 (AFB1) through the evaluation of kidney function tests and histopathological examination of renal tissues, targeting the therapeutic role of Marjoram (*Origanum vulgare* essential oil-OEO) in improving health status. Forty-eight New Zealand Whites growing rabbits (four weeks old) weighing on average 660.5 ± 2.33 g were randomly and equally distributed into four groups, each of which had four replicas of three animals as the following: Control group (only basal diet), AFB1 group (0.3 mg AFB1/kg diet), OEO group (1 g OEO/kg diet) and co-exposed group (1 g OEO/kg + 0.3 mg AF/kg diet). Our study lasted eight weeks and was completed at 12 weeks of age. The results revealed that OEO decreased the toxic effects of AFB1 in rabbit kidneys by substantially reducing the cystatin C levels in the AFB1 group. Additionally, OEO decreased oxidative stress and lipid peroxidation levels in the co-exposed group. Moreover, OEO reduced DNA damage and inflammatory response in addition to the down-regulation of stress and inflammatory cytokines-encoding genes. Besides, OEO preserved the cytoarchitecture of rabbits' kidneys treated with AFB1. In conclusion, *O. vulgare* essential oil supplementation ameliorated the deleterious effects of AFB1 on the rabbits' kidneys by raising antioxidant levels, decreasing inflammation, and reversing oxidative DNA damage.

Keywords: Marjoram essential oil; AFB1; rabbits; kidney; oxidative damage

Key Contribution: AFB1 can cause DNA damage and downregulation of stress and inflammatory-related genes. However; OEO can significantly reduce oxidative stress and inflammation.

1. Introduction

Mycotoxins are fungi's corrosive and cytotoxic metabolites and are considered widespread pollutants in the feed materials [1,2]. According to statistics, one-fourth of the feed components produced globally contain mycotoxins [3]. Mycotoxin contamination of forage and feed components is a serious issue that requires more attention. Mycotoxins are produced mainly by *Aspergillus flavus*, *Aspergillus parasiticus*, and *Aspergillus fumigatus* and Aflatoxins are thought to be the most harmful ones [4–6].

In developing counties like Egypt, there is a significant potential for the production of rabbits. The rabbit requires little area and little capital investment, and it grows quickly and has a high capacity for reproduction. Since rabbits have a functional caecum, they can naturally use the plentiful fibrous agricultural byproducts and herbaceous plants as food [7]. The presence of aflatoxins in feeds is a hazard to rabbit performance.

Despite the infrequent outbreaks of acute aflatoxins toxicity, animals are the most susceptible population, exposed to high accumulations of aflatoxins through the intake of contaminated feedstuffs [8].

Due to its broad toxicity, aflatoxin B1 (AFB1) has attained great attention. AFB1 was graded as a Group A carcinogen by the WHO in 1993, and its consumption in rabbits is related to a high hazard for liver and kidney cancer [9]. Additionally, AFB1 can potentially cause growth impairment, malnutrition, and immunosuppression, particularly in young animals [10].

Essential oils (EOs) are aromatic compounds derived from plant sources and are known for their high volatility and complicated structure. They are heavily used in pharmaceutical, agricultural, nutritional, and cosmetic industries because of their well-known antibacterial, antifungal, and antioxidant properties [11–13]. Concerning antifungal properties, various types of research have shown that essential oils can prevent the spread of a variety of fungi in foodstuffs, including *Aspergillus* [14]. Due to its antifungal characteristics, oregano essential oil is one of the most researched essential oils [15,16].

Due to its flavor-adding capacity, *Origanum vulgare* L. (*Lamiaceae*) is one of the most popular condiments used in food and pet food preparation [16–19]. Additionally, it is widely used in the production of essential oils due to its antimicrobial characteristics, which include the presence of carvacrol and thymol, which are effective against different types of fungi [20]. Moreover, Origanum can be used as a detoxifying substance against aflatoxins, and this point should be investigated [21]. The current study's objective is to ascertain if *O. vulgare* essential oil (OEO) may have any protective benefits against aflatoxins by investigating the harmful effects of aflatoxins on renal function, oxidative stress, inflammatory responses, and the histological structure of rabbits' kidneys.

2. Results

2.1. Effects on Kidney Function Markers

Urea and creatinine levels in the serum of the AFB1-exposed group were considerably higher than those in the control group, as indicated in Table 1. Urea and creatinine levels did not significantly differ from the control after using OEO supplements. Additionally, the urea levels were still higher than the control values in the AFB1 + OEO group, while the creatinine levels were stable.

Table 1. Effect of supplementation of OEO (1 g/kg diet) on kidney function markers of rabbits exposed to AFB1 (0.3 mg/kg diet) for eight weeks.

Items	Kidney Function Markers					
	Urea (mg/dL)	**Creatinine (mg/dL)**	**Total Bilirubin (mg/dL)**	**Direct Bilirubin (mg/dL)**	**B2 Microglobulin (ng/mL)**	**Cystatin C (ng/mL)**
Control	31.47 ± 0.93^{c}	1.20 ± 0.01^{b}	1.01 ± 0.00^{c}	0.61 ± 0.01^{b}	0.14 ± 0.00^{a}	0.11 ± 0.00^{b}
OEO	31.18 ± 0.63^{c}	1.20 ± 0.01^{b}	1.01 ± 0.01^{c}	0.54 ± 0.03^{b}	0.14 ± 0.00^{a}	0.10 ± 0.00^{b}
AFB1	64.26 ± 1.63^{a}	1.84 ± 0.02^{a}	1.82 ± 0.03^{a}	0.87 ± 0.1^{a}	0.14 ± 0.00^{a}	0.39 ± 0.01^{a}
AFB1 + OEO	43.59 ± 1.85^{b}	1.20 ± 0.01^{b}	1.34 ± 0.06^{b}	0.64 ± 0.09^{b}	0.14 ± 0.00^{a}	0.12 ± 0.00^{b}
p-value	<0.001	<0.001	<0.001	<0.001	0.859	<0.001

$^{a–c}$ Different superscripts within each column are significantly different ($p < 0.05$).

The total and direct bilirubin concentrations in the AFB1-supplemented group were significantly greater than in control. In the OEO group, total and direct bilirubin levels did not alter significantly. In the AFB1 + OEO group, total bilirubin was higher than the control, but direct bilirubin remained steady.

Regarding glomerular filtration rate (GFR), Beta-2 microglobulin levels were not significantly altered in any treatment group. The findings of cystatin C indicated that aflatoxins increase their level compared to the control. At the same time, it remains constant in the OEO and AFB1 + OEO groups compared to the control (Table 1).

2.2. Effects on Oxidative Stress Biomarkers in Kidney

Table 2 lists the values of several antioxidant enzymes and oxidative stress indicators in the kidney of rabbits. Superoxide-dismutase (SOD) activity levels did not significantly alter in any treatment group relative to the control group.

Table 2. Effect of supplementation of OEO (1 g/kg diet) on antioxidants and oxidative stress biomarkers in the kidney of rabbits exposed to AFB1 (0.3 mg/kg diet) for eight weeks.

Items	Antioxidant and Oxidative Stress Parameters					
	SOD (μg/g tissue)	CAT (μg/g tissue)	GSH (μg/g tissue)	MDA (nmol/g tissue)	PC (nmol/g tissue)	8-OHdG (ng/mL)
Control	0.20 ± 0.01	0.54 ± 1.90 ^{b}	0.21 ± 0.00 ^{a}	0.17 ± 0.00 ^{b}	5.58 ± 0.08 ^{b}	0.23 ± 0.01 ^{b}
OEO	0.20 ± 0.01	0.80 ± 1.01 ^{a}	0.20 ± 0.00 ^{a}	0.11 ± 0.00 ^{c}	5.35 ± 0.07 ^{b}	0.19 ± 0.01 ^{b}
AFB1	0.20 ± 0.03	0.30 ± 0.01 ^{c}	0.15 ± 0.00 ^{b}	0.46 ± 0.00 ^{a}	8.92 ± 0.15 ^{a}	0.35 ± 0.00 ^{a}
AFB1 + OEO	0.20 ± 0.01	0.28 ± 0.01 ^{c}	0.07 ± 0.01 ^{c}	0.17 ± 0.01 ^{b}	5.57 ± 0.04 ^{b}	0.24 ± 0.01 ^{b}
p-value	0.992	<0.001	<0.001	<0.001	<0.001	<0.001

$^{a–c}$ Different superscripts within each column are significantly different ($p < 0.05$). SOD: superoxide dismutase, CAT: catalase, GSH: reduced glutathione, MDA: malondialdehyde, PC: protein carbonyl, 8-OHdG: 8-hydroxy-2-deoxyguanosine.

Catalase activity was considerably higher in the OEO group but lower in the AFB1 and AFB1 + OEO-exposed groups compared to the control. GSH levels in the AFB1 and AFB1 + OEO groups significantly decreased, remaining constant in the OEO groups compared to the control.

Concerning oxidative stress biomarkers, lipid peroxidation, as demonstrated by the formation of MDA, was significantly elevated in the AFB1 group relative to the control. At the same time, there were no significant changes in MDA levels among the OEO and AFB1 + OEO groups.

Compared to the control, the AF-treated group showed a substantially (*** $p < 0.001$) higher level of protein carbonyl (PC) production, a marker of oxidative protein damage. However, there were no significant differences between the control, OEO, or AFB1 + OEO groups.

Regarding DNA oxidation biomarkers, our findings showed no observable differences in the serum level of 8-OHdG between control, OEO, or AFB1 + OEO groups. The 8-OHdG content was significantly higher in the AFB1 group than in the control and experimental groups.

2.3. Effects on Inflammatory Biomarkers

Table 3 illustrates the effect of AFB1 and OEO supplementation on inflammatory biomarkers. TNF-α levels increased dramatically in response to AFB1 administration alone compared to the control group. Co-supplementation of OEO with AF significantly decreased TNF-α levels to the control value. Nitric oxide (NO) levels did not significantly alter in none of the treatment groups compared to the control. The levels of TGF- β1 and VEGF were significantly (*** $p < 0.001$) increased in the AFB1 group. In contrast, co-treatment of OEO with AFB1 significantly decreased the TGF-β1 and VEGF levels more than AFB1 alone; however, it did not reach the control values.

Table 3. Effect of supplementation of OEO (1 g/kg diet) on inflammatory biomarkers in rabbits exposed to AFB1 (0.3 mg/kg diet) for eight weeks.

Items.	Inflammatory Biomarkers			
	TNFα (pg/mL)	NO (nmol/μL)	VEGF (pg/mL)	TGF-β1 (pg/mL)
Control	174.29 ± 0.69 [b]	87.11 ± 0.62 [a]	4.42 ± 0.02 [c]	12.29 ± 0.13 [c]
OEO	169.50 ± 0.89 [b]	87.19 ± 0.53 [a]	4.03 ± 0.09 [c]	9.64 ± 0.22 [d]
AFB1	237.87 ± 2.46 [a]	87.51 ± 0.17 [a]	15.85 ± 0.27 [a]	27.82 ± 1.09 [a]
AFB1 + OEO	173.19 ± 0.29 [b]	87.54 ± 0.48 [a]	5.58 ± 0.36 [b]	19.84 ± 0.83 [b]
p-value	<0.001	0.891	<0.001	<0.001

[a–d] Different superscripts within each column are significantly different ($p < 0.05$). TNFα: tumor necrosis factor α, NO: nitric oxide, VEGF: vascular endothelial growth factor, TGF-β1: transforming growth factor beta 1.

2.4. Effects on Stress and Inflammatory Cytokines-Encoding Genes

As depicted in Figure 1, our findings demonstrated that most investigated genes had their expression levels significantly rise due to exposure to AFB1-contaminated feed.

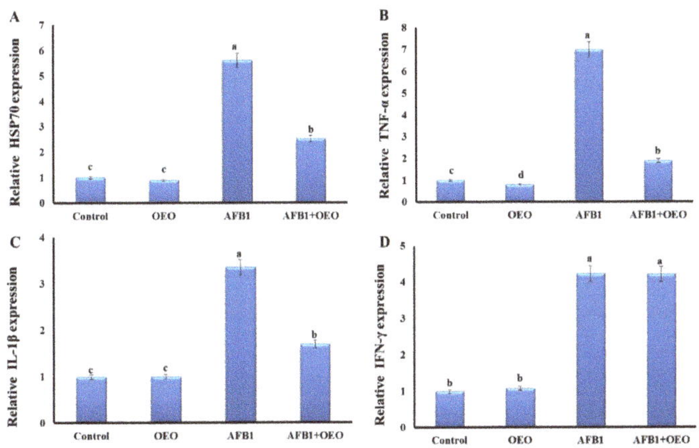

Figure 1. Effect of supplementation of OEO on the expression pattern of stress-related genes (**A**) and pro-inflammatory cytokines (**B**): TNFα; (**C**): IL-1β, and (**D**): IFN-γ) in the kidney of rabbits exposed to AFB1. Values are mean ± SE; bars that are not sharing a common superscript letter (a,b,c,d) differ significantly at $p < 0.05$.

Inflammatory markers expression significantly increased as a result of AFB1 exposure. There is an upregulation in the TNF-α gene expression in the AFB1 group, while there is downregulation in the OEO group. In the AFB1 + OEO group, a significant increase in TNF-α gene expression was observed.

Expression levels of IL-1β and HSP70 genes were significantly increased in the AFB1 group compared to the control. OEO alone did not affect these markers but caused a decrease in the expression of IL-1β and HSP70 genes compared to the AFB1 group.

Regarding IFN-γ gene expression, there is a significant increase in the AFB1 and AFB1 + OEO groups, while there is no change in the OEO group compared to the control.

2.5. Bioaccumulation of AFB1 Residues in Kidney Tissue

Analysis of AFB1 residues in the kidney tissue of control and treated rabbits are represented in Figure 2.

The highest concentration of aflatoxin residues was detected in the kidney of AFB1-exposed rabbits. OEO caused a significant decrease in the residual accumulation of AFB1 in the co-exposed group. However, the OEO group showed the lowest level of AFB1 residues among all the groups.

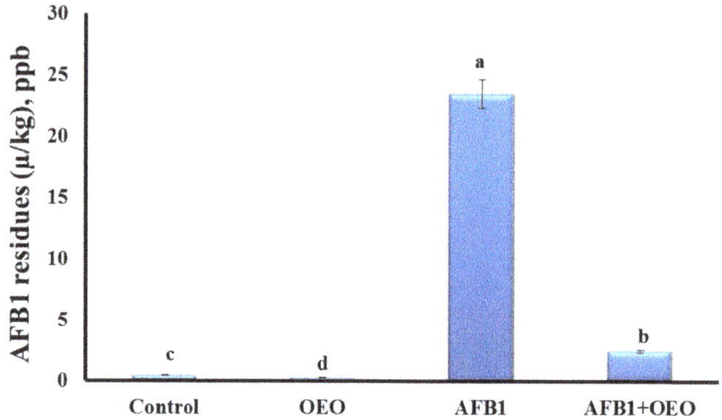

Figure 2. Effect of supplementation of OEO on the bioaccumulation of AFB1 residues in the kidney of rabbits. Values are mean ± SE; bars that are not sharing a common superscript letter (a,b,c,d) differ significantly at $p < 0.05$.

2.6. Histopathological Findings

The kidney from the control and OEO rabbits showed normal histological structure. The rabbit's kidney in the AFB1 group showed the disappearance of Bowman's spaces in glomeruli, degeneration of renal tubules epithelium renal blood vessels fibrosis with vacuolated media. The kidneys in the combination group showed shrunken glomeruli (Figure 3).

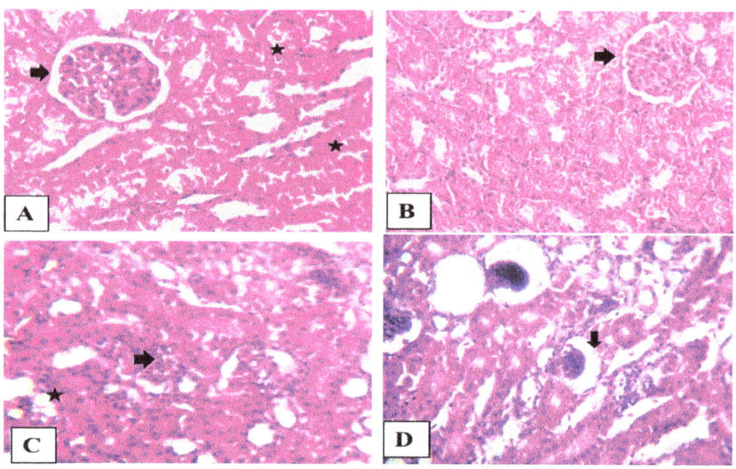

Figure 3. (**A**) Photomicrograph of the rabbit's kidney (Control) and (**B**) (OEO) showing normal glomerular tufts and cellularity (arrow) besides normal proximal renal tubules (stars). (**C**) Photomicrograph of the rabbit's kidney (Aflatoxin) showing disorder of glomeruli (arrows) characterized by the disappearance of Bowman's spaces beside degenerated renal tubule epithelium (star). (**D**) Photomicrograph of the rabbit's kidney (Aflatoxin + OEO) showing shrunken glomeruli (arrow). Scale bar is 100 μm. H&E stain.

3. Discussion

As a result of the ongoing growth in the incidence of aflatoxicosis in developing nations, it is currently a global public health issue [5,22].

The laboratory findings usually examine urea and creatinine levels to determine the integrity and functionality of the nephrons. We observed that the AFB1 group had higher urea and creatinine levels than the control group, demonstrating the adverse effects of aflatoxin on the kidneys. These observations concur with prior studies [23]. However, co-supplementation with OEO restored kidney functional biomarker activities and levels, indicating that OEO acted as a safeguard against aflatoxicosis in the rabbits' kidneys. The abovementioned results are consistent with those of Zowail et al. [24].

We assessed the glomerular filtration rate, a valuable marker for evaluating renal functioning. It is crucial to determine the levels of several freely filtered low molecular weight proteins, which are more pertinent than assessing the concentration of creatinine alone, to determine the impairments in GFR. These proteins include beta-2 microglobulin and cystatin C [25]. In our study, The AF-treated rabbits had considerably higher blood levels of urea, creatinine, and cystatin C than the control. This suggested the presence of kidney and glomerular filtration dysfunction. This result was in accordance with Orsi et al. [26] and Abdou et al. [27].

Several studies have demonstrated that oxidative stress plays a critical pathogenic role in the tissue damage caused by AFB1 that results in inflammatory damage [28]. AFB1 generated free radicals that can cause oxidative damage to cells after being bio-transformed into the electrophilic, intermediate metabolite AFB1 8,9-epoxide [29]. The results of the current investigation show that exposure to AFB1 causes oxidative stress to be induced in the kidney of rabbits. This was demonstrated by the dramatically decreased activities of the antioxidant molecules CAT and GSH and the significantly elevated levels of MDA and PC activities. In prior findings, it was revealed that the administration of AFB1 to rats and mouse models lowered the activity of antioxidant enzymes [30]. Catalase and GSH levels were markedly elevated when OEO was concurrently administered to rabbits. Additionally, MDA and PC levels in the serum have significantly decreased. These findings imply that OEO is increasing the levels and activity of antioxidant molecules. Additionally, recent research has demonstrated that OEO is beneficial in delaying lipid oxidation [31,32] and acts as a good antioxidant [33].

Following exposure to mycotoxins, oxidative stress and inflammation are closely related [34]. AFB1 exposure has been associated with the release of pro-inflammatory cytokines (including TNF-α, IL-1, and IL-6) and inflammation-related tissue damage [35]. In our study, there is upregulation in the expression of TNF-α in the kidney tissue of rabbits in addition to increased expression levels of IL-1β, INF, and HSP70 genes. The first and most significant inflammatory mediator is TNF-α, which controls the metabolic reactions of another tissue and encourages the production of different cytokines during the inflammation process [36].

The alterations in the inflammatory cytokine expression levels in the rabbits' kidneys showed that the AFB1 exposure caused inflammatory damage. OEO supplementation dramatically reduced TNF-α, TGF, and TGF-β1, suggesting that OEO has anti-inflammatory potential. For instance, Arranz et al. showed that OEO decreased the production of IL-1, TNF-α, and IL-6 in LPS-activated THP-1 macrophage cells of humans [37]. Furthermore, Han and Parker demonstrated that, on activated-primary human newborn fibroblasts, OEO dramatically reduced the levels of different inflammatory biomarkers such as MCP-1, ICAM-1, and VCAM-1 [38]. These observations indicated that the OEO might have anti-inflammatory potential.

The activation of programmed cell death by oxidative DNA damage is another crucial mechanism of AFB1-induced kidney injury [39–41]. DNA damage inside a tissue or an organ is linked to the production of 8-OHdG, which may cause the activation of p53 and apoptosis.

In the current study, DNA oxidation biomarker 8-OHdG was significantly elevated in the AFB1 group compared to the control, and co-supplementation with OEO has an anti-genotoxic effect. Previous studies in mice exposed to cyclophosphamide [42] and people exposed to radioiodine [43] found that using *Origanum vulgare* extract in the diet reduces genotoxicity.

The results of the histological analysis of the kidney tissue in the AFB1 group had deteriorated renal tubule epithelium as well as glomeruli abnormalities characterized by the loss of Bowman's spaces, demonstrating fibrosis of the major renal blood arteries, provided additional validation for the development of inflammation and apoptosis by AFB1.

4. Conclusions

The present study's findings established that AFB1 negatively affected kidney tissue structure and function and induced its accumulation in renal tissues, while the use of natural feed additives such as OEO ameliorated physiological functions and health aspects by reducing oxidative stress, increasing the levels of antioxidants, resolving inflammation, and reversing DNA damage.

Moreover, OEO could reduce the AFB1 residues suggesting that besides its powerful antioxidant activity, it could also act as a chelator and this dual beneficial action makes OEO a powerful candidate for protecting rabbits from environmental contaminants.

5. Materials and Methods

5.1. Chemicals

O. vulgare essential oil (OEO) was obtained from Elhawag Company for Natural Oils, Nasr City, and Cairo, Egypt (LOT#: 150/2/159). The bioactive components of OEO were identified by gas chromatography-mass spectroscopy (GC–MS) analysis [44] and are represented in Table 4.

Table 4. Retention time and peak area (%) of the different compounds found in *O. vulgare* analyzed by GC-MS.

No.	Bioactive Chemical Constituents	RT (min)	Peak Area, %
1	19-Octanol	9.61	0.27
2	ci10s-trans Sabinene hydrate	10.33	3.47
3	Cycl11ohexanol	10.33	3.47
4	Terpinen-4-ol	12.03	1.24
5	Estragole	12.36	2.61
6	Thymol methyl ether	12.37	2.3
7	Carvacrol methyl ether	12.41	5.6
8	2,4-Decadienal	14.32	3.70
9	2-Pentene	15.72	1.77
10	Methyleugenol	16.45	0.06
11	Epiglobulol	18.48	0.48
12	Junipene	18.48	0.48
13	Naphthalene	18.62	0.37
14	Nerolidyl acetate	18.76	0.12
15	l-(+)-Ascorbic acid	29.11	16.11
16	5-Amino-2-methyl-2-phenyl-2	29.66	6.20
17	cis-13-Octadecenoic acid	32.63	32.63
18	Oleic Acid	32.63	48.57
19	Oleic acid, eicosyl ester	32.85	1.15

AFB1 preparation and selection of the used concentration were based on the previous observations [6]. To obtain AFB1, *Aspergillus flavus* MD 341 was retrieved from the Central Lab. of Residues of Agricultural Products, Dokki, Egypt and incubated for 8 days in liquid media with 2% of yeast extract and 20% of sucrose. A reversed-phase column was utilized for the extraction, filtration, and quantitative HPLC analysis of aflatoxins. 45% methanol served as the mobile phase and was inoculated into the apparatus at a flow rate of 1 mL per minute. The analyses were performed using a fluorescence detector, and the column temperature was set to 40 °C. Aflatoxin standard was bought from Sigma-Aldrich (St. Louis, MO, USA). The media was found to contain only AFB1.

5.2. Animals and Experimental Design

Forty-eight male New Zealand Whites growing rabbits (4 weeks old) weighing on average 660.5 ± 2.33 g were obtained from the laboratory animal farm, Faculty of Veterinary Medicine, University, Zagazig, Egypt. Rabbits were randomly and equally distributed into four groups, each of which had four replicas of 3 animals. The duration of this study was eight weeks, so it was completed at 12 weeks of age. The investigated groups were as the following: control group (only basal diet), AFB1 group (0.3 mg AFB1/kg diet), OEO group (1 g OEO/kg diet), and Combination group (1 g OEO/kg + 0.3 mg AF/kg diet). Rabbits were housed in proper cages (50 cm × 40 cm × 30 cm). Every rabbit was raised under the same management and sanitary settings, and the nutritional requirements were fulfilled by the feed they were given (Table 5).

Table 5. Ingredients and composition of commercial growing rabbit diet (% of the as-fed diet).

Items	Basal Diet
Ingredient	
Soybean meal	15
Wheat bran	25
Berseem hay	30
Barely grain	28
Limestone	1
NaCl	0.5
Premix *	0.5
Total	100
Calculated composition, %	
Digestible energy, MJ/kg	10.85
Crude protein	17.29
Calcium	0.87
Phosphorus	0.54

* Each one kg of premix (minerals and vitamins mixture) contains vit. A, 20,000 IU; vit. D3, 15,000 IU; vit. E, 8.33 g; vit. K, 0.33 g; vit. B1, 0.33 g; vit. B2, 1.0 g; vit. B6, 0.33 g; vit. B5, 8.33 g; vit. B12, 1.7 mg; pantothenic acid, 3.33 g; biotin, 33 mg; folic acid, 0.83 g; choline chloride, 200 g.

The animal study was reviewed and approved by the institutional Ethics Committee of Zagazig University, Zagazig, Egypt (ZU-IACUC/2/F/387/2022).

5.3. Sampling and Analysis

After the last treatment, rabbits (6 per group) were euthanized, and blood samples were taken in sterile tubes, allowed to be clotted, and after that, centrifuged at 4000 rpm for 10 min. The collected serum was kept at −20 °C until they were tested. We collect serum to detect kidney function markers. Other blood samples were collected on heparinized tubes for plasma collection for measuring inflammatory markers. Specimens from the kidney tissue were excised, washed with normal saline (0.9% NaCl), and preserved in formalin for histological analysis. For further biochemical analysis, another kidney sample set was frozen at −20 °C. After being instantly frozen in liquid nitrogen, the third group of kidney specimens was kept at −80 °C until the RNA extraction.

5.4. Antioxidant Biomarkers

Kidney tissues were homogenized (10% *w*/*v*) in potassium phosphate buffer solution (pH 7.4) for antioxidant analyses, followed by 15 min centrifugation at 3000 rpm. Following the manufacturer's guidelines, the obtained supernatant was used to test the activity of superoxide dismutase (SOD), catalase (CAT), and the level of reduced glutathione (GSH) using commercial bio diagnostic kits from BioMérieux, Marcy-l'Etoile, France.

5.5. Oxidative Stress Biomarkers

Malondialdehyde (MDA), a biomarker of lipid peroxidation, and Protein Carbonyls (PC), a marker of oxidative protein damage, were assessed in renal tissues, while 8-hydroxy-

2-deoxyguanosine (8-OHdG), a biomarker of DNA oxidative damage was measured in serum, using ELISA kits from MyBiosource.com, San Diego, CA, USA (Cat No., MBS040484, MBS1601647, and MBS726394, respectively) following the manufacturer's instructions.

5.6. Glomerular Filtration and Kidney Function Markers

Beta-2 microglobulin (B2M) and cystatin C were determined by ELISA kits (Rabbit ELISA kits; MBS2602011 and MBS285072, respectively) from MyBioSource, San Diego, California. While kits for measuring urea, creatinine, and total and direct bilirubin were purchased from (Bio Med Diagnostic, Giza, Egypt).

5.7. Inflammatory Biomarkers

Using ELISA kits from MyBioSource, San Diego, California, Nitric oxide (NO), transforming growth factor-1 (TGF-1), and vascular endothelial growth factor (VEGF) were measured in plasma while tumor necrosis factor (TNF-α) was assessed in serum (Cat No. MBS8243214, MBS704933, MBS751261, and MBS2500169, respectively).

5.8. Vicam AflaTest Fluorometer Technique

Using the mycotoxin calibration standards, AflaTest was calibrated. The purified water (2 mL) and the plain reagent (1 mL methanol and 1 mL developer) should both read 0 ppb on the calibrated fluorometer to assure accuracy. Each sample (25 g) was blended at high speed for 1 min with 125 mL of methanol: water (60:40) and 5 g of sodium chloride. The fluted filter paper was used to filter the mixture, and the extract (20 mL) was diluted with purified water (20 mL). A 1.5 mL glass microfiber filter was used to filter the diluted extract into a clean container. Filtered extract (10 mL, represents 1 g sample) at a flow rate of nearly two drops/s was passed through the affinity column, followed by water (10 mL, 2 drops/s). The AFB1 was eluted with HPLC methanol (1 mL, 1 drop/s) and then collected in a glass cuvette (VICAM part # 34000). The eluate and developer (1 mL) were well combined before being added to the calibrated fluorometer.

5.9. Gene's Transcriptional Analysis (qRT-PCR Study)

Total RNA was extracted from frozen kidney samples using the TRIzol reagent (easyREDTM, iNtRON Biotechnology, Gyeonggi-do, South Korea). The first-strand cDNA was produced from the isolated RNA using the Quantitect® Reverse Transcription kit from Qiagen, Hilden, Germany. Following the instructions provided by the kit's manufacturer, RNA extraction and cDNA synthesis were carried out. The forward and reverse sequences of the primers for the investigated genes, including the housekeeping gene β-actin, the heat shock protein-70 (HSP-70), interleukin-1β (IL-1β), interferon-(IFN-γ), and tumor necrosis factor (TNF-α), are listed in Table 6.

Table 6. Primers sequences (forward and reverse) used for Real-Time qPCR studies.

Accession No.		Genes Sequence
NM_001101683.1	F: 5′-CTGGAACGGTGAAGGTGACA-3′ R: 5′-CGGCCACATTGCAGAACTTT-3′	β-actin
XM_002719559.3	F: 5′-CGTGGAGTCCTACACCTACAAC-3′ R: 5′-ACTCGTCTTTCTCGGCCATC-3′	HSP70
NM_001082263	F: 5′-CTGCACTTCAGGGTGATCG-3′ R: 5′-CTACGTGGGCTAGAGGCTTG-3′	TNF-α
NM_001082201	F: 5′-TTGAAGAAGAACCCGTCCTCTG-3′ R: 5′-CTCATACGTGCCAGACAACACC-3′	IL-1β
NM_001081991	F: 5′-TGCCAGGACACACTAACCAGAG-3′ R: 5′-TGTCACTCTCCTCTTTCCAATTCC-3′	IFN-γ

The qPCR analysis was carried out using the Rotor-Gene Q instrument and a QuantiTect® SYBR® Green PCR kit (Qiagen, Hilden, Germany) under the following thermocycler conditions: 10 min at 95 °C, then 40 cycles at 95 °C for 15 s, 60 °C for 30 s, and 72 °C for 30 s. To confirm the specificity of the PCR, melt-curve analysis was carried out. The comparative $2^{-\Delta\Delta Ct}$ methodology recommended by Livak and Schmittgen was used to determine the relative mRNA expression pattern for each gene [45].

5.10. Histopathological Examination

Specimens of the kidney were obtained and fixed in buffered formalin 20% for two days. The samples were embedded in paraffin following gradual dehydration in 70% to 100% ethanol. Subsequently cut into sections (5 μm), stained by hematoxylin and eosin (H&E), and the histopathological analysis was conducted with the aid of the Olympus BX51 light microscope, Tokyo, Japan [46].

5.11. Statistical Analysis

The general linear models' approach, modified by SPSS for the user's handbook using one-way ANOVA, was used to evaluate the data statistically. The post-hoc Newman-Keuls test (* $p < 0.05$) was used to assess whether or not there were differences between the treatments.

Author Contributions: Conceptualization, M.A.H., M.A. and E.E.-H.; methodology, A.M.A.A.-E. and A.W.Z.; software, S.A.M.M.; validation, A.D.C., E.E.-H., M.A. and M.R.F.; formal analysis, S.M.A.-Z. and R.A.; investigation, M.A. and A.M.A.A.-E.; resources, E.E.-H. and M.A.; data curation, S.A.M.M., M.M.A. and A.W.Z.; writing—original draft preparation, M.A.H., M.A., A.D.C., E.E.-H., S.A.M.M. and M.R.F.; writing—review and editing, M.A.H., M.A., A.D.C., E.E.-H. and M.R.F.; visualization, E.E.-H., M.R.F. and M.A.; project administration, M.A., E.E.-H. and M.A.H.; funding acquisition, M.A. and E.E.-H. All authors have read and agreed to the published version of the manuscript.

Funding: Researchers Supporting Project (RSPD2023R581), King Saud University (Riyadh, Saudi Arabia).

Institutional Review Board Statement: The animal study was reviewed and approved by the institutional Ethics Committee of Zagazig University, Egypt (ZU-IACUC/2/F/387/2022). 27 November 2022 to 27 November 2024.

Informed Consent Statement: Not applicable.

Data Availability Statement: The data presented in this study are available on request from the corresponding authors.

Acknowledgments: Researchers Supporting Project (RSPD2023R581), King Saud University (Riyadh, Saudi Arabia).

Conflicts of Interest: The authors declare no conflict of interest.

References

1. Jedziniak, P.; Panasiuk, Ł.; Pietruszka, K.; Posyniak, A. Multiple mycotoxins analysis in animal feed with LC-MS/MS: Comparison of extract dilution and immunoaffinity clean-up. *J. Sep. Sci.* **2019**, *42*, 1240–1247. [CrossRef]
2. Tahir, M.A.; Abbas, A.; Muneeb, M.; Bilal, R.M.; Hussain, K.; Abdel-Moneim, A.-M.E.; Farag, M.R.; Dhama, K.; Elnesr, S.S.; Alagawany, M. Ochratoxicosis in poultry: Occurrence, environmental factors, pathological alterations and amelioration strategies. *World's Poult. Sci. J.* **2022**, *78*, 727–749. [CrossRef]
3. Hassan, A.A.; Mogda, K.M.; Ibrahim, E.M.; Naglaa, M.A.; Flourage, A.M.A.; Rady, M.; Darwish, A.S. Aflatoxicosis in Rabbits with particular reference to their control by N. Acetyl Cysteine and Probiotic. *Int. J. Curr. Res.* **2016**, *8*, 25547–25560.
4. Barati, M.; Chamani, M.; Mousavi, S.N.; Hoseini, S.A.; Taj Abadi Ebrahimi, M. Effects of biological and mineral compounds in aflatoxin-contaminated diets on blood parameters and immune response of broiler chickens. *J. Appl. Anim. Res.* **2018**, *46*, 707–713. [CrossRef]
5. Reda, F.M.; Ismail, I.E.; El-Mekkawy, M.M.; Farag, M.R.; Mahmoud, H.K.; Alagawany, M. Dietary supplementation of potassium sorbate, hydrated sodium calcium almuniosilicate and methionine enhances growth, antioxidant status and immunity in growing rabbits exposed to aflatoxin B1 in the diet. *J. Anim. Physiol. Anim. Nutr.* **2020**, *104*, 196–203. [CrossRef] [PubMed]
6. Ismail, I.E.; Farag, M.R.; Alagawany, M.; Mahmoud, H.K.; Reda, F.M. Efficacy of some feed additives to attenuate the hepato-renal damage induced by aflatoxin B1 in rabbits. *J. Anim. Physiol. Anim. Nutr.* **2020**, *104*, 1343–1350. [CrossRef] [PubMed]
7. Dalle Zotte, A. Rabbit farming for meat purposes. *Anim. Front.* **2014**, *4*, 62–67. [CrossRef]

8. Farombi, O.E. Aflatoxin contamination of foods in developing countries: Implications for hepatocellular carcinoma and chemopreventive strategies. *Afr. J. Biotechnol.* **2006**, *5*, 1–14.
9. El-Nahla, S.M.; Imam, H.M.; Moussa, E.A.; Ibrahim, A.M.; Ghanam, A.R. Teratogenic effects of aflatoxin in Rabbits (*Oryctolagus cuniculus*). *J. Vet. Anat.* **2013**, *6*, 67–85. [CrossRef]
10. Mohsenzadeh, M.S.; Hedayati, N.; Riahi-Zanjani, B.; Karimi, G. Immunosuppression following dietary aflatoxin B1 exposure: A review of the existing evidence. *Toxin Rev.* **2016**, *35*, 121–127. [CrossRef]
11. Bakkali, F.; Averbeck, S.; Averbeck, D.; Idaomar, M. Biological effects of essential oils—A review. *Food Chem. Toxicol.* **2008**, *46*, 446–475. [CrossRef]
12. Wang, L.; Liu, B.; Jin, J.; Ma, L.; Dai, X.; Pan, L.; Liu, Y.; Zhao, Y.; Xing, F. The Complex Essential Oils Highly Control the Toxigenic Fungal Microbiome and Major Mycotoxins During Storage of Maize. *Front. Microbiol.* **2019**, *10*, 1643. [CrossRef] [PubMed]
13. Iseppi, R.; Di Cerbo, A.; Aloisi, P.; Manelli, M.; Pellesi, V.; Provenzano, C.; Camellini, S.; Messi, P.; Sabia, C. In Vitro Activity of Essential Oils Against Planktonic and Biofilm Cells of Extended-Spectrum beta-Lactamase (ESBL)/Carbapenamase-Producing Gram-Negative Bacteria Involved in Human Nosocomial Infections. *Antibiotics* **2020**, *9*, 272. [CrossRef] [PubMed]
14. Carson, C.F.; Hammer, K.A. Chemistry and Bioactivity of Essential Oils. In *Lipids and Essential Oils as Antimicrobial Agents*; John Wiley & Sons: Hoboken, NJ, USA, 2011; pp. 203–238. [CrossRef]
15. Carmo, E.S.; de Oliveira Lima, E.; de Souza, E.L. The potential of *Origanum vulgare* L. (Lamiaceae) essential oil in inhibiting the growth of some food-related Aspergillus species. *Braz. J. Microbiol.* **2008**, *39*, 362–367. [CrossRef]
16. Alagawany, M.; Farag, M.R.; Salah, A.S.; Mahmoud, M.A. The role of oregano herb and its derivatives as immunomodulators in fish. *Rev. Aquac.* **2020**, *12*, 2481–2492. [CrossRef]
17. Di Cerbo, A.; Morales-Medina, J.C.; Palmieri, B.; Pezzuto, F.; Cocco, R.; Flores, G.; Iannitti, T. Functional foods in pet nutrition: Focus on dogs and cats. *Res. Vet. Sci.* **2017**, *112*, 161–166. [CrossRef]
18. Di Cerbo, A.; Palmieri, B.; Chiavolelli, F.; Guidetti, G.; Canello, S. Functional Foods in Pets and Humans. *Int. J. Appl. Res. Veterianry Med.* **2014**, *12*, 192–199.
19. Guidetti, G.; Di Cerbo, A.; Giovazzino, A.; Rubino, V.; Palatucci, A.T.; Centenaro, S.; Fraccaroli, E.; Cortese, L.; Bonomo, M.G.; Ruggiero, G.; et al. In Vitro Effects of Some Botanicals with Anti-Inflammatory and Antitoxic Activity. *J. Immunol. Res.* **2016**, *2016*, 5457010. [CrossRef]
20. Araújo, D.L.; Machado, B.A.d.S.; Mascarenhas, J.M.F.; Alves, S.P.; Sousa, S.L.F.d.; Moura, L.C.d.; Melo, B.C.; Luz, T.E.M.; Loures, L.L.d.S.; Apolinário, J.M.d.S.d.S.; et al. Analysis of the antimicrobial activity of the essential oil of oregano (*Origanum vulgare*): A review study on the main effects on pathogens. *Res. Soc. Dev.* **2021**, *10*, e36810212584. [CrossRef]
21. Mirza Alizadeh, A.; Golzan, S.A.; Mahdavi, A.; Dakhili, S.; Torki, Z.; Hosseini, H. Recent advances on the efficacy of essential oils on mycotoxin secretion and their mode of action. *Crit. Rev. Food Sci. Nutr.* **2022**, *62*, 4726–4751. [CrossRef]
22. Tahir, N.I.; Hussain, S.; Javed, M.; Rehman, H.; Shahzady, T.G.; Parveen, B.; Ali, K.G. Nature of aflatoxins: Their extraction, analysis, and control. *J. Food Saf.* **2018**, *38*, e12561. [CrossRef]
23. Yilmaz, S.; Kaya, E.; Karaca, A.; Karatas, O. Aflatoxin B(1) induced renal and cardiac damage in rats: Protective effect of lycopene. *Res. Vet. Sci.* **2018**, *119*, 268–275. [CrossRef]
24. Zowail, M.; El-Balshy, R.; Asran, A.; Khidr, F.; Amer, N. Protective effect of Origanum majorana against the toxicity of bromadialone on adult male albino rat. *Egypt. J. Exp. Biol. (Zool.)* **2019**, *15*, 263. [CrossRef]
25. Filler, G.; Bokenkamp, A.; Hofmann, W.; Le Bricon, T.; Martinez-Bru, C.; Grubb, A. Cystatin C as a marker of GFR–history, indications, and future research. *Clin. Biochem.* **2005**, *38*, 1–8. [CrossRef]
26. Orsi, R.B.; Oliveira, C.A.; Dilkin, P.; Xavier, J.G.; Direito, G.M.; Correa, B. Effects of oral administration of aflatoxin B1 and fumonisin B1 in rabbits (*Oryctolagus cuniculus*). *Chem. Biol. Interact.* **2007**, *170*, 201–208. [CrossRef] [PubMed]
27. Abdou, I.M.; Abdel-Hamied, M.N.; Mohamady, S.N.; Abdullah, S.H.; Gehan, N.A.G. The Positive effects of vitamin E and selenium with activated charcoal against aflatoxin induced oxidative stress and cytotoxic damage in rabbits. *Egypt. J. Chem. Environ. Health* **2015**, *1*, 798–820. [CrossRef]
28. Zhao, W.; Wang, L.; Liu, M.; Jiang, K.; Wang, M.; Yang, G.; Qi, C.; Wang, B. Transcriptome, antioxidant enzyme activity and histopathology analysis of hepatopancreas from the white shrimp Litopenaeus vannamei fed with aflatoxin B1(AFB1). *Dev. Comp. Immunol.* **2017**, *74*, 69–81. [CrossRef]
29. Shen, H.M.; Ong, C.N.; Shi, C.Y. Involvement of reactive oxygen species in aflatoxin B1-induced cell injury in cultured rat hepatocytes. *Toxicology* **1995**, *99*, 115–123. [CrossRef]
30. Akinrinde, A.S.; Adebiyi, O.E.; Asekun, A. Amelioration of Aflatoxin B1-induced gastrointestinal injuries by Eucalyptus oil in rats. *J. Complement. Integr. Med.* **2019**, *17*, 20190002. [CrossRef]
31. Rodriguez-Garcia, I.; Silva-Espinoza, B.A.; Ortega-Ramirez, L.A.; Leyva, J.M.; Siddiqui, M.W.; Cruz-Valenzuela, M.R.; Gonzalez-Aguilar, G.A.; Ayala-Zavala, J.F. Oregano Essential Oil as an Antimicrobial and Antioxidant Additive in Food Products. *Crit. Rev. Food Sci. Nutr.* **2016**, *56*, 1717–1727. [CrossRef]
32. Alagawany, M.; Abd El-Hack, M.E.; Farag, M.R.; Shaheen, H.M.; Abdel-Latif, M.A.; Noreldin, A.E.; Patra, A.K. The usefulness of oregano and its derivatives in poultry nutrition. *World's Poult. Sci. J.* **2018**, *74*, 463–474. [CrossRef]
33. Zhang, L.Y.; Peng, Q.Y.; Liu, Y.R.; Ma, Q.G.; Zhang, J.Y.; Guo, Y.P.; Xue, Z.; Zhao, L.H. Effects of oregano essential oil as an antibiotic growth promoter alternative on growth performance, antioxidant status, and intestinal health of broilers. *Poult. Sci.* **2021**, *100*, 101163. [CrossRef]

34. Marin, D.E.; Bulgaru, C.V.; Anghel, C.A.; Pistol, G.C.; Dore, M.I.; Palade, M.L.; Taranu, I. Grape Seed Waste Counteracts Aflatoxin B1 Toxicity in Piglet Mesenteric Lymph Nodes. *Toxins* **2020**, *12*, 800. [CrossRef] [PubMed]
35. Dey, D.K.; Chang, S.N.; Kang, S.C. The inflammation response and risk associated with aflatoxin B1 contamination was minimized by insect peptide CopA3 treatment and act towards the beneficial health outcomes. *Environ. Pollut.* **2021**, *268*, 115713. [CrossRef]
36. Cruceriu, D.; Baldasici, O.; Balacescu, O.; Berindan-Neagoe, I. The dual role of tumor necrosis factor-alpha (TNF-alpha) in breast cancer: Molecular insights and therapeutic approaches. *Cell. Oncol.* **2020**, *43*, 1–18. [CrossRef] [PubMed]
37. Arranz, E.; Jaime, L.; López de las Hazas, M.C.; Reglero, G.; Santoyo, S. Supercritical fluid extraction as an alternative process to obtain essential oils with anti-inflammatory properties from marjoram and sweet basil. *Ind. Crops Prod.* **2015**, *67*, 121–129. [CrossRef]
38. Han, X.; Parker, T.L. Anti-inflammatory, tissue remodeling, immunomodulatory, and anticancer activities of oregano (*Origanum vulgare*) essential oil in a human skin disease model. *Biochim. Open* **2017**, *4*, 73–77. [CrossRef] [PubMed]
39. Li, H.; Li, S.; Yang, H.; Wang, Y.; Wang, J.; Zheng, N. l-Proline Alleviates Kidney Injury Caused by AFB1 and AFM1 through Regulating Excessive Apoptosis of Kidney Cells. *Toxins* **2019**, *11*, 226. [CrossRef]
40. Gallo, A.; Landi, R.; Rubino, V.; Di Cerbo, A.; Giovazzino, A.; Palatucci, A.T.; Centenaro, S.; Guidetti, G.; Canello, S.; Cortese, L.; et al. Oxytetracycline induces DNA damage and epigenetic changes: A possible risk for human and animal health? *PeerJ* **2017**, *5*, e3236. [CrossRef]
41. Di Cerbo, A.; Rubino, V.; Morelli, F.; Ruggiero, G.; Landi, R.; Guidetti, G.; Canello, S.; Terrazzano, G.; Alessandrini, A. Mechanical phenotyping of K562 cells by the Micropipette Aspiration Technique allows identifying mechanical changes induced by drugs. *Sci. Rep.* **2018**, *8*, 1219. [CrossRef]
42. Habibi, E.; Shokrzadeh, M.; Ahmadi, A.; Chabra, A.; Naghshvar, F.; Keshavarz-Maleki, R. Genoprotective effects of *Origanum vulgare* ethanolic extract against cyclophosphamide-induced genotoxicity in mouse bone marrow cells. *Pharm. Biol.* **2015**, *53*, 92–97. [CrossRef] [PubMed]
43. Arami, S.; Ahmadi, A.; Haeri, S.A. The radioprotective effects of *Origanum vulgare* extract against genotoxicity induced by (131)I in human blood lymphocyte. *Cancer Biother. Radiopharm.* **2013**, *28*, 201–206. [CrossRef] [PubMed]
44. Smigielski, K.; Dolot, M.; Raj, A. Composition of the Essential Oils of Ginseng Roots of Panax quinquefolium L. and Panax ginseng C.A. Meyer. *J. Essent. Oil Bearing Plants* **2006**, *9*, 261–266. [CrossRef]
45. Livak, K.J.; Schmittgen, T.D. Analysis of relative gene expression data using real-time quantitative PCR and the 2(-Delta Delta C(T)) Method. *Methods* **2001**, *25*, 402–408. [CrossRef]
46. Suvarna, K.S.; Layton, C.; Bancroft, J.D. *Bancroft's Theory and Practice of Histological Techniques E-Book*; Elsevier Health Sciences: Amsterdam, The Netherlands, 2018.

toxins

MDPI

Article

Fe_3O_4@COF(TAPT–DHTA) Nanocomposites as Magnetic Solid-Phase Extraction Adsorbents for Simultaneous Determination of 9 Mycotoxins in Fruits by UHPLC–MS/MS

Jie Wang [1,2,†], Qingwen Huang [2,†], Wenbo Guo [2], Dakai Guo [2], Zheng Han [2] and Dongxia Nie [1,2,*]

1 School of Health Science and Engineering, University of Shanghai for Science and Technology, Shanghai 200093, China
2 Institute for Agro-Food Standards and Testing Technology, Shanghai Academy of Agricultural Sciences, Shanghai 201403, China
* Correspondence: niedongxia@saas.sh.cn; Tel.: +86-21-37196975
† These authors contributed equally to this work.

Abstract: In this study, a simple and efficient magnetic solid-phase extraction (MSPE) strategy was developed to simultaneously purify and enrich nine mycotoxins in fruits, with the magnetic covalent organic framework nanomaterial Fe_3O_4@COF(TAPT–DHTA) as an adsorbent. The Fe_3O_4@COF(TAPT–DHTA) was prepared by a simple template precipitation polymerization method, using Fe_3O_4 as magnetic core, and 1,3,5-tris-(4-aminophenyl) triazine (TAPT) and 2,5-dihydroxy terephthalaldehyde (DHTA) as two building units. Fe_3O_4@COF(TAPT–DHTA) could effectively capture the targeted mycotoxins by virtue of its abundant hydroxyl groups and aromatic rings. Several key parameters affecting the performance of the MSPE method were studied, including the adsorption solution, adsorption time, elution solvent, volume and time, and the amount of Fe_3O_4@COF(TAPT–DHTA) nanomaterial. Under optimized MSPE conditions, followed by analysis with UHPLC–MS/MS, a wide linear range (0.05–200 μg kg^{-1}), low limits of detection (0.01–0.5 μg kg^{-1}) and satisfactory recovery (74.25–111.75%) were achieved for the nine targeted mycotoxins. The established method was further successfully validated in different kinds of fruit samples.

Keywords: covalent organic framework; magnetic solid-phase extraction; mycotoxins; UHPLC–MS/MS; fruits

Key Contribution: In this paper; Fe_3O_4@COF(TAPT–DHTA) was used for the first time as an adsorbent for the simultaneous enrichment of nine mycotoxins. Under optimized MSPE conditions, followed by analysis with UHPLC–MS/MS, good linearity, sensitivity and recovery were achieved. The established method can be used to monitor the contamination of fruit by the nine mycotoxins.

1. Introduction

Mycotoxins are toxic secondary metabolites produced by toxigenic fungi under suitable environmental conditions. Fruits increasingly favored by consumers, owing to their high moisture content, rich nutrition and improper harvest or storage conditions, are highly susceptible to various fungi such as *Alternaria, Aspergillus, Fusarium*, etc. [1,2]. Among these metabolites, aflatoxin B_1(AFB_1), ochratoxin A (OTA), zearalenone (ZEN) and *Alternaria* toxins, mainly including tentoxin (TEN), altenuene (ALT), altenusin (ALS), alternariol monomethyl ether (AME), alternariol (AOH) and tenuazonic acid (TeA), are the most frequently found in various fruits [3,4]. Those mycotoxins can cause acute and chronic toxic effects (teratogenicity, cytotoxicity, reproductive and developmental toxicity, etc) on animals and human [5,6]. Owing to mycotoxins' high toxicity and widespread contamination of food, the European Food Safety Authority (EFSA) has established a threshold of toxicological concern (TTC) values of 2.5 ng/kg bw/day for AOH and AME, and 1.5 μg/kg bw/day for TeA [7]. Maximum levels of AFB_1, OTA and ZEN in different types of food have been set in the European Union, USA, Canada, China and other countries [8,9]. Given the widely occurring co-contamination of mycotoxins in various fruits, it is imperative to establish an efficient analytical method to simultaneously determine multiple mycotoxins in different kinds of fruits, such as watermelon, hawthorn, melon, tomato, strawberry, etc. [10,11].

Conventional methods for the analysis of mycotoxins in foodstuffs commonly rely on high-performance liquid chromatography (HPLC) coupled with ultraviolet, fluorescence or mass spectrum detectors [12,13]. Nevertheless, the matrix of fruit samples contains large amounts of pigments, cellulose and minerals, which could dramatically impede the detection of trace mycotoxins in food. Therefore, efficient enrichment and purification of multiple mycotoxins in fruit samples is crucial before instrumental analysis. Recently, the most commonly used sample pretreatment methods have included liquid-liquid extraction (LLE) [14], solid-phase extraction (SPE) [15] and QuEChERS methods [16]. Among these methods, immunoaffinity SPE approach has high specificity, but it is also expensive, time consuming and, in particular, not applicable to the simultaneous purification of multiple mycotoxins. Magnetic solid phase extraction (MSPE), as a new kind of SPE, has attracted great attention by virtue of its easy separation, convenient operation and time-saving qualities [17]. In the MSPE process, the magnetic sorbents are directly dispersed in the sample solution for rapid and efficient extraction of analytes, and then quickly separated by an external magnetic field. Many kinds of magnetic nanomaterials have already been developed to enrich targeted analytes and eliminate matrix interferences [18–20].

Covalent organic frameworks (COFs) are a new class of crystalline material, which can be constructed with organic building units by covalent bonds of elements (C, O, N, H, etc.) [21]. COFs, which have been shown to exhibit many unique characteristics such as large specific surface areas, permanent porosity, rich functional groups, and good thermal and chemical stability, have received increasing attention in the field of sample pretreatment [22,23]. Given that the structure and surface properties of COFs has been mainly dependent on covalent linkage topology schemes and organic monomers, considerable attention has been focused on the exploration of various synthetic strategies in recent years [24,25]. For instance, Xu et al. demonstrated the superiority of triazine-based COF for the extraction of phenoxy carboxylic acid pesticide residue [26]. Li and co-workers synthesized Fe_3O_4@COF(TpDA) material to enrich plant growth regulators from fruits and vegetables through π-π and hydrogen bonding interactions [27]. Nevertheless, owing to the low concentrations, different functional groups and polarity of various mycotoxins, it is still challenging to achieve the simultaneous and efficient extraction of multiple mycotoxins from complex matrices. To date, the COF-based MSPE methods for the determinations of mycotoxins have rarely been studied [28].

In this study, using Fe_3O_4 as magnetic core, and 1,3,5-tris-(4-aminophenyl) triazine (TAPT) and 2,5-dihydroxy terephthalaldehyde (DHTA) as two building units, Fe_3O_4@COF(TAPT–DHTA) was firstly designed and applied as an MSPE adsorbent to extract nine mycotoxins (Table S1) from fruits. The targeted mycotoxins were then analyzed by UHPLC–MS/MS (Table S2). The prepared Fe_3O_4@COF(TAPT–DHTA) adsorbents were expected to effectively enrich mycotoxins by virtue of their abundant hydroxyl groups and aromatic rings. The features of Fe_3O_4@COF(TAPT–DHTA) were characterized and several key factors affecting MSPE were optimized. In addition, the possible adsorption mechanism was also discussed. The proposed method was further validated and applied to the analysis of mycotoxins in fruits including watermelon, hawthorn, melon, tomato and strawberry. The schematic of the fabrication and application of Fe_3O_4@COF(TAPT–DHTA) is shown in Figure 1.

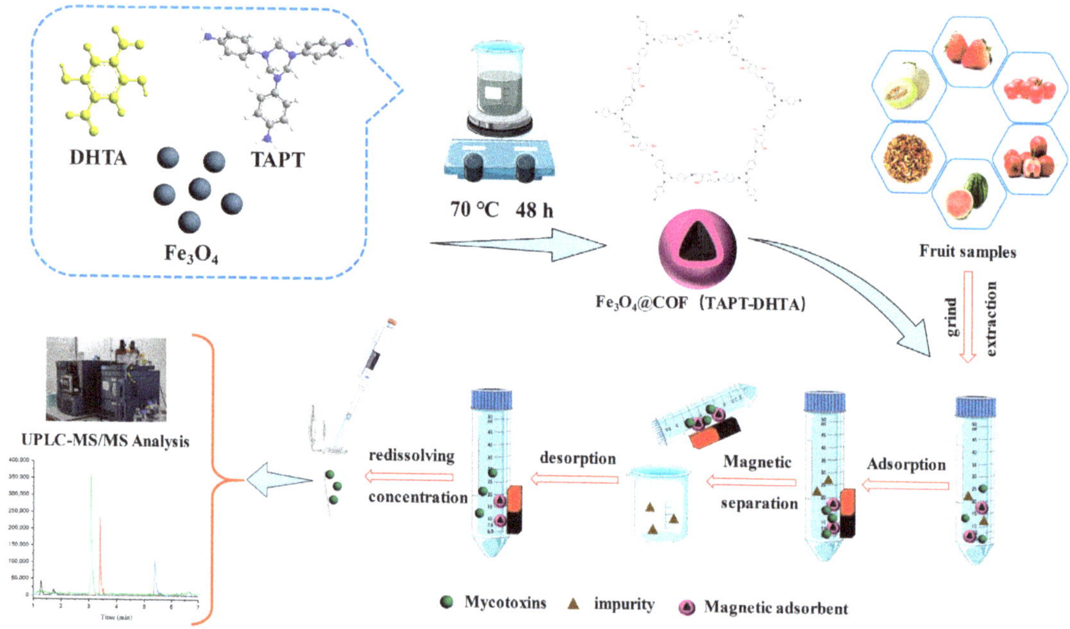

Figure 1. Schematic illustration of the fabrication of Fe_3O_4@COF(TAPT–DHTA) and the established MSPE procedure.

2. Results and Discussion

2.1. Characterization of Fe_3O_4@COF(TAPT–DHTA)

The morphology of the prepared nanomaterial was characterized by SEM (Figure 2A,B). It can be seen that Fe_3O_4 has a regular spherical structure with a diameter of approximately 200 nm. However, Fe_3O_4@COF(TAPT–DHTA) exhibited a dense surface with a significantly larger particle size of nearly 850 nm, proving that COFs were successfully grafted onto Fe_3O_4 nanoparticles.

The formation of Fe_3O_4@COF(TAPT–DHTA) was verified by FT-IR spectroscopy (Figure 2C). For Fe_3O_4, the characteristics peak of the Fe–O–Fe vibration was observed at 588 cm^{-1}. For Fe_3O_4@COF(TAPT–DHTA), the peaks at 1670 cm^{-1} and 3329 cm^{-1} were assigned to the C=O stretching vibration of DHTA and N–H stretching of TAPT, respectively [26]. The additional typical peaks appeared at 1367 cm^{-1} and 1501 cm^{-1} owing to the presence of the triazine ring [29]. Meanwhile, the stretching bands at 1620 cm^{-1} might be ascribed to the formation of imine bonds, proving the successful condensation of formyl linker and amine node [26,29].

Figure 2D shows the UV–VIS absorption spectra of Fe_3O_4 and Fe_3O_4@COF(TAPT–DHTA), respectively. Compared with Fe_3O_4, Fe_3O_4@COF(TAPT–DHTA) showed an absorption peak at 345 nm, which was attributed to the existence of the conjugated double bond in COF. The strong conjugated double bond made Fe_3O_4@COF(TAPT–DHTA) nanocomposites ideal adsorbents for the extraction of benzene ring-containing mycotoxins. In fact, after capturing nine mycotoxins (Figure S1), the absorption peak of Fe_3O_4@COF(TAPT–DHTA) at 345 nm was red-shifted to 351–358 nm, indicating the strong π-π-stacking interaction between the nanocomposites and targeted mycotoxins [30].

Furthermore, the energy dispersive X-ray spectroscopy (EDAX) line-scanning method was conducted to analyze the elements contained in Fe_3O_4@COF(TAPT–DHTA) (Figure 2E). The results demonstrated that the prepared nanomaterial contains four elements, Fe, O, C and N, which also proved the successful synthesis of the nanomaterial.

The magnetic properties of Fe_3O_4 and Fe_3O_4@COF(TAPT–DHTA) were also investigated, and the magnetization curves are shown in Figure 2F. The saturation magnetization values of Fe_3O_4 and Fe_3O_4@COF(TAPT–DHTA) were 46.542 and 18.457 emu g^{-1}, respectively. Although the magnetization values of Fe_3O_4@COF(TAPT–DHTA) decreased compared with pure particles, it could still provide sufficient magnetism for MSPE. As shown in the inset of Figure 2F, Fe_3O_4@COF(TAPT–DHTA) was well-dispersed in the solvent, and could be collected within 30 s by an external magnetic field.

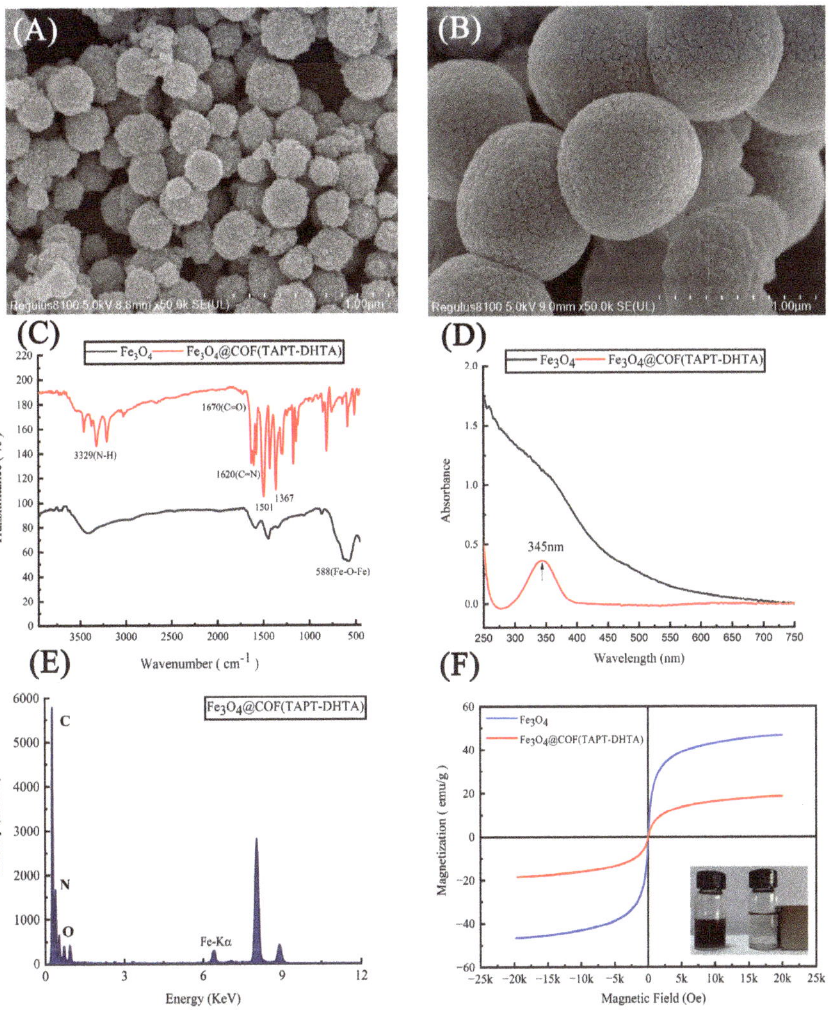

Figure 2. SEM images of Fe_3O_4 (**A**) and Fe_3O_4@COF(TAPT–DHTA) (**B**); FT-IR (**C**) and UV–VIS (**D**) spectra of the Fe_3O_4 and Fe_3O_4@COF(TAPT–DHTA); EDAX spectra of Fe_3O_4@COF(TAPT–DHTA) (**E**); and hysteresis loops of the Fe_3O_4 and Fe_3O_4@COF(TAPT–DHTA) (**F**).

2.2. Optimization of MSPE Conditions

Fe_3O_4@COF(TAPT–DHTA) nanomaterial, with rich hydroxyl groups and aromatic rings, was utilized as competent adsorbent in MSPE for the selective enrichment of multiple mycotoxins. To achieve excellent MSPE procedure, several important parameters were studied by using spiked tomato samples (20 μg kg^{-1}), including adsorption solution, pH, ionic strength, adsorption time, elution solvent, elution volume and time, and Fe_3O_4@COF(TAPT–DHTA) amount.

2.2.1. MSPE Adsorption Solution

The contents of organic solvent in the adsorption solution played a crucial role in the adsorption process. Therefore, the effect of different contents of acetonitrile (0%, 1%, 2%, 3%, 4% and 5%, *v*/*v*) were examined. As shown in Figure 3A, when 1% acetonitrile in water was applied, the highest recoveries of targeted mycotoxins except for AFB_1 were obtained (83–104%). For AFB_1, there was no significant difference in recovery between 1% acetonitrile (79%) and 2% acetonitrile (81%) ($p > 0.05$, Table S3). To ensure adsorption efficiency of the Fe_3O_4@COF(TAPT–DHTA) composite for all the targeted analytes, 1% acetonitrile in water was used as the adsorption solution.

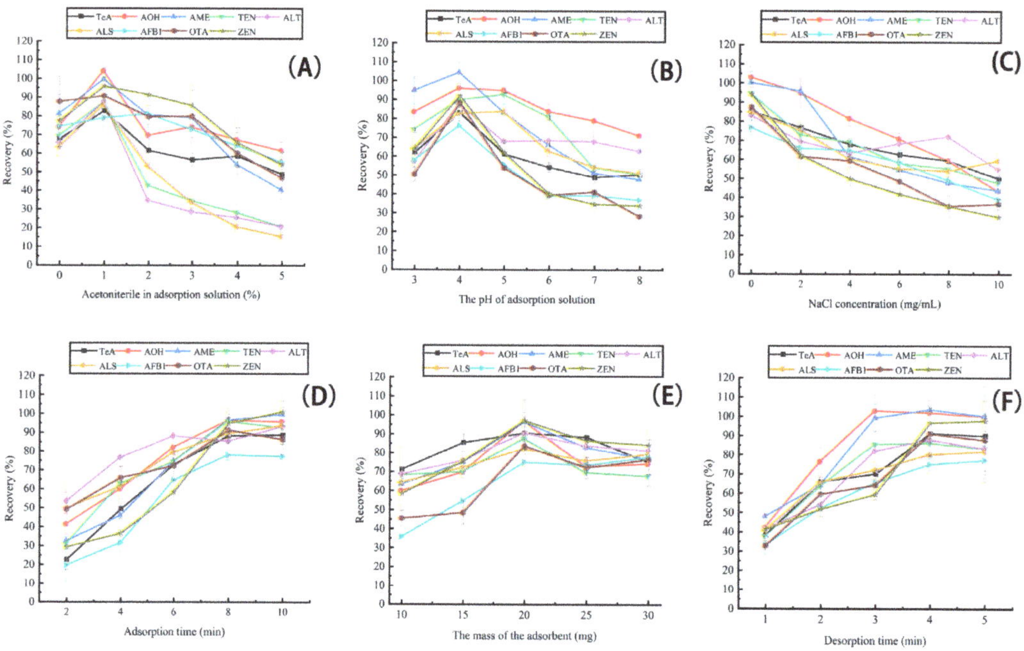

Figure 3. Effects of the key parameters on the performance of Fe_3O_4@COF(TAPT–DHTA) MSPE procedure, including (**A**) percentage of acetonitrile in adsorption solution, (**B**) the pH of adsorption solution, (**C**) the concentration of NaCl in adsorption solution, (**D**) adsorption time, (**E**) the amount of adsorbent and (**F**) desorption time. The concentrations of mycotoxins tested were 20 μg kg^{-1} ($n = 3$).

The pH value of the sample solution had a great influence on the charge property of the surface of the Fe_3O_4@COF(TAPT–DHTA), as well as on the stability and existing forms of the targeted mycotoxins. Figure 3B shows the recoveries of the nine mycotoxins in different pH values. The recoveries first increased and then decreased when the pH varied from acidic to basic. When pH was set as 4.0, satisfactory recoveries (76–104%) were achieved for all nine mycotoxins. Although the recoveries for ALS and TEN obtained at pH 5.0 were a little higher than those at pH 4.0, no significant difference was found for them ($p > 0.05$, Table S4). This result might be due to the characteristics of analytes and nanomaterial [26]. As can be seen in Table S1, the mycotoxins had polar hydroxyl groups and carboxyl groups with pKa values between 3.08 and 7.58, which were not conducive to the formation of hydrogen bonds with Fe_3O_4@COF (TAPT–DHTA) under alkaline conditions. Therefore, the optimal pH was set to 4.0.

To explore the effect of ionic strength on the extraction efficiency of mycotoxins, different sample solutions with various amounts of sodium chloride (0, 2, 4, 6, 8, and 10 mg mL^{-1}) were evaluated (Figure 3C). It was observed that the recoveries of the targeted mycotoxins significantly decreased with the increased salt concentration. Only in the case of no salt addition were the highest recoveries (76–103%) obtained for all nine mycotoxins. In particular, the recoveries for six mycotoxins (TeA, TEN, ALT, AFB_1, OTA and ZEN) were much higher than those obtained when the adsorption solution contained sodium chloride ($p < 0.05$, Table S5). This phenomenon might be explained by the fact that the intermolecular aggregation of mycotoxins was enhanced owing to the addition of salt ions, which inhibited the further adsorption of isolated mycotoxin molecules onto the Fe_3O_4@COF(TAPT–DHTA) [28]. Hence, NaCl was not used for subsequent experiments.

The adsorption time could also affect the extraction efficiency. A series of experiments was conducted to optimize the adsorption times (2, 4, 6, 8 and 10 min). It was observed that the recoveries of the nine mycotoxins gradually increased in the range of 2–8 min (Figure 3D). When 8 min was used, satisfactory recoveries (78–96%) were obtained for all mycotoxins. When the time was further prolonged, no significant improvement could be observed ($p > 0.05$, Table S6). Therefore, 8 min was finally determined for further experiments.

2.2.2. The Amount of Fe_3O_4@COF(TAPT–DHTA)

The amount of adsorbent is a critical factor in the MSPE process. The effects of various amounts (10, 15, 20, 25, 30 mg) of Fe_3O_4@COF(TAPT–DHTA) nanomaterial on the recoveries of the nine mycotoxins were compared (Figure 3E). When 20 mg Fe_3O_4@COF(TAPT–DHTA) was used, the adsorption and desorption of the nine mycotoxins achieved equilibrium, and the highest recoveries were achieved within the acceptable range (75–97%) except for AFB_1. Although the recovery of AFB_1 was a little lower than that in the case of 30 mg, there was no significant difference between them ($p > 0.05$, Table S7). When the amounts were 10 and 15 mg, Fe_3O_4@COF(TAPT–DHTA) could not provide adequate adsorption sites for these mycotoxins. Conversely, when the amounts of adsorbent were excessive, such as 25 and 30 mg, it might not have been easy to elute these mycotoxins from Fe_3O_4@COF(TAPT–DHTA). Thus, 20 mg of Fe_3O_4@COF(TAPT–DHTA) was selected for study.

2.2.3. Elution Solvent and Time

Depending on the effect of the adsorption mechanism of the targeted mycotoxins on the Fe_3O_4@COF(TAPT–DHTA) nanomaterial, the π-π, hydrogen bonding and hydrophobic interactions between adsorbent and analytes could be destroyed by organic solvents. Therefore, elution solvent with different polarities was optimized. As indicated in Figure S2, when methanol/acetonitrile/formic acid (80/19/1) was used, the recoveries of seven mycotoxins (TeA, AOH, AME, TEN, ALT, ALS and OTA) were acceptable in the range of 83–107%. However, the recoveries of AFB_1 and ZEN were much lower compared to methanol ($p < 0.05$, Table S8). Only pure methanol used as elution solvent could achieve satisfactory recoveries for all nine of the mycotoxins (81–99%). The best elution performance of methanol is attributed to the strong polarity, which could compete for the hydrogen binding sites with mycotoxins and lead to the destruction of the formed hydrogen bonding between adsorbent and mycotoxins.

The volume of the elution solvent is crucial in the MSPE procedure. Different volumes (3, 9, 15 mL) of elution solvent were investigated (Figure S3). The results demonstrated that 3 mL elution solvent was enough to achieve satisfactory recoveries (78–99%). With the increase of the elution volume (9 and 15 mL), the recoveries of the targeted mycotoxins were not obviously improved ($p > 0.05$, Table S9). In addition, different elution time (1, 2, 3, 4 and 5 min) was studied, as shown in Figure 3F. The recoveries of these mycotoxins increased with an increase of elution time ranging from 1 to 4 min, and then remained stable ($p > 0.05$, Table S10), suggesting the complete elution of the targeted mycotoxins from Fe_3O_4@COF(TAPT–DHTA) nanomaterial. Consequently, 3.0 mL of pure methanol for 4 min were the optimal elution conditions.

Therefore, based on the optimization of these key parameters, the optimal MSPE conditions were determined as follows: 20 mg Fe_3O_4@COF(TAPT–DHTA) as magnetic adsorbent, extraction time of 8 min using 3 mL water containing 1% acetonitrile as adsorption solution (pH 4.0), and 3 mL methanol as elution solvent with desorption time of 4 min.

2.3. Method Validation

For further characterization of the MSPE performance of Fe_3O_4@COF(TAPT–DHTA), the purification efficiency of the synthesized nanomaterials was investigated in five kinds of fruits including watermelon, hawthorn, melon, tomato and strawberry. The sample solutions treated with Fe_3O_4@COF(TAPT–DHTA) were almost colorless and transparent (Figure 4A) in comparison with the untreated solutions, indicating that Fe_3O_4@COF(TAPT–DHTA) could eliminate the interferences. The matrix effects of the targeted mycotoxins in the five kinds of fruits were in the range of 80.04–106.87% except for TeA (127.19–144.88%), ALS (46.31–102.74%) and ZEN (55.01–83.33%), as shown in Figure 4B. To achieve accurate quantification, matrix-matched calibration curves were applied to compensate for the matrix effects in this study.

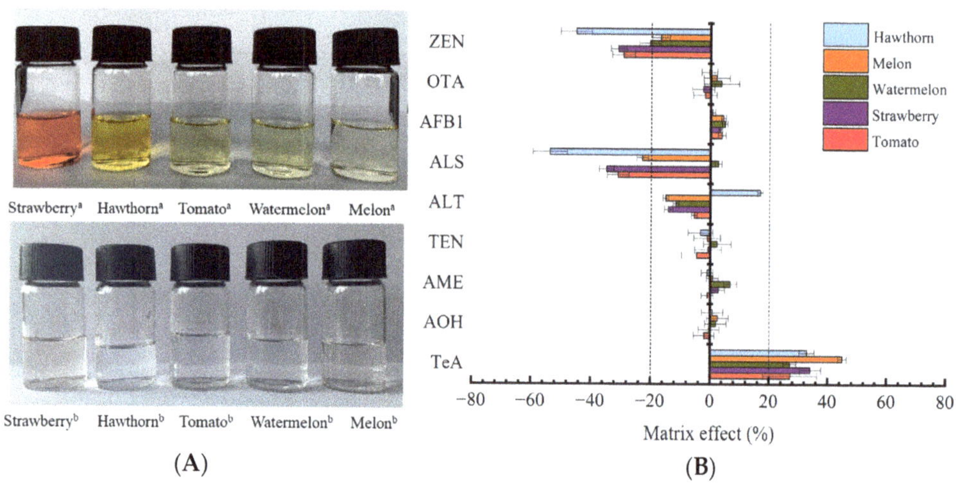

Figure 4. Visual appearance (**A**) and matrix effects (**B**) of 9 mycotoxins in hawthorn, melon, watermelon, strawberry and tomato matrices treated with Fe_3O_4@COF(TAPT–DHTA), a without treatment by Fe_3O_4@COF(TAPT–DHTA) and b treated with Fe_3O_4@COF (TAPT–DHTA).

The current analytical method demonstrated an optimal selectivity, because there were no apparent interfering peaks presenting near the retention time of all the targeted mycotoxins in the tomato matrix (Figure S4). Excellent linearity was obtained for all the analytes, with correlation coefficients (R^2) > 0.99 over the concentration range of 0.05–200 μg kg^{-1} in the matrices for all five kinds of fruits (Table 1). The LODs of the nine mycotoxins ranged from 0.01 to 0.5 μg kg^{-1}, and the LOQs ranged from 0.05 to 1.0 μg kg^{-1}. Acceptable recoveries in the range of 74.25–111.75% were also obtained. The intra- and inter-day precisions were in the range of 2.08–9.01% and 2.22–12.92%, respectively (Table 2). All the above results indicated that the established UHPLC–MS/MS method had high sensitivity, accuracy and precision, and could be applied for the simultaneous quantitative analysis of multiple mycotoxins in different fruits. Compared with previously reported methods, the proposed method possesses equal or even greater sensitivities and recoveries (Table S11).

Table 1. Linear range, detection limit and quantification limit of MSPE combined with UHPLC–MS/MS in different matrices.

Matrix	Mycotoxin	Formula	Linear Range (μg kg−1)	LOD [a] (μg kg−1)	LOQ [b] (μg kg−1)	R^2
Neat solution	TeA	y = 2477.96x + 3394.45	0.5–200			0.9973
	AOH	y = 1411.49x + 941.08	0.5–200			0.9959
	AME	y = 1817.16x + 322.29	0.5–200			0.9989
	TEN	y = 5284.07x−58.31	0.05–200			0.9969
	ALT	y = 524.42x + 638.12	1–200			0.9928
	ALS	y = 143.63x + 11.17	0.5–200			0.9982
	AFB_1	y = 13387.40x + 2131.25	0.05–200			0.9978
	OTA	y = 12142.70x + 1612.83	0.1–200			0.9998
	ZEN	y = 2592.33x − 45.41	0.2–200			0.9956
Tomato	TeA	y = 3151.74x + 2661.33	0.5–200	0.20	0.50	0.9975
	AOH	y = 1381.90x + 88.99	0.5–200	0.20	0.50	0.9961
	AME	y = 1802.95x + 139.73	0.5–200	0.10	0.50	0.9968
	TEN	y = 5037.77x − 1339.19	0.05–200	0.01	0.05	0.9975
	ALT	y = 495.76x + 2523.89	1–200	0.30	1.00	0.9973
	ALS	y = 99.21x − 44.54	1–200	0.50	1.00	0.9933
	AFB_1	y = 13910.30x − 3759.31	0.5–200	0.20	0.50	0.9957
	OTA	y = 11913.80x − 3011.72	0.1–200	0.05	0.10	0.9969
	ZEN	y = 1836.13x − 408.39	0.5–200	0.20	0.50	0.9966
Watermelon	TeA	y = 3147.83x + 3812.73	0.5–200	0.20	0.50	0.9988
	AOH	y = 1439.20x + 166.73	0.5–200	0.20	0.50	0.9998
	AME	y = 1942.00x + 660.55	0.5–200	0.10	0.50	0.9995
	TEN	y = 5411.94x−86.24	0.1–200	0.05	0.10	0.9995
	ALT	y = 463.38x + 2883.52	1–200	0.30	1.00	0.9952
	ALS	y = 147.57x − 105.27	1–200	0.50	1.00	0.9985
	AFB_1	y = 14048.90x − 407.44	0.05–200	0.01	0.05	0.9998
	OTA	y = 12596.50x − 1507.82	0.5–200	0.20	0.50	0.9991
	ZEN	y = 2064.16x − 407.63	0.5–200	0.10	0.50	0.9923
Melon	TeA	y = 3590.11x + 3091.17	0.5–200	0.20	0.50	0.9981
	AOH	y = 1447.78x + 35.81	0.5–200	0.20	0.50	0.9973
	AME	y = 1832.77x + 632.65	0.5–200	0.10	0.50	0.9993
	TEN	y = 5228.12x−492.67	0.1–200	0.05	0.10	0.9993
	ALT	y = 446.22x + 6640.54	1–200	0.30	1.00	0.9976
	ALS	y = 110.85x − 74.61	1–200	0.50	1.00	0.9968
	AFB_1	y = 13980.80x − 1061.58	0.1–200	0.02	0.10	0.9998
	OTA	y = 12393.30x − 462.63	0.5–200	0.10	0.50	0.9994
	ZEN	y = 2160.20x − 575.37	0.5–200	0.10	0.50	0.9931
Strawberry	TeA	y = 3323.10x + 3464.16	0.5–200	0.20	0.50	0.9972
	AOH	y = 1406.51x − 68.41	0.5–200	0.20	0.50	0.9983
	AME	y = 1869.42x + 594.67	0.5–200	0.10	0.50	0.9992
	TEN	y = 5245.70x − 460.86	0.1–200	0.05	0.10	0.9996
	ALT	y = 450.92x + 2765.14	1–200	0.30	1.00	0.9922
	ALS	y = 93.84x + 58.02	1–200	0.50	1.00	0.9940
	AFB_1	y = 13843.00x − 1246.62	0.1–200	0.02	0.10	0.9996
	OTA	y = 11841.30x − 470.11	0.5–200	0.10	0.50	0.9993
	ZEN	y = 1792.03x − 54.64	0.5–200	0.10	0.50	0.9983
Hawthorn	TeA	y = 3292.18x + 5601.93	0.5–200	0.20	0.50	0.9958
	AOH	y = 1423.11x + 12363.80	0.5–200	0.20	0.50	0.9943
	AME	y = 1799.34x + 365.52	0.5–200	0.20	0.50	0.9996
	TEN	y = 5082.12x + 1587.85	0.05–200	0.02	0.05	0.9996
	ALT	y = 419.79x + 9686.70	1–200	0.30	1.00	0.9910
	ALS	y = 66.53x + 34.26	1–200	0.50	1.00	0.9918
	AFB_1	y = 13483.20x − 304.85	0.1–200	0.02	0.10	0.9992
	OTA	y = 12097.60x − 1068.53	0.5–200	0.10	0.50	0.9993
	ZEN	y = 1426.21x − 169.96	0.5–200	0.10	0.50	0.9972

[a] Limit of detection (S/N = 3). [b] Limit of quantification (S/N = 10).

Table 2. Spike recoveries, intra-day and inter-day precision of 9 mycotoxins in different matrices based on MSPE and HPLC-MS/MS methods.

Mycotoxins	Spiked Levels ($\mu g\ kg^{-1}$)	Tomato			Watermelon			Melon			Strawberry			Hawthorn		
		Recovery (X ± SD, %) n = 3	Intra-Day Precision (Intra-RSD, %) n = 5	Inter-Day Precision (Inter-RSD, %) n = 5	Recovery (X ± SD, %) n = 3	Intra-Day Precision (Intra-RSD, %) n = 5	Inter-Day Precision (Inter-RSD, %) n = 5	Recovery (X ± SD, %) n = 3	Intra-Day Precision (Intra-RSD, %) n = 5	Inter-Day Precision (Inter-RSD, %) n = 5	Recovery (X ± SD, %) n = 3	Intra-Day Precision (Intra-RSD, %) n = 5	Inter-Day Precision (Inter-RSD, %) n = 5	Recovery (X ± SD, %) n = 3	Intra-Day Precision (Intra-RSD, %) n = 5	Inter-Day Precision (Inter-RSD, %) n = 5
TeA	2	95.8 ± 0.2	2.4	8.7	84.4 ± 0.2	4.9	9.2	94.6 ± 0.2	6.9	8.2	99.2 ± 0.2	5.0	10.6	96.3 ± 0.2	6.6	11.4
	50	90.8 ± 1.5	3.1	6.2	79.2 ± 0.4	5.4	3.7	84.5 ± 1.4	7.5	11.1	78.5 ± 1.4	4.2	11.7	96.4 ± 2.0	3.8	8.8
	100	85.6 ± 3.1	2.6	6.1	74.8 ± 3.7	4.9	8.6	81.1 ± 3.8	5.4	7.8	80.4 ± 5.7	7.6	10.7	88.7 ± 6.2	3.9	11.7
AOH	2	82.0 ± 0.1	3.8	4.3	88.3 ± 0.1	4.9	6.0	93.1 ± 0.2	3.8	12.4	82.0 ± 0.1	3.2	3.7	103.6 ± 0.1	2.1	9.64
	50	84.5 ± 1.3	4.3	12.9	92.4 ± 2.2	5.4	9.7	100.9 ± 4.5	3.4	7.4	101.9 ± 4.3	5.2	8.6	95.9 ± 4.3	6.9	8.7
	100	111.7 ± 3.4	3.8	5.1	94.2 ± 2.3	4.2	4.0	82.5 ± 4.9	7.7	4.6	83.2 ± 4.6	4.8	9.2	76.3 ± 4.5	4.3	9.7
AME	2	98.1 ± 0.2	3.1	7.5	89.1 ± 0.2	2.3	8.4	104.4 ± 0.2	3.9	5.3	87.7 ± 0.1	3.4	7.0	109.9 ± 0.2	6.2	5.81
	50	82.5 ± 1.1	2.8	8.4	78.4 ± 1.3	3.0	5.7	91.2 ± 2.6	4.3	9.6	96.5 ± 1.6	4.5	10.8	107.9 ± 1.3	5.2	8.3
	100	78.2 ± 3.9	5.4	8.4	81.5 ± 4.8	3.0	4.7	88.8 ± 2.9	4.1	5.4	91.7 ± 5.6	3.8	10.1	98.9 ± 3.1	5.4	5.2
TEN	2	89.7 ± 0.1	4.4	7.9	75.2 ± 0.1	5.7	2.2	82.8 ± 0.1	5.2	5.8	92.9 ± 0.1	7.2	11.9	88.1 ± 0.1	6.7	6.7
	50	91.5 ± 0.7	5.2	5.1	83.7 ± 1.5	6.4	11.7	96.6 ± 1.6	5.5	6.2	91.9 ± 1.3	6.7	9.3	92.0 ± 1.3	5.6	9.7
	100	75.4 ± 1.8	7.3	3.9	90.1 ± 2.3	6.2	4.3	75.7 ± 4.3	3.9	9.4	79.8 ± 4.5	5.8	9.5	82.5 ± 2.3	6.1	4.7
ALT	2	86.1 ± 0.1	7.9	7.2	80.4 ± 0.2	8.6	7.6	81.1 ± 0.1	3.3	9.1	99.9 ± 0.2	4.0	12.7	96.8 ± 0.2	4.6	10.7
	50	90.8 ± 1.4	6.2	4.4	81.5 ± 1.4	7.5	11.8	75.6 ± 1.1	5.5	9.3	78.5 ± 1.4	7.2	11.7	99.4 ± 2.0	5.3	9.3
	100	85.6 ± 3.1	6.1	11.5	76.9 ± 2.7	7.5	6.2	75.6 ± 2.4	6.8	5.2	74.9 ± 1.7	8.1	3.8	77.6 ± 2.2	4.6	4.8
ALS	2	93.0 ± 0.1	3.6	5.9	85.7 ± 0.2	4.3	3.0	87.8 ± 0.1	2.7	4.2	83.6 ± 0.2	8.6	6.6	84.7 ± 0.1	4.8	11.0
	50	86.8 ± 1.9	4.1	6.8	90.3 ± 1.3	2.2	4.2	88.2 ± 1.7	7.5	10.4	88.0 ± 1.8	3.0	11.4	88.4 ± 1.3	5.1	8.9
	100	88.71 ± 5.84	4.7	6.1	83.1 ± 4.0	5.4	6.1	91.3 ± 5.3	5.7	9.1	90.7 ± 4.9	3.8	8.9	89.9 ± 5.6	4.4	4.2
AFB_1	2	81.1 ± 0.1	2.9	8.4	81.3 ± 0.1	7.9	10.6	96.6 ± 0.1	6.1	5.1	102.1 ± 0.2	7.1	8.7	97.1 ± 0.1	3.6	8.4
	50	83.6 ± 0.9	3.9	7.6	79.0 ± 1.8	4.4	8.9	79.6 ± 1.9	8.6	9.5	85.4 ± 1.7	5.9	10.9	85.3 ± 1.7	5.8	8.9
	100	77.2 ± 2.6	4.6	6.1	82.5 ± 5.9	5.4	11.5	74.3 ± 4.7	5.5	8.9	80.6 ± 2.9	6.3	5.9	77.4 ± 5.1	5.0	10.9
OTA	2	108.1 ± 0.1	6.2	5.4	100.6 ± 0.1	4.5	3.5	89.8 ± 0.2	6.7	12.3	75.6 ± 0.1	5.5	8.5	90.6 ± 0.2	3.3	9.2
	50	111.7 ± 1.3	5.7	7.5	97.3 ± 1.6	7.4	11.0	99.6 ± 1.8	4.2	11.9	93.6 ± 1.7	3.8	11.80	98.5 ± 1.6	6.9	11.0
	100	84.1 ± 4.4	6.0	8.7	76.3 ± 5.6	6.9	12.3	75.7 ± 1.9	7.7	4.2	83.1 ± 2.5	2.1	4.9	83.8 ± 1.9	7.8	3.8
ZEN	2	78.8 ± 0.2	4.3	9.6	91.2 ± 0.1	5.2	5.0	107.5 ± 0.1	3.8	4.6	84.8 ± 0.2	7.2	9.9	78.2 ± 0.4	6.2	3.2
	50	99.7 ± 1.5	7.1	10.0	84.8 ± 1.6	9.0	8.4	101.7 ± 1.5	4.2	9.9	76.6 ± 1.8	6.7	11.2	79.2 ± 0.7	5.0	5.8
	100	85.3 ± 2.8	6.2	5.4	91.2 ± 4.0	4.8	7.4	87.1 ± 2.3	3.8	4.4	82.0 ± 4.9	5.7	3.8	87.1 ± 3.3	5.9	6.4

2.4. Method Application

A validated approach was employed to investigate mycotoxin contamination of a total of 100 samples of the five kinds of fruits, i.e., tomato, strawberry, watermelon, melon and hawthorn. As shown in Table S12, ALT was the most frequently detected mycotoxin with incidences (concentration ranges) of 40% (2.2–44.5 μg kg^{-1}), 55% (3.4–54.8 μg kg^{-1}), 25% (29.5–56.3 μg kg^{-1}), 50% (43.4–123.7 μg kg^{-1}) and 45% (38.8–190.4 μg kg^{-1}) in the tomato, strawberry, watermelon, melon and hawthorn samples, respectively. TeA was detected in the tomato and strawberry samples, with incidences (concentration ranges) of 50% (3.8–6.5 μg kg^{-1}) and 45% (1.9–5.6 μg kg^{-1}), respectively. The incidence (concentration ranges) of AOH in the tomato, strawberry and hawthorn samples was 10% (3.05–4.0 μg kg^{-1}), 15% (4.9–20.0 μg kg^{-1}) and 25% (3.7–14.2 μg kg^{-1}), respectively. These results are similar to previous reports, with pollution levels at the same level as other regions [31,32]. In addition, concentrations of AME, TEN and ALS were detected in the range of 1.4–16.8 μg kg^{-1}, 0.6–18.2 μg kg^{-1} and 1.3–8.1 μg kg^{-1}, respectively. OTA, AFB_1 and ZEN were not detected in the five kinds of fruit samples. Therefore, the survey results demonstrated that the fruits were mainly susceptible to contamination by *Alternaria* mycotoxins, which might impose health risks to the consumer.

To demonstrate the accuracy of the method, a comparison between the current established approach and the reference method [33] was performed on typical tomato samples (nos. 7, 14 and 20). There was no significant difference between the results obtained by the two methods, with the RSDs lower than 10% ($p > 0.05$, Table S13), verifying the accuracy and applicability of the developed UHPLC–MS/MS method.

3. Conclusions

In summary, Fe_3O_4@COF(TAPT–DHTA) was synthesized and utilized as an MSPE adsorbent for the first time to enrich and determine trace levels of nine mycotoxins in fruits by coupling with UHPLC–MS/MS. Owing to the presence of abundant aromatic rings and the carbonyl group in the structure of the adsorbent, the effective enrichment of the targeted mycotoxins was achieved by virtue of a strong π-π interaction and hydrogen bonding between the mycotoxins and Fe_3O_4@COF(TAPT–DHTA). The established method showed a wide linear range, high sensitivity, satisfactory recoveries and good precision, and was successfully employed for the analysis of mycotoxins in real fruit samples. The fruits were found to be easily contaminated with different mycotoxins, i.e., TeA, AOH, ALT, etc. Therefore, continuous monitoring of the occurrence of multiple mycotoxins is essential to ensure of the safe consumption of fruits.

4. Materials and Methods

4.1. Chemicals and Materials

All the organic solvents, acids, alkalis and salts were HPLC or analytical grade. Acetonitrile and methanol were purchased from Merck (Darmstadt, Germany). Ammonium acetate, formic acid, aqueous ammonia, $FeCl_3 \cdot 6H_2O$ (≥99.5%) and $FeCl_2 \cdot 4H_2O$ (≥99%) were provided by Sinopharm Chemical Reagent Co. (Shanghai, China).1,2-dichlorobenzene, 1-Butanol and 1,4-Dioxane were supplied by Macklin Co. (Shanghai, China). DHTA and TAPT were obtained from Yuanye Bio-Technology Co. (Shanghai, China). Nylon filters (0.22 μm) were obtained from Navigator Lab Instrument Co. Ltd. (New York, NY, USA). Deionized water was prepared by a Milli-Q water purification system (Millipore, Bedford, MA, USA).

High purity (≥98%) standards of TeA, AOH, AME, TEN, ALT, ALS, AFB_1, OTA and ZEN were obtained from Sigma-Aldrich (St. Louis, MO, USA). Standard stock solutions of the nine mycotoxins (10 μg mL^{-1}) were prepared in acetonitrile and stored at −20 °C in the dark. Their chemical structures and physicochemical parameters are shown in Table S1.

A total of 100 random fruit samples were provided by local markets and supermarkets in Shanghai. The samples were ground into powder or pulp and stored at −20 °C.

4.2. Apparatus and Characterization

Scanning electron microscopy (SEM) images were achieved on a ZEISS Gemini SEM 300 electron microscope (Oberkochen, Baden-Warburg, Germany) operated at 3.0 kV. Fourier transform infrared (FTIR) spectroscopy with a recording range of 500–4000 cm^{-1} was performed with an FTIR spectrometer (FTIR, Thermo Scientific Nicolet iS20, Waltham, MA, USA). Ultraviolet-visible (UV–VIS) absorption spectra were obtained using a JENA2010 spectrophotometer (JENA, Turingia, Germany). A Super-X Spectrometer (Waltham, MA, USA) used for elemental analysis and energy spectrum analysis. The magnetic property was studied with a LakeShore7404 vibrating sample magnetometer (MI, USA).

4.3. Preparation of Fe_3O_4@COF(TAPT–DHTA)

The preparation procedures of Fe_3O_4@COF(TAPT–DHTA) are shown in Figure 1. Firstly, Fe_3O_4 magnetic nanoparticles (MNPs) were fabricated according to the previously reported Massart method with minor changes [34,35]. Briefly, 2.6 g $FeCl_3{\cdot}6H_2O$ and 1.59 g $FeCl_2$ $4H_2O$ were dissolved in 12.5 mL water containing 0.43 mL of 30% HCl under nitrogen atmosphere. Afterwards, the obtained mixture was added dropwise into 125 mL 1.5 M NaOH aqueous solution under vigorous stirring for 40 min. The synthesized Fe_3O_4 product was collected using a strong magnet and washed several times with water, then dried at 65 °C for further use. Secondly, Fe_3O_4@COF(TAPT–DHTA) was fabricated using a simple synthesis method [36]. The prepared Fe_3O_4 MNPs (200 mg), DHTA (105 mg) and TAPT (85 mg) were mixed in the 40 mL 1,4-dioxane/butanol (1/1, *v*/*v*) solvent system, and then sonicated for 5 min. This was followed by adding 0.5 mL of 36% acetic acid, and the mixture was then stirred for 2 h at 25 °C. Then, 4.5 mL acetic acid/deionized water (2/1, *v*/*v*) was dropped into the mixture and refluxed at 75 °C for 48 h. Finally, the obtained Fe_3O_4@COF(TAPT–DHTA) was separated by magnet, washed several times with methanol and dried for further use.

4.4. Sample Preparation

A fruit sample of 2.0 g was vortexed in 10 mL acetonitrile/formic acid (99/1, *v*/*v*) for 1 min, and then ultrasonicated for 60 min at room temperature (25 °C). After centrifugation at 8000 rpm for 5 min, a 3 mL aliquot of the supernatant was evaporated to dryness under nitrogen flow and dissolved to 3 mL with an aqueous solution containing 1% acetonitrile (pH 4.0). This was followed by the addition of 20 mg of Fe_3O_4@COF(TAPT–DHTA), the mixture was vortexed for 8 min to fully adsorb the mycotoxins onto the nanocomposites. Afterwards, the supernatant was discarded by a magnet and 3 mL methanol was added to elute the targeted mycotoxins for 4 min under ultrasonic conditions. The elution was collected and dried with nitrogen at 50 °C. Finally, the residues were re-dissolved with 1 mL acetonitrile/water containing 5 mmol L^{-1} ammonium acetate (20/80, *v*/*v*), passed through a 0.22 μm nylon filter before analysis. The schematic illustration of MSPE is shown in Figure 1. To avoid introducing other impurities, Fe_3O_4@COF(TAPT–DHTA) nanocomposites were used as disposable purification adsorbents.

4.5. UHPLC–MS/MS Analysis

Chromatographic analysis of the nine mycotoxins was accomplished by an Acquity UHPLC (Waters, Milford, MA, USA) with an analytical column EC-C18 (100 mm × 3.0 mm, i.d. 2.7 μm) (Agilent, Santa Clara, CA, USA) maintained at 40 °C. The binary gradient mixture consisted of (A) methanol and (B) water containing 5 mmol L^{-1} ammonium acetate used as a mobile phase, with a flow rate of 0.4 mL min^{-1}. The gradient elution program was set as follows: 0–5 min, 70% B; 5–7 min, 10% B; 7.5–8.5 min, 70% B. The injection volume was 3.0 μL.

Waters XEVO TQ-S mass spectrometer system (Waters, Milford, MA, USA) was performed to detect separated mycotoxins both in positive (ESI^+) and negative (ESI^-) mode. The capillary voltages were set at 2.5 kV (ESI^+) and 1.5 kV (ESI^-). The source and desolvation temperatures were 150 °C and 500 °C, respectively. The desolvation gas flow and cone gas flow were 800 L h^{-1} and 150 L h^{-1}, respectively. Multiple reaction monitoring (MRM) mode was established and is shown in Table S2, and data analysis was performed using MassLynx v4.1 and Targetlynx (Waters).

4.6. Method Validation

The performance of the established method was carefully validated according to the recommendations of European Commission Decision 2002/657/EC by determination of linearity, matrix effect, sensitivity, accuracy (recovery), and precision (%RSD) [37]. Different concentrations (0.5, 1, 2, 5, 10, 20, 50, 100 and 200 ng mL^{-1}) of mixed standard solutions of the nine mycotoxins were freshly prepared both in blank matrix solution and standard solutions (acetonitrile/water containing 5 mmol L^{-1} ammonium acetate, 20/80, v/v), respectively. The calibration curves were obtained by plotting the responses (y) versus analyte concentrations (x). The limit of detection (LOD) and limit of quantification (LOQ) were applied to evaluate the sensitivity of the method, which were defined by the signal-to-noise ratio (S/N) of 3 and 10, respectively. The recoveries and intra- and inter-day precision were measured using blank samples spiked with three different concentrations of mycotoxins (2, 50, and 100 μg kg^{-1}). The relative standard deviations (RSDs) in a single day and five consecutive days were devoted to estimation of the intra-day precision and inter-day precision, respectively. The matrix effect (ME) (%) was calculated according to the following equation:

$$ME\ (\%) = (\text{Slope matrix spiked} - \text{Slope standard solution})/\text{Slope standard solution} \times 100\%$$

where Slope matrix spiked and Slope standard solution represented the slope of the calibration curve in the matrix and reagent solution, respectively.

4.7. Statistical Analysis

Data were analyzed using SPSS Statistics 18 (SPSS, Inc., Chicago, IL, USA) and presented as mean ± SD. Significant tests were conducted by the analysis of variance (ANOVA) followed by Duncan's least significant test. The Shapiro–Wilk test was used to check the normality of the data before using ANOVA.

Supplementary Materials: The following supporting information can be downloaded at: https://www.mdpi.com/article/10.3390/toxins15020117/s1. References [16,38–43] are cited in the supplementary materials. Table S1. Structure and physicochemical parameters of 9 mycotoxins; Table S2. Mass spectrometry parameters of 9 mycotoxins; Table S3. Significance analysis of adsorbed solution; Table S4. Significance analysis of pH value of adsorbed solution; Table S5. Significance analysis of NaCl concentration in adsorbed solution; Table S6. Significance analysis of adsorption time; Table S7. Significance analysis of adsorbent amount; Table S8. Significance analysis of elution kinds; Table S9. Significance analysis of elution volume; Table S10. Significance analysis of elution time; Table S11. Comparison of the Fe_3O_4@COF(TAPT–DHTA)-based MSPE method developed in this study with the sample pretreatment approaches reported in previous studies; Table S12. Contamination levels of 9 mycotoxins in five fruits (tomato, strawberry, watermelon, melon, hawthorn); Table S13. Comparison between the developed UHPLC–MS/MS method and the reference methods by using tomato samples nos. 7, 14 and 20; Figure S1. UV–VIS spectra of targeted mycotoxins, Fe_3O_4@COF(TAPT–DHTA) before and after extraction of target mycotoxins; Figure S2. Comparison of the purification efficiency of 9 mycotoxins in the spiked tomato sample by 8 candidate elution solutions; Figure S3. Effects of eluent volume on Fe_3O_4@COF (TAPT–DHTA) MSPE process; Figure S4. MRM chromatography of 9 mycotoxins in (A) solvent standard solution and (B) matrix standard solution (100 μg kg^{-1}).

Author Contributions: Conceptualization, J.W., D.N. and Z.H.; methodology, J.W., Q.H., W.G. and D.G.; validation, J.W., Q.H., W.G. and D.G.; formal analysis, J.W., W.G. and D.G.; investigation, Q.H., W.G. and D.G.; data curation, Q.H.; writing—original draft preparation, J.W.; writing—review and editing, D.N. and Z.H.; visualization, J.W.; supervision, D.N. and Z.H.; project administration, D.N. and Z.H.; funding acquisition, D.N., Z.H. and W.G. All authors have read and agreed to the published version of the manuscript.

Funding: This work was funded by the Shanghai Agriculture Applied Technology Development Program, China [Grant No.X20210302], the National Natural Science Foundation of China [Grant No. 31972178] and Shanghai Rising-Star Program [Grant No.21QB1403300].

Institutional Review Board Statement: Not applicable.

Informed Consent Statement: Not applicable.

Data Availability Statement: Not applicable.

Conflicts of Interest: The authors declare no conflict of interest.

References

1. Drusch, S.; Ragab, W. Mycotoxins in Fruits, Fruit Juices, and Dried Fruits. *J. Food Prot.* **2003**, *66*, 1514–1527. [CrossRef]
2. Zhao, K.; Shao, B.; Yang, D.J.; Li, F.Q. Natural occurrence of four Alternaria mycotoxins in tomato- and citrus-based foods in China. *J. Agric. Food Chem.* **2015**, *63*, 343–348. [CrossRef] [PubMed]
3. Juan, C.; Oueslati, S.; Manes, J. Evaluation of alternaria mycotoxins in strawberries: Quantification and storage condition. *Food Addit. Contam.* **2016**, *33*, 861–868. [CrossRef] [PubMed]
4. Meng, J.J.; Guo, W.B.; Zhao, Z.H.; Zhang, Z.Q.; Nie, D.X.; Tangni, E.K.; Han, Z. Production of alternaria toxins in yellow peach (*Amygdalus persica*) upon artificial inoculation with *Alternaria alternate*. *Toxins* **2021**, *13*, 656. [CrossRef] [PubMed]
5. Ahmad, A.; Yu, J.H. Toxicity, and analysis of major mycotoxins in Food. *Int. J. Environ. Res. Public Health* **2017**, *14*, 632. [CrossRef]
6. Chen, A.Q.; Mao, X.; Sun, Q.H.; Wei, Z.X.; Li, J.; You, Y.L.; Zhao, J.Q.; Jiang, G.B.; Wu, Y.N.; Wang, L.P.; et al. Alternaria mycotoxins: An overview of toxicity, metabolism, and analysis in food. *J. Agric. Food Chem.* **2021**, *69*, 7817–7830. [CrossRef]
7. Fan, K.; Guo, W.B.; Huang, Q.W.; Meng, J.J.; Yao, Q.; Nie, D.X.; Han, Z.; Zhao, Z.H. Assessment of human exposure to five alternaria mycotoxins in China by biomonitoring approach. *Toxins* **2021**, *13*, 762. [CrossRef]
8. *GB 2761-2017*; National Food Safety Standard-Maximum Levels of Mycotoxins in Foods. National Standard of People's Republic of China: Beijing, China, 2017.
9. EC. Commission Regulation (EC) No 1881/2006 of 19 December 2006 setting maximum levels for certain contaminants in foodstuffs. *Off. J. Eur. Union* **2007**, *364*, 5–25.
10. Wang, M.; Jiang, N.; Xian, H.; Wei, D.Z.; Shi, L.; Feng, X.Y. A single-step solid phase extraction for the simultaneous determination of 8 mycotoxins in fruits by ultra-high performance liquid chromatography tandem mass spectrometry. *J. Chromatogr. A* **2016**, *1429*, 22–29. [CrossRef]
11. Van de Perre, E.; Deschuyffeleer, N.; Jacxsens, L.; Vekeman, F.; Van Der Hauwaert, W.; Asam, S.; Rychlik, M.; Devlieghere, F.; De Meulenaer, B. Screening of moulds and mycotoxins in tomatoes, bell peppers, onions, soft red fruits and derived tomato products. *Food Control* **2014**, *37*, 165–170. [CrossRef]
12. Gao, S.Q.; Wu, Y.Q.; Xie, S.Y.; Shao, Z.C.; Bao, X.M.; Yan, Y.M.; Wu, Y.Y.; Wang, J.X.; Zhang, Z.E. Determination of aflatoxins in milk sample with ionic liquid modified magnetic zeolitic imidazolate frameworks. *J. Chromatogr. B* **2019**, *1128*, 121778. [CrossRef] [PubMed]
13. La Barbera, G.; Capriotti, A.L.; Cavaliere, C.; Foglia, P.; Montone, C.M.; Chiozzi, R.Z.; Lagana, A. A rapid magnetic solid phase extraction method followed by Liquid Chromatography-Tandem Mass spectrometry analysis for the determination of mycotoxins in cereals. *Toxins* **2017**, *9*, 147. [CrossRef] [PubMed]
14. Qiao, X.T.; Zhang, J.; Yang, Y.J.; Yin, J.; Li, H.; Xing, Y.; Shao, B. Development of a simple and rapid LC-MS/MS method for the simultaneous quantification of five Alternaria mycotoxins in human urine. *J. Chromatogr. B* **2020**, *1144*, 122096. [CrossRef] [PubMed]
15. Goncalves, C.; Tolgyesi, A.; Bouten, K.; Robouch, P.; Emons, H.; Stroka, J. Determination of alternaria toxins in tomato, wheat, and sunflower seeds by SPE and LC-MS/MS-A method validation through a collaborative trial. *J. AOAC Int.* **2022**, *105*, 80–94. [CrossRef]
16. De Berardis, S.; De Paola, E.L.; Montevecchi, G.; Garbini, D.; Masino, F.; Antonelli, A.; Melucci, D. Determination of four alternaria alternata mycotoxins by QuEChERS approach coupled with liquid chromatography-tandem mass spectrometry in tomato-based and fruit-based products. *Food Res. Int.* **2018**, *106*, 677–685. [CrossRef]
17. Targuma, S.; Njobeh, P.B.; Ndungu, P.G. Current applications of magnetic nanomaterials for extraction of mycotoxins, pesticides, and pharmaceuticals in food commodities. *Molecules* **2021**, *26*, 4284. [CrossRef]

18. Moreno, V.; Zougagh, M.; Rios, A. Hybrid nanoparticles based on magnetic multiwalled carbon nanotube-nano C18SiO2 composites for solid phase extraction of mycotoxins prior to their determination by LC-MS. *Mikrochim. Acta* **2016**, *183*, 871–880. [CrossRef]
19. Ma, K.X.; Zhang, M.X.; Miao, S.C.; Gu, X.Y.; Li, N.; Cui, S.H.; Yang, J. Magnetic solid-phase extraction of pyrethroid pesticides in environmental water samples with $CoFe_2O_4$-embedded porous graphitic carbon nanocomposites. *J. Sep. Sci.* **2018**, *41*, 3441–3448. [CrossRef]
20. Jia, Y.Q.; Wang, Y.H.; Yan, M.; Wang, Q.; Xu, H.J.; Wang, X.S.; Zhou, H.Y.; Hao, Y.L.; Wang, M.M. Fabrication of iron oxide@MOF-808 as a sorbent for magnetic solid phase extraction of benzoylurea insecticides in tea beverages and juice samples. *J. Chromatogr. A* **2020**, *1615*, 460766. [CrossRef]
21. Lin, X.P.; Wang, X.Q.; Wang, J.; Yuan, Y.W.; Di, S.S.; Wang, Z.W.; Xu, H.; Zhao, H.Y.; Zhao, C.S.; Ding, W.; et al. Magnetic covalent organic framework as a solid-phase extraction absorbent for sensitive determination of trace organophosphorus pesticides in fatty milk. *J. Chromatogr. A* **2020**, *1627*, 461387. [CrossRef]
22. Wen, L.; Liu, L.; Wang, X.; Wang, M.L.; Lin, J.M.; Zhao, R.S. Spherical mesoporous covalent organic framework as a solid-phase extraction adsorbent for the ultrasensitive determination of sulfonamides in food and water samples by liquid chromatography-tandem mass spectrometry. *J. Chromatogr. A* **2020**, *1625*, 461275. [CrossRef] [PubMed]
23. Li, N.; Du, J.J.; Wu, D.; Liu, J.C.; Li, N.; Sun, Z.W.; Li, G.L.; Wu, Y.N. Recent advances in facile synthesis and applications of covalent organic framework materials as superior adsorbents in sample pretreatment. *TrAC. Trends Anal. Chem.* **2018**, *108*, 154–166. [CrossRef]
24. Emmerling, S.T.; Schuldt, R.; Bette, S.; Yao, L.; Dinnebier, R.E.; Kastner, J.; Lotsch, B.V. Interlayer interactions as design tool for Large-Pore COFs. *J. Am. Chem. Soc.* **2021**, *143*, 15711–15722. [CrossRef]
25. Vardhan, H.; Hou, L.X.; Yee, E.; Nafady, A.; Al-Abdrabalnabi, M.A.; Al-Enizi, A.M.; Pan, Y.X.; Yang, Z.Y.; Ma, S.Q. Vanadium docked covalent-organic frameworks: An effective heterogeneous catalyst for modified mannich-type reaction. *ACS Sustain. Chem. Eng.* **2019**, *7*, 4878–4888. [CrossRef]
26. Xu, G.J.; Hou, L.F.; Li, B.Y.; Wang, X.L.; Liu, L.; Li, N.; Wang, M.L.; Zhao, R.S. Facile preparation of hydroxyl bearing covalent organic frameworks for analysis of phenoxy carboxylic acid pesticide residue in plant-derived food. *Food Chem.* **2021**, *345*, 128749. [CrossRef]
27. Li, N.; Wu, D.; Li, X.T.; Zhou, X.X.; Fan, G.S.; Li, G.L.; Wu, Y.N. Effective enrichment and detection of plant growth regulators in fruits and vegetables using a novel magnetic covalent organic framework material as the adsorbents. *Food Chem.* **2020**, *306*, 125455. [CrossRef]
28. Wang, Y.F.; Mu, G.D.; Wang, X.J.; Zhang, F.; Li, Y.L.; Lu, D.J.; Chen, F.M.; Yang, M.L.; He, M.Y.; Liu, T. Fast construction of core-shell structured magnetic covalent organic framework as sorbent for solid-phase extraction of zearalenone and its derivatives prior to their determination by UHPLC-MS/MS. *Mikrochim. Acta* **2021**, *188*, 246. [CrossRef] [PubMed]
29. Mu, M.M.; Wang, Y.W.; Qin, Y.T.; Yan, X.L.; Li, Y.; Chen, L.G. Two-Dimensional Imine-Linked covalent organic frameworks as a platform for selective oxidation of olefins. *ACS Appl. Mater. Interfaces* **2017**, *9*, 22856–22863. [CrossRef]
30. Song, B.; Wei, H.; Wang, Z.Q.; Zhang, X.; Smet, M.; Dehaen, W. Supramolecular Nanofibers by Self-Organization of Bola-amphiphiles through a Combination of Hydrogen Bonding and π-π Stacking Interactions. *Adv. Mater.* **2007**, *19*, 416–420. [CrossRef]
31. Fan, Y.Y.; Liu, F.J.; He, W.Z.; Qin, Q.M.; Hu, D.Q.; Wu, A.B.; Jiang, W.B.; Wang, C. Screening of multi-mycotoxins in fruits by ultra-performance liquid chromatography coupled to ion mobility quadrupole time-of-flight mass spectrometry. *Food Chem.* **2022**, *368*, 130858. [CrossRef]
32. Tang, Y.; Mu, L.; Cheng, J.X.; Du, Z.X.; Yang, Y.Y. Determination of multi-class mycotoxins in apples and tomatoes by combined use of QuEChERS method and Ultra-High-Performance Liquid Chromatography Tandem mass spectrometry. *Food Anal. Method* **2020**, *13*, 1381–1390. [CrossRef]
33. *SN-T 4295-2015*; Determination of Alternaria Toxins Residues in Fruit and Vegetables for Export-LC-MS/MS Method. General Administration of Quality Supervision, Inspection and Quarantine of the People's Republic of China: Beijing, China, 2015.
34. Massart, R. Preparation of aqueous magnetic liquids in alkaline and acidic media. *IEEE Trans. Magn.* **1981**, *17*, 1247–1248. [CrossRef]
35. Runowski, M.; Lis, S. Synthesis, surface modification/decoration of luminescent–magnetic core/shell nanomaterials, based on the lanthanide doped fluorides ($Fe_3O_4/SiO_2/NH_2/PAA/LnF_3$). *J. Lumin.* **2016**, *170*, 484–490. [CrossRef]
36. Yin, S.J.; Zhao, C.P.; Jiang, H.; Lu, M.; Wang, Y.; Chen, H.; Yang, F.Q. Preparation of amino-functionalized covalent organic framework modified Fe_3O_4 nanoparticles for the selective enrichment of flavonoid glycosides. *Microchem. J.* **2021**, *164*, 105990. [CrossRef]
37. EC. Commission Decision 2002/657/EC Implementing Council Directive 96/23/EC concerning the performance of analytical methods and the interpretation of results. *Off. J. Eur. Communities* **2002**, *L221*, 8–36. Available online: https://www.legislation.gov.uk/eudn/2002/657 (accessed on 20 May 2022).
38. Zhao, Y.; Wan, L.H.; Bai, X.L.; Liu, Y.M.; Zhang, F.P.; Liu, Y.M.; Liao, X. Quantification of mycotoxins in vegetable oil by UPLC-MS/MS after magnetic solid-phase extraction. *Food Addit. Contam.* **2017**, *34*, 1201–1210. [CrossRef]
39. Xu, H.W.; Sun, J.D.; Wang, H.M.; Zhang, Y.Z.; Sun, X.L. Adsorption of aflatoxins and ochratoxins in edible vegetable oils with dopamine-coated magnetic multi-walled carbon nanotubes. *Food Chem.* **2021**, *365*, 130409. [CrossRef] [PubMed]

40. Li, W.K.; Zhang, H.X.; Shi, Y.P. Simultaneous determination of aflatoxin B_1 and zearalenone by magnetic nanoparticle filled amino-modified multi-walled carbon nanotubes. *Anal. Methods* **2018**, *10*, 3353–3363. [CrossRef]
41. Tang, Z.T.; Han, Q.R.; Yu, G.; Liu, F.; Tan, Y.Z.; Peng, C. Fe_3O_4@PDA/MIL-101(Cr) as magnetic solid-phase extraction sorbent for mycotoxins in licorice prior to ultrahigh-performance liquid chromatography-tandem mass spectrometry analysis. *Food Sci. Nutr.* **2022**, *10*, 2224–2235. [CrossRef] [PubMed]
42. Wang, C.; Fan, Y.; He, W.; Hu, D.; Wu, A.; Wu, W. Development and Application of a QuEChERS-Based Liquid Chromatography Tandem Mass Spectrometry Method to Quantitate Multi-Component Alternaria Toxins in Jujube. *Toxins* **2018**, *10*, 382. [CrossRef]
43. Xing, L.J.; Zou, L.J.; Luo, R.F.; Wang, Y. Determination of five Alternaria toxins in wolfberry using modified QuEChERS and ultra-high performance liquid chromatography-tandem mass spectrometry. *Food Chem.* **2020**, *311*, 125975. [CrossRef] [PubMed]

toxins

MDPI

Article

Development of Acid Hydrolysis-Based UPLC–MS/MS Method for Determination of *Alternaria* Toxins and Its Application in the Occurrence Assessment in Solanaceous Vegetables and Their Products

Hongxia Tang [1], Wei Han [1], Shaoxiang Fei [1], Yubo Li [1], Jiaqing Huang [1], Maofeng Dong [1,2,*], Lei Wang [3], Weimin Wang [1] and Ying Zhang [3]

1 Pesticide Safety Evaluation Research Center, Shanghai Academy of Agricultural Sciences, Shanghai 201403, China
2 Key Laboratory for Safety Assessment (Environment) of Agricultural Genetically Modified Organisms, Ministry of Agriculture and Rural Affairs, Institute for Agro-Food Standards and Testing Technology, Shanghai Academy of Agricultural Sciences, Shanghai 201403, China
3 School of Resources and Environment, Northeast Agricultural University, Harbin 150030, China
* Correspondence: dmf385@163.com; Tel.: +86-21-62203612; Fax: +86-21-62203612

Abstract: In this work, we proposed an acid hydrolysis-based analytical method for the detection of *Alternaria* toxins (ATs) in solanaceous vegetables and their products with solid-phase extraction (SPE) and ultrahigh-performance liquid chromatography–tandem mass spectrometry (UPLC–MS/MS). This study was the first to reveal that some compounds in the eggplant matrix bind to altenusin (ALS). Validation under optimal sample preparation conditions showed that the method met the EU criteria, exhibiting good linearity (R^2 > 0.99), matrix effects (−66.6–−20.5%), satisfying recovery (72.0–107.4%), acceptable precision (1.5–15.5%), and satisfactory sensitivity (0.05–2 μg/kg for limit of detection, 2–5 μg/kg for limit of quantification). Out of 393 marketed samples, only 47 samples were detected, ranging from 0.54–806 μg/kg. Though the occurrence ratio (2.72%) in solanaceous vegetables could be negligible, the pollution status in solanaceous vegetable products was much more serious, and the incidences were 41.1%. In the 47 contaminated samples, the incidences were 4.26% for alternariol monomethyl ether (AME), 6.38% for alternariol (AOH) and altenuene (ALT), 42.6% for tentoxin (TEN), and 55.3% for tenuazonic acid (TeA).

Keywords: *Alternaria* toxins; modified analytical method; solanaceous vegetables; solanaceous vegetable products; UPLC–MS/MS

Key Contribution: An acid hydrolysis-based analytical method had been development for the analysis of six *Alternaria* toxins and this proposed method was applied for their occurrence in solanaceous vegetables and solanaceous vegetable products in Shanghai, China.

1. Introduction

As secondary metabolites of *Alternaria* strains, *Alternaria* toxins (ATs) are readily produced during unfavorable climatic conditions, insect attacks, and the transport, processing, and storage of agricultural products, and are widely found in various foods, posing great potential risks to human and animal health due to their carcinogenic, teratogenic, and cytotoxic properties [1–4]. Although more than 70 ATs had been confirmed to be poisonous, contamination and dietary safety assessments had focused on only some ATs, including alternariol (AOH), alternariol monomethyl ether (AME), altenuene (ALT), altenusin (ALS), tentoxin (TEN), and tenuazonic acid (TeA) [5–7]. Previous studies had revealed that AT contamination in vegetables [8], fruits [8–10], tomato products [11], wine [12,13], cereal grains [14,15], dried fruits [16], and juices [12] was widespread, and a high incidence had been observed in processed products, ranging from 62–100% [11,12,17–19]. Among these, vegetables and their products were also considered to be some of the most highly AT-contaminated foods [20]. Therefore, due to the considerable consumption of these products and their high susceptibility to *Alternaria* infection, the occurrence of ATs on vegetables and their products in China had been studied, revealing the contamination status and providing data regarding dietary safety risks, which have rarely been reported in China compared to other countries.

As described in these reports, many analytical methods have been proposed for the detection of ATs or ATs and other mycotoxins in various food matrixes. Compared with gas chromatography-mass spectrometry (GC-MS) and liquid chromatography–diode array detection (LC–DAD), liquid chromatography–tandem mass spectrometry (LC–MS/MS) was the best choice for detecting ATs due to its excellent sensitivity and high efficiency [2,6,8, 9,21–23]. However, for sample pretreatment, acetonitrile or formic acid-acetonitrile were the most common extraction solvents following solid-phase extraction (SPE), the QuEChERS (quick, easy, inexpensive, effective, rugged, and safe) method, or direct injection with the isotope internal standard method [4,19,24,25]. Given the availability of different matrixes, SPE cartridges (HLB) and adsorption materials (GCB and C_{18}) had been optimized and utilized to clean up the samples [1,5,10,22]. However, among these studies, only a few had achieved the simultaneous detection of six ATs (AOH, AME, ALT, ALS, TEN, and TeA) in wine [12], herbs [6], and juices [12] via the QuEChERS method or direct analysis. It is well known that the adjustment and optimization of the sample pretreatment are necessary for accuracy when the analytes and matrixes have been changed. As shown in this work, the existing analytical methods would not be applicable because of the binding residue of ALS. Therefore, a new method needed to be established to ensure the accuracy of the monitoring results.

Among the various vegetable species, including cucumber, tomato, eggplant, and pepper, solanaceous vegetables undoubtedly play an important role in vegetable production, household consumption, and processed products [26]. Solanaceous vegetables are well known to be susceptible to *Alternaria* infection, which occurs more widely in processed products [11]. Therefore, tomato and its products are the main vegetables used for AT contamination monitoring and safety evaluation. However, in China, relevant reports have tended to focus more on the development of methods than on the investigation of contamination status, and due to differences in consumption habits, it has been more realistic and in line with the requirements for food safety assessment to carry out pollution monitoring of ATs in the main foods consumed in China, such as chili paste and cherry tomato. In this work, using solanaceous vegetables and their products as the subject, an AT occurrence assessment was first carried out to determine the contamination level in Shanghai, which is often overlooked and has become a monitoring gap in food safety evaluations in China.

Based on validation via the reported analytical methods, a new and reliable analytical method was developed and validated for the detection of six ATs (AOH, AME, ALT, ALS, TEN, and TeA) in solanaceous vegetables and their products using acid hydrolysis, SPE, and ultrahigh-performance liquid chromatography–tandem mass spectrometry (UPLC–MS/MS). To ensure the reliability and credibility of the research results, 939 commercial samples were collected, including chili paste, eggplant, ketchup, pepper, and tomato samples. Finally, a survey of AT contamination in solanaceous vegetables and their production in the Shanghai area was proposed, representing the first time that the regional contamination levels of six ATs in solanaceous vegetables and their products had been studied in China.

2. Results and Discussion

2.1. Optimization of Extraction

2.1.1. Validation via Reported Methods

By using previously reported methods [10,16], validation was carried out via extraction with acetonitrile, but the matrix was replaced with chili paste, eggplant, and tomato. As shown in Figure 1A, although the results showed that most recoveries were acceptable (74.3–104.7%), the value for ALS in eggplant was only 16.2% when the spiked level was 1000 μg/kg. Then, regardless of whether the volume of the extraction solution was expanded, or whether the extraction solvent (such as acetone and methanol) was adjusted and some substances (such as formic acid and ammonium acetate) were added, there was no obvious improvement in the recovery of ALS in eggplant.

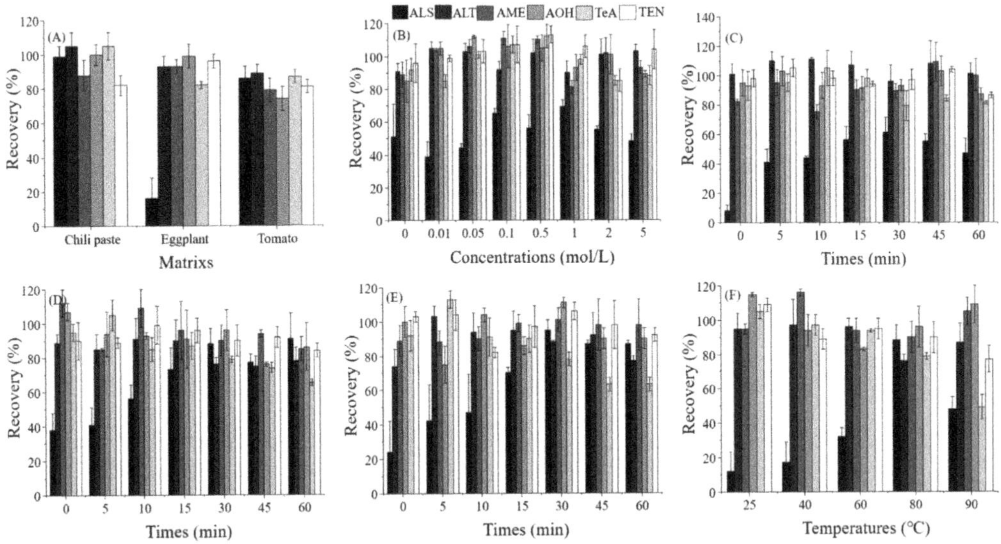

Figure 1. Optimization of the extraction conditions (1000 μg/kg, *n* = 5). (**A**) Recoveries of *Alternaria* toxins from chili paste, eggplant, and tomato by extraction with acetonitrile. (**B**) Optimization of the hydrochloric acid concentration at 80 °C for 15 min in eggplant. (**C**) Optimization of the heating time with 0.1 mol/L hydrochloric acid at 80 °C in eggplant. (**D**) Optimization of the heating time with 0.5 mol/L hydrochloric acid at 80 °C in eggplant. (**E**) Optimization of the heating time with 1 mol/L hydrochloric acid at 80 °C in eggplant. (**F**) Optimization of the heating temperatures with 0.5 mol/L hydrochloric acid for 30 min in eggplant.

In line with the structural formula of ALS, there were two hydroxyl groups and one carboxyl group, and the bound residues may have formed because of the presence of alkaline compounds in eggplant; in comparison, good results were obtained in tomato due to its acidity. Additionally, this result was confirmed by spiking 2 mL of ALS standard solution in acetonitrile (1000 μg/L) into 2 g of eggplant. After vortexing for 15 min, the concentration of ALS in acetonitrile was only 380 μg/L, showing the existence of a binding residue. Therefore, as the first step, it was necessary to eliminate the binding residue between ALS and the eggplant matrix to improve the recovery of ALS.

2.1.2. Optimization of Extraction Conditions

According to the relevant literature, hydrochloric acid and calefaction were selected to remove the binding residue, which was a conventional and effective approach used in some analytical methods [23,27–30]. Focusing on the concentration of hydrochloric acid, heating time, and heating temperature, the optimization of extraction conditions was carried out using the recovery test in 2.0 g of eggplant (1000 μg/kg) and direct analysis after extraction in 8 mL of acetonitrile.

After exposure to 80 °C for 15 min, the effect of the different concentrations of hydrochloric acid (2 mL) on AT recovery was first estimated. The volume of hydrochloric acid solvent was 2 mL, and the concentrations of hydrochloric acid selected were 0, 0.01, 0.05,

0.1, 0.5, 1, 2, and 5 mol/L. Figure 1B showed that the recovery of ALS was closely related to the concentration of hydrochloric acid, but for other ATs, the values were not significantly changed, ranging from 81.1–113.4%. When the concentration of hydrochloric acid was 0.1, 0.5, or 1 mol/L, the recoveries (56.1–68.9%) of ALS were distinctly higher than the values (38.9–55.1%) under other conditions reported for the analytical method, which was a significant improvement over the reported analytical methods. Therefore, at three hydrochloric acid concentrations (0.1, 0.5 and 1 mol/L) and 80 °C, the recoveries were determined while varying the heating time. Figure 1C–E showed the results at the three different hydrochloric acid concentrations. At heating times ranging from 0–60 min, the recoveries of ALT, AME, AOH, and TEN were stable and acceptable, with values of 74.1–111.7%. Although the range (77.8–113.3%) for TeA was satisfactory, the values decreased when increasing the hydrochloric acid concentration or heating time, especially when the heating time was 45 min or 60 min. However, for ALS, the heating time had a significant effect on the results, and all the recoveries increased as the heating time was extended under the three hydrochloric acid concentrations. At hydrochloric acid concentrations of 0.5 mol/L and 1 mol/L and heating times of 30–60 min, the recoveries of ALS were fairly good, ranging from 77.4–94.6%. Therefore, 0.5 mol/L and 30 min were selected as the best hydrochloric acid concentration and heating time, respectively.

Finally, based on the above tests, the heating temperature was also investigated, and five heating temperatures (25, 40, 60, 80, and 90 °C) were included in the experiment. The results (Figure 1F) showed that the recoveries (75.8–115.5%) of ALT, AME, AOH, and TEN were all acceptable and were consistent with the results of previous experiments. For TeA, when the heating temperature was 90 °C, the recovery was only 49.0%, but in the lower temperature conditions, the values were much more stable and acceptable (77.0–108.9%). The recoveries of ALS increased when increasing the heating temperature (25–80 °C), ranging from 12.4% to 87.8%. However, when the heating temperature was 90 °C, similar to TeA, the value decreased to 48.4%, revealing that both ALS and TeA were degraded at higher temperatures or longer heating times.

Therefore, the following sample pretreatment conditions were established: 2 mL of 0.5 mol/L hydrochloric acid, 30 min of heating time, and a heating temperature of 80 °C.

2.2. Optimization of Clean-Up Conditions

To ensure the accuracy of the method and based on other papers, acetonitrile was used to extract the target compounds from the hydrochloric acid mixture, and the extraction solvent volume was optimized. With the volume of acetonitrile set to 2, 4, 6, 8, 12, 16, or 20 mL, the recoveries of ATs in eggplant were determined. As shown in Figure 2A, when the volume of acetonitrile was in the range of 4–20 mL, the recoveries of ALS, ALT, AME, AOH, and TEN were 72.2–91.0%, but for TeA, the volume of acetonitrile needed to be at least 8 mL, attaining recoveries of 81.7–87.3%. Therefore, 8 mL was selected as the volume of acetonitrile for the extraction solvent.

As shown in the literature and for the standards, SPE is a powerful clean-up tool in analytical methods [9,13,16,22]. Although some papers have described the application of SPE for the detection of ATs, to date, simultaneous analysis of all six major toxins via SPE has not been reported. Therefore, in this work, we attempted to select a suitable SPE column due to the change in the target object. Eleven kinds of commercial SPE columns were tested. As shown in Figure 2B, after clean-up using the SPE columns and elution with 2 mL of acetonitrile, except for the C_8 columns, none of the 10 columns could achieve the simultaneous purification of six target compounds, attaining recoveries of 0.2–123.1%, which did not meet the requirements of the EU guidelines (70–120%). However, with the C_8 columns, the recoveries (77.7–112.7%) of the six ATs were much better and met the requirements of analytical methods. Therefore, the C_8 columns were selected as the clean-up tool.

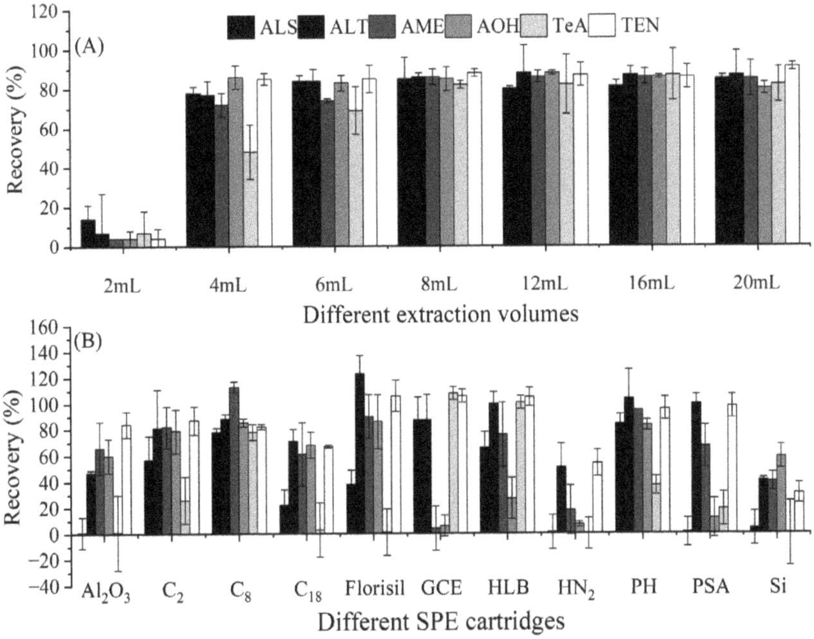

Figure 2. Optimization of the extraction and SPE columns in eggplant (1000 μg/kg, $n = 3$). (**A**) Optimization of the extraction volume. (**B**) Selection of the SPE column.

2.3. Validation Method

For validation, five typical solanaceous vegetables and vegetable products, namely chili paste, eggplant, ketchup, pepper, and tomato, were selected as the evaluation matrixes. Moreover, six ATs were isolated through chromatographic separation (Figure 3). As shown in Table 1, in different concentration ranges and matrixes, the target compounds exhibited a good linear relationship, with correlation coefficients exceeding 0.998. Then, for the MEs, as shown in Figure 4, the values in these matrixes indicated obvious matrix inhibition effects, and the recoveries in unpurified samples and purified samples were −94.7–−47.1% and −66.6–−20.5%, respectively. All the absolute values for the ME were significantly lower than those of the unpurified samples, proving that the proposed method had an obvious impurity removal effect. However, because the values exceeded ±20%, the matrix standard solution was indispensable for accurate quantitative analysis. The accuracy and precision were determined via the spiked recovery test. Table 1 showed that in the five matrixes, all the recoveries and RSDs were acceptable, reaching 72.0–107.4% and below 15.5%, respectively. Calculated according to three signal/noise ratios, the LODs of the six ATs ranged from 0.05–2 μg/kg in the five matrixes. The LOQ of AME was 5 μg/kg, and for other ATs, the LOQ was 2 μg/kg. Therefore, the validation results demonstrated that the proposed analytical method was acceptable and satisfactory, complying with EU criteria [31].

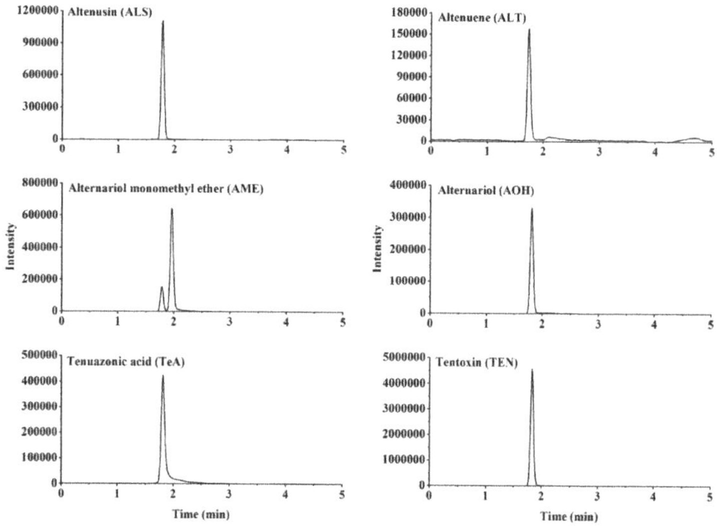

Figure 3. Chromatograms of *Alternaria* toxins in the mixed standard solution. The concentration is 100 μg/L.

Table 1. Method validation for *Alternaria* toxins in solanaceous vegetables and their products.

Compound	Matrix	Linear Range (μg/L)	Calibration Curve	Correlation Coefficient (R^2)	LOD (μg/kg)	Average Recovery Rate (%) (RSD (%)) (n = 6)		
						2 μg/kg	10 μg/kg	1000 μg/kg
ALS	Chili paste	1–1000	y = 4802x + 14064	1.000	0.5	77.2 (5.7)	76.0 (2.6)	82.8 (2.8)
	Eggplant	1–1000	y = 4548x + 39771	0.999	0.5	75.2 (3.4)	81.0 (8.2)	75.6 (3.0)
	Ketchup	1–1000	y = 5456x + 26143	1.000	0.5	75.6 (3.6)	72.0 (2.8)	80.0 (6.4)
	Pepper	1–1000	y = 3766x + 21551	1.000	0.5	72.4 (2.3)	76.0 (4.5)	81.6 (4.6)
	Tomato	1–1000	y = 5304x + 47966	0.999	0.5	76.8 (3.4)	76.6 (3.5)	82.8 (3.4)
ALT	Chili paste	1–1000	y = 1400x + 5590	1.000	0.5	100.4 (15.5)	96.2 (7.5)	91.2 (3.5)
	Eggplant	1–1000	y = 1963x − 1128	0.999	0.5	85.4 (6.7)	85.8 (5.5)	91.2 (4.9)
	Ketchup	1–1000	y = 1780x + 1184	1.000	0.5	97.4 (13.0)	82.6 (5.4)	98.0 (12.5)
	Pepper	1–1000	y = 1219x + 6487	1.000	0.5	96.4 (13.0)	91.0 (4.2)	92.6 (6.6)
	Tomato	1–1000	y = 1329x + 9554	1.000	0.5	103.6 (11.6)	89.0 (4.3)	90.8 (5.9)
AME	Chili paste	5–1000	y = 290x + 2029	0.999	2	93.0 (3.1) [a]	87.8 (12.8)	92.6 (8.0)
	Eggplant	5–1000	y = 541x + 5632	0.999	2	105.2 (13.7) [a]	91.0 (1.3)	89.8 (4.1)
	Ketchup	5–1000	y = 627x + 6041	0.999	2	85.4 (10.1) [a]	86.2 (3.3)	92.8 (10.8)
	Pepper	5–1000	y = 484x + 1091	1.000	2	103.2 (12.2) [a]	93.2 (6.3)	89.2 (2.9)
	Tomato	5–1000	y = 607x + 13240	0.998	2	87.6 (14.2) [a]	96.6 (3.2)	96.0 (2.4)
AOH	Chili paste	5–1000	y = 840x + 6452	1.000	2	86.0 (14.1)	93.8 (7.2)	92.2 (3.5)
	Eggplant	1–1000	y = 1291x + 9660	1.000	0.5	86.8 (5.3)	101.4 (2.6)	95.2 (3.6)
	Ketchup	1–1000	y = 1559x + 11605	0.998	0.5	107.4 (4.7)	89.2 (6.6)	95.8 (10.8)
	Pepper	1–1000	y = 875x + 8618	1.000	0.5	85.4 (14.8)	92.6 (5.3)	92.4 (4.2)
	Tomato	1–1000	y = 1056x + 12184	0.999	0.5	84.2 (15.1)	89.8 (7.1)	93.0 (3.8)
TeA	Chili paste	0.5–1000	y = 9007x + 59591	1.000	0.2	73.2 (3.5)	75.4 (4.8)	78.6 (3.6)
	Eggplant	0.1–1000	y = 12988x + 70681	1.000	0.05	74.8 (2.4)	77.2 (2.8)	76.8 (6.4)
	Ketchup	0.1–1000	y = 13465x + 36066	1.000	0.05	75.4 (4.5)	76.2 (3.3)	82.0 (4.6)
	Pepper	0.5–1000	y = 10670x + 22379	1.000	0.2	72.6 (3.3)	75.2 (1.5)	77.8 (3.7)
	Tomato	0.5–1000	y = 10876x + 70959	1.000	0.2	75.8 (2.9)	75.8 (3.4)	81.8 (7.7)
TEN	Chili paste	0.1–1000	y = 12846x + 23279	1.000	0.05	87.6 (8.2)	80.6 (5.5)	96.0 (2.4)
	Eggplant	0.1–1000	y = 14980x + 31438	1.000	0.05	86.8 (1.5)	84.0 (2.4)	93.6 (3.8)
	Ketchup	0.1–1000	y = 15575x + 12590	1.000	0.05	94.0 (11.2)	78.2 (4.2)	96.0 (10.7)
	Pepper	0.1–1000	y = 12366x + 25385	1.000	0.05	90.2 (2.1)	94.6 (4.1)	93.4 (3.5)
	Tomato	0.1–1000	y = 14814x + 26054	1.000	0.05	94.4 (10.6)	92.2 (5.9)	92.4 (3.9)

[a]: The lowest spiked level of AME was 5 μg/kg.

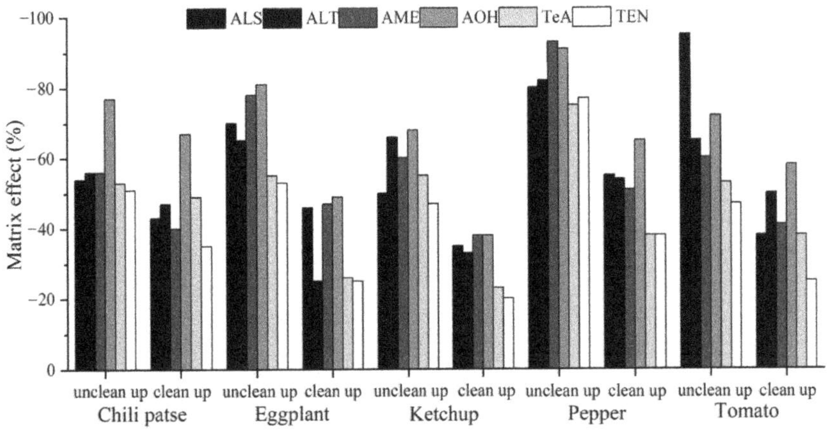

Figure 4. Matrix effects of solanaceous vegetables and their products.

2.4. Method Application

In 2019, a total of 939 commercialized samples were randomly collected from supermarkets, online vendors and farmer's markets in Shanghai, China. The collected samples comprised 33 chili paste, 244 eggplant, 23 ketchup, 450 pepper, and 189 tomato (including cherry tomato) samples. As shown in Table 2, overall, the results showed that only 5.00% were (47/939) positive samples among the collected samples, and except for ALS, the other five target compounds were detected, with levels ranging from 0.54–806 μg/kg. Although the AT contamination range in the solanaceous vegetables was low (only 2.72%; 24/883), the rate in the solanaceous vegetable products was much higher, reaching 41.1% (23/56). Among the samples, the rates of positive samples, from high to low, were as follows: 65.2% (15/23) for ketchup, 24.2% (8/33) for chili paste, 3.70% (7/189) for tomato, 2.46% (6/244) for eggplant, and 2.44% (11/450) for pepper. In general, the contamination rates (2.44–3.70%) in solanaceous vegetables were not significantly different among eggplant, pepper, and tomato, but the values in ketchup and chili paste reached up to 65.2% and 24.2%, respectively, proving that AT contamination in solanaceous vegetable products was greater than that in solanaceous vegetables. Although there had been reports on AT contamination in vegetables, some studies had focused on other foods [5,10,12–14]. Among them, some studies in several countries revealed the AT contamination in ketchup, and the positive rates of the ketchup samples were much higher (40–92%), and the pollution level of ATs ranged from 42–814 μg/kg, which was similar to the results of our study [11,18,19,32,33].

In contrast, AOH, AME, ALT, TEN, and TeA were detected in solanaceous vegetables and solanaceous vegetable products, but the degree of contamination varied with the matrix and AT. Among them, TeA and TEN were the two toxins with the highest contamination frequencies, accounting for 55.3% (26/47) and 42.6% (20/47) of the positive samples, respectively. TeA and TEN were recognized as the most widespread of these toxins, present in materials such as ketchup [11,32], wolfberry [5], grain [17], food commodities [34], and grain-based products [14]. As the most frequently detected toxin, the concentrations of TeA in the five matrixes were in the range not detected (ND) to 19.3 μg/kg in chili paste, ND–214 μg/kg in eggplant, ND–337 μg/kg in ketchup, ND-806 μg/kg in pepper, and ND in tomato. The concentrations of TEN in the five matrixes were in the range of ND–0.56 μg/kg in chili paste, ND–6.44 μg/kg in eggplant, ND–2.81 μg/kg in ketchup, ND–13.1 μg/kg in pepper, and ND–6.43 μg/kg in tomato. The TeA-positive samples were solanaceous vegetable products (four chili paste and fourteen ketchup) and related solanaceous vegetables (two eggplant and six pepper samples). This was very different from the result for TEN, as the numbers of TEN-positive samples of solanaceous vegetables and solanaceous vegetable products were fifteen (four eggplant, five pepper, and six tomato samples) and five (two chili paste and three ketchup samples), respectively.

Table 2. Contamination by *Alternaria* toxins in solanaceous vegetables and their products in Shanghai, China.

Matrix	N^{pos}/N	N^{qual}	N^{quant}	Avg^{quan} (μg/kg)	Min^{quan} (μg/kg)	Max^{quan} (μg/kg)
Chili paste (N = 33)						
ALS	0/33	0	0	-	-	-
ALT	0/33	0	0	-	-	-
AME	0/33	0	0	-	-	-
AOH	0/33	0	0	-	-	-
TeA	4/33	0	4	12.1	3.44	19.3
TEN	2/33	2	0		-	-
Eggplant (N = 244)						
ALS	0/244	0	0	-	-	-
ALT	0/244	0	0	-	-	-
AME	0/244	0	0	-	-	-
AOH	0/244	0	0	-	-	-
TeA	2/244	0	2	128	41.8	214
TEN	4/244	2	2	4.27	2.10	6.44
Ketchup (N = 23)						
ALS	0/23	0	0	-	-	-
ALT	0/23	0	0	-	-	-
AME	2/23	0	2	12.3	11.9	12.6
AOH	3/23	0	3	8.06	5.75	10.3
TeA	14/23	0	14	85.1	5.61	337
TEN	3/23	2	1	2.81	2.81	2.81
Pepper (N = 450)						
ALS	0/450	0	0	-	-	-
ALT	2/450	0	2	18.2	16.3	20.0
AME	0/450	0	0	-	-	-
AOH	0/450	0	0	-	-	-
TeA	6/450	0	6	157	7.29	806
TEN	5/450	2	3	8.07	2.28	13.1
Tomato (N = 189)						
ALS	0/189	0	0	-	-	-
ALT	1/189	0	1	5.29	5.29	5.29
AME	0/189	0	0	-	-	-
AOH	0/189	0	0	-	-	-
TeA	0/189	0	0	-	-	-
TEN	6/189	4	2	4.58	2.72	6.43

N^{pos}/N: Number of positive samples/number of samples analyzed, N^{qual}: number of samples above LOD but below LOQ, N^{quant}: number of quantified samples, Avg^{quan}: mean concentrations of *Alternaria* toxins in quantified samples, Min^{quan}: minimum concentrations of *Alternaria* toxins in quantified samples, Max^{quan}: maximum concentrations of *Alternaria* toxins in quantified samples.

However, for AOH, AME, and ALT, there were very few positive samples, with proportions of 6.38% (3/47), 4.26% (2/47), and 6.38% (3/47), respectively. Additionally, the concentrations were also much lower than those of TeA or TEN, with values of 5.29–20 μg/kg for AOH, 11.9–12.6 μg/kg for AME, and 5.75–10.3 μg/kg for ALT. All the samples positive for AOH and AME were ketchup samples; in contrast, ALT was detected only in solanaceous vegetables (one tomato and one pepper).

Therefore, from the above survey results, AT contamination in solanaceous vegetable products is higher and more widespread than that in solanaceous vegetables. Compared with the reported papers, the rates of positive samples or the pollution level in ketchup were prominent, which was similar to the results of our study [32,33]. The low occurrence rate of *Alternaria* infection may be due to the shorter preservation and transportation time of the commercialized solanaceous vegetables. The other reason may be the concentration factors in the processing of the vegetable products. For example, the dry matter of tomatoes is 5–6%, while in ketchup it is 28–30%; this difference may cause the higher detection rates of ATs in the vegetable products. Anyways, the current situation of severe contamination in solanaceous vegetable products led to concerns regarding food safety, and it prompted us to study the contamination in processed foods.

3. Conclusions

In light of the ALS binding residue present in the alkaline matrix, the strategy involving the addition of acid and heating at an optimal temperature and time was proposed. The development method was first established for detecting six ATs in eggplant and was also validated in other solanaceous vegetables and their products for monitoring the six ATs in accordance with the EU criteria [31].

Furthermore, a survey of AT contamination was carried out in Shanghai, China. The experimental results for a total of 939 commercialized samples indicated that the AT levels in solanaceous vegetables were negligible with (2.72%, 24/883), but the incidence in solanaceous vegetable products was much higher (41.1%, 23/56). Among the six toxins detected (AOH, AME, ALT, TEN, and TeA), as the most frequently detected ATs, TEN and TeA were present at similar levels, with a detection rate of 42.6–55.3% in the positive samples, which was much higher than the values for AOH, AME, and ALT (4.26–6.38%). This study was the first to describe the issue of mycotoxin-bound residues in a plant origin matrix, and the proposed analytical method also led to the creation of a new preprocessing approach. Additionally, this study revealed AT contamination in solanaceous vegetables and their products in China, which revealed the high detection rate in the solanaceous vegetable products.

4. Materials and Methods

4.1. Chemicals, Reagents, and Materials

The AOH standard (95% purity) was purchased from Cayman Chemical (Ann Arbor, MI, USA) and the ALT standard (98% purity) was purchased from Toronto Research Chemicals (North York, ON, Canada). The AME (99.5% purity), TeA (99.5% purity), and ALS (99.4% purity) standards were purchased from AdipoGen Life Sciences (San Diego, CA, USA), and the TEN standard was purchased from LKT Laboratories (Sao Paulo, Brazil). HPLC-grade solvents (acetonitrile and methanol) were purchased from Merck (Darmstadt, Germany). Octylsilane (C_8) (100 mg, 1 mL), ethylsilane (C_2) (100 mg, 1 mL), phenyl (PH) (100 mg, 1 mL), primary secondary amine (PSA) (100 mg, 1 mL), graphitized carbon black (GCB) (100 mg, 1 mL), and alumina (Al_2O_3) (100 mg, 1 mL) columns were obtained from Shimadzu Corp. (Kyoto, Japan). Aminopropyl (HN_2) (500 mg, 6 mL) and florisil (500 mg, 3 mL) columns were obtained from Agilent Technologies Inc. (Wilmington, DE, USA). Octadecylsilane (C_{18}) (200 mg, 3 mL), hydrophilic–lipophilic balance (HLB) (60 mg, 3 mL), and silica (Si) (1 g, 6 mL) columns were obtained from ANPEL Laboratory Technologies Inc. (Shanghai, China).

4.2. Sample Preparation

Approximately 2.0 g of homogenized sample were weighed into 50 mL polytetrafluoroethylene (PTFE) centrifuge tubes. After adding 2 mL of 0.5 mol/L hydrochloric acid solvent, the tube was heated in an 80 °C water bath for 30 min. Then 8 mL of acetonitrile were added to the tube at room temperature. After shaking vigorously for 3 min, 2 g of NaCl were added to the tube, and the tube was shaken for 30 s and centrifuged at 4500 rpm for 5 min in preparation for cleanup.

The supernatant was transferred into the SPE cartridge for the cleanup. In the SPE cleanup procedure, the C8 cartridge was pre-eluted with 1 mL of acetonitrile, then 4 mL of the extraction solution were loaded into the C_8 cartridges, and 2 mL of acetonitrile were loaded onto the cartridge for elution. The collected eluate was dried in a 60 °C water bath under nitrogen. The sample was redissolved in 1 mL of acetonitrile and passed through a 0.22 μm PTFE filter for analysis.

4.3. Apparatus and UPLC–MS/MS Conditions

Detection was performed with a UPLC-MS/MS instrument (triple–quadrupole mass spectrometer) from Shimadzu (Kyoto, Japan) equipped with an electrospray ionization source. Separation of the ATs was achieved using a Waters ACQUITY UPLC BEH C18 (2.1 mm × 50 mm, 1.7 μm) column (Waters, Milford, MA, USA). The injection volume, column temperature, and flow rate were set at 5 μL, 40 °C, and 0.4 mL/min, respectively. The mobile phase included acetonitrile (A) and 0.1% formic acid water, with the following elution gradient: 0–0.5 min (10% A), 0.5–1.5 min (10–90% A), 1.5–3.0 min (90% A), 3.0–3.5 min (90–10% A), and 3.5–5.0 min (10% A).

The low-energy collision dissociation tandem–mass spectrometric analysis (CID–MS/MS) conditions were set using the following parameters: 1.5 mL/min nebulizing gas flow (99.9%, N_2), 400 °C heat block temperature, 15 mL/min drying gas flow (99.9%, N_2), 3.5 kV interface voltage, and 230 kPa collision-induced dissociation gas pressure (99.999%, Ar). After automatic optimization, the MS parameters of the ATs were determined and are shown in Table 3. Quantification was performed using multiple reaction monitoring (MRM) of selected precursor ion → product ions transitions.

Table 3. MS/MS parameters for analyzing *Alternaria* toxins.

Compound	Structure	Retention Time (min)	Precursor Ion (*m/z*)	Product Ion (*m/z*)	Q1 Pre Bias (v)	CE (v)	Q3 Pre Bias (v)
ALS	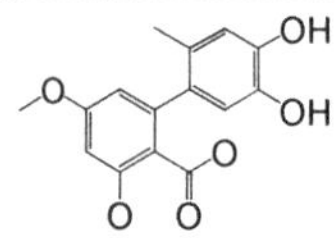 $C_{15}H_{12}O_6$	1.77	288.80	230.10 * 245.10	21.0 21.0	21 15	15.0 11.0
ALT	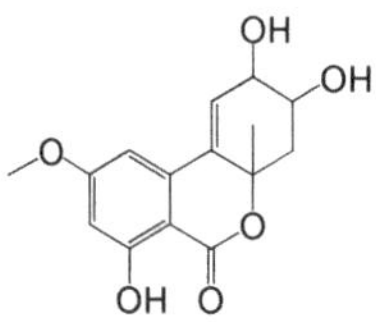 $C_{15}H_{16}O_6$	1.73	293.00	257.10 * 275.10	−15.0 −15.0	−16.0 −9.0	−30.0 −19.0
AME	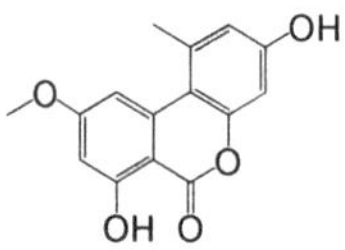 $C_{15}H_{12}O_5$	1.95	273.00	128.10 * 258.00	−15.0 −15.0	−46.0 −27.0	−25.0 −18.0
AOH	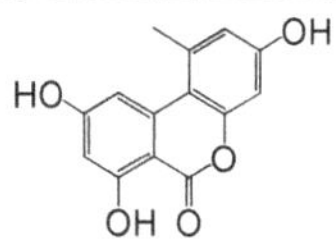 $C_{14}H_{10}O_5$	1.80	259.00	185.10 * 213.10	−14.0 −14.0	−32.0 −26.0	−12.0 −22.0
TeA	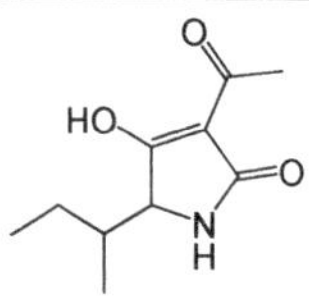 $C_{10}H_{15}O_3N$	1.80	198.10	125.00 * 153.10	−11.0 −11.0	−17.0 −14.0	−24.0 −16.0
TEN	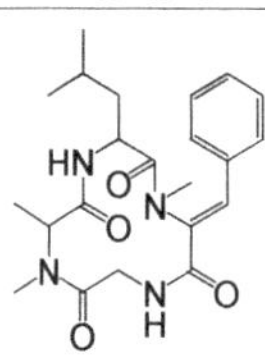 $C_{22}H_{30}N_4O_4$	2.12	415.20	199.20 * 312.20	−13.0 −12.0	−15.0 −23.0	−12.0 −12.0

*: Quantitative ion.

4.4. Method Validation

As an essential step, validation of the analytical method was carried out by determining a series of parameters, including the linear curve, matrix effect (ME), accuracy, precision (relative standard deviations, RSDs), limit of detection (LOD), and limit of quantitation (LOQ). A series of standard solutions with five concentrations were prepared, and the linear curve was described using the concentrations and the instrument response. The ME was obtained via the following formula: ME = (the slope of calibration curves in the matrix—the slope of calibration curves in solvent)/the slope of calibration curves in solvent $\times$ 100%. If the ME was within $\pm$20%, it could be ignored; otherwise, there was an obvious ME, and a matrix standard solution was necessary to ensure the accuracy of the analytical methods [35]. After spiking with three different levels and five repetitions in five matrixes (chili paste, eggplant, ketchup, pepper, and tomato), the accuracy and precision of the proposed method were determined. The LOD was the concentration corresponding to three times the signal/noise ratio [7], and the LOQ was considered the lowest spiked level [36].

Author Contributions: Conceptualization, M.D.; methodology, H.T. and M.D.; validation, S.F.; formal analysis, J.H., Y.Z. and L.W.; data curation, H.T. and W.H.; writing—review and editing, M.D.; language polishing, Y.L.; funding acquisition, W.W. All authors have read and agreed to the published version of the manuscript.

Funding: This work was supported by the Shanghai Agriculture Applied Technology Development Program, China [Grant No. X20210302].

Institutional Review Board Statement: Not applicable.

Informed Consent Statement: Not applicable.

Data Availability Statement: Not applicable.

Conflicts of Interest: The authors declare no conflict of interest.

References

1. Dong, H.; Xian, Y.; Xiao, K.; Wu, Y.; Zhu, L.; He, J. Development and comparison of single-step solid phase extraction and QuEChERS clean-up for the analysis of 7 mycotoxins in fruits and vegetables during storage by UHPLC-MS/MS. *Food Chem.* **2019**, *274*, 471–479. [CrossRef] [PubMed]
2. Rodriguez-Carrasco, Y.; Molto, J.C.; Manes, J.; Berrada, H. Development of a GC-MS/MS strategy to determine 15 mycotoxins and metabolites in human urine. *Talanta* **2014**, *128*, 125–131. [CrossRef] [PubMed]
3. Hessel-Pras, S.; Kieshauer, J.; Roenn, G.; Luckert, C.; Braeuning, A.; Lampen, A. In vitro characterization of hepatic toxicity of Alternaria toxins. *Mycotoxin Res.* **2019**, *35*, 157–168. [CrossRef]
4. Chen, A.; Mao, X.; Sun, Q.; Wei, Z.; Li, J.; You, Y.; Zhao, J.; Jiang, G.; Wu, Y.; Wang, L.; et al. Alternaria Mycotoxins: An Overview of Toxicity, Metabolism, and Analysis in Food. *J. Agric. Food Chem.* **2021**, *69*, 7817–7830. [CrossRef] [PubMed]
5. Xing, L.; Zou, L.; Luo, R.; Wang, Y. Determination of five Alternaria toxins in wolfberry using modified QuEChERS and ultra-high performance liquid chromatography-tandem mass spectrometry. *Food Chem.* **2020**, *311*, 125975. [CrossRef]
6. Zhao, X.; Liu, D.; Yang, X.; Zhang, L.; Yang, M. Detection of seven Alternaria toxins in edible and medicinal herbs using ultra-high performance liquid chromatography-tandem mass spectrometry. *Food Chem. X* **2022**, *13*, 100186. [CrossRef]
7. Qiao, X.; Zhang, J.; Yang, Y.; Yin, J.; Li, H.; Xing, Y.; Shao, B. Development of a simple and rapid LC-MS/MS method for the simultaneous quantification of five Alternaria mycotoxins in human urine. *J. Chromatogr. B Anal. Technol. Biomed. Life Sci.* **2020**, *1144*, 122096. [CrossRef]
8. Van de Perre, E.; Deschuyffeleer, N.; Jacxsens, L.; Vekeman, F.; Van Der Hauwaert, W.; Asam, S.; Rychlik, M.; Devlieghere, F.; De Meulenaer, B. Screening of moulds and mycotoxins in tomatoes, bell peppers, onions, soft red fruits and derived tomato products. *Food Control* **2014**, *37*, 165–170. [CrossRef]
9. Wang, M.; Jiang, N.; Xian, H.; Wei, D.; Shi, L.; Feng, X. A single-step solid phase extraction for the simultaneous determination of 8 mycotoxins in fruits by ultra-high performance liquid chromatography tandem mass spectrometry. *J. Chromatogr. A* **2016**, *1429*, 22–29. [CrossRef]
10. Guo, W.; Fan, K.; Nie, D.; Meng, J.; Huang, Q.; Yang, J.; Shen, Y.; Tangni, E.K.; Zhao, Z.; Wu, Y.; et al. Development of a QuEChERS-Based UHPLC-MS/MS Method for Simultaneous Determination of Six Alternaria Toxins in Grapes. *Toxins* **2019**, *11*, 87. [CrossRef]

11. Noser, J.; Schneider, P.; Rother, M.; Schmutz, H. Determination of six Alternaria toxins with UPLC-MS/MS and their occurrence in tomatoes and tomato products from the Swiss market. *Mycotoxin Res.* **2011**, *27*, 265–271. [CrossRef]
12. Zwickel, T.; Klaffke, H.; Richards, K.; Rychlik, M. Development of a high performance liquid chromatography tandem mass spectrometry based analysis for the simultaneous quantification of various Alternaria toxins in wine, vegetable juices and fruit juices. *J. Chromatogr. A* **2016**, *1455*, 74–85. [CrossRef]
13. Fan, C.; Cao, X.; Liu, M.; Wang, W. Determination of Alternaria mycotoxins in wine and juice using ionic liquid modified countercurrent chromatography as a pretreatment method followed by high-performance liquid chromatography. *J. Chromatogr. A* **2016**, *1436*, 133–140. [CrossRef] [PubMed]
14. Walravens, J.; Mikula, H.; Rychlik, M.; Asam, S.; Ediage, E.N.; Di Mavungu, J.D.; Van Landschoot, A.; Vanhaecke, L.; De Saeger, S. Development and validation of an ultra-high-performance liquid chromatography tandem mass spectrometric method for the simultaneous determination of free and conjugated Alternaria toxins in cereal-based foodstuffs. *J. Chromatogr. A* **2014**, *1372*, 91–101. [CrossRef]
15. Ji, X.; Xiao, Y.; Wang, W.; Lyu, W.; Wang, X.; Li, Y.; Deng, T.; Yang, H. Mycotoxins in cereal-based infant foods marketed in China: Occurrence and risk assessment. *Food Control* **2022**, *138*, 108998. [CrossRef]
16. Wei, D.; Wang, Y.; Jiang, D.; Feng, X.; Li, J.; Wang, M. Survey of Alternaria Toxins and Other Mycotoxins in Dried Fruits in China. *Toxins* **2017**, *9*, 200. [CrossRef]
17. López, P.; Venema, D.; de Rijk, T.; de Kok, A.; Scholten, J.M.; Mol, H.G.J.; de Nijs, M. Occurrence of Alternaria toxins in food products in The Netherlands. *Food Control* **2016**, *60*, 196–204. [CrossRef]
18. Hickert, S.; Bergmann, M.; Ersen, S.; Cramer, B.; Humpf, H.U. Survey of Alternaria toxin contamination in food from the German market, using a rapid HPLC-MS/MS approach. *Mycotoxin Res.* **2016**, *32*, 7–18. [CrossRef]
19. De Berardis, S.; De Paola, E.L.; Montevecchi, G.; Garbini, D.; Masino, F.; Antonelli, A.; Melucci, D. Determination of four Alternaria alternata mycotoxins by QuEChERS approach coupled with liquid chromatography-tandem mass spectrometry in tomato-based and fruit-based products. *Food Res. Int.* **2018**, *106*, 677–685. [CrossRef] [PubMed]
20. Qiao, X.; Yin, J.; Yang, Y.; Zhang, J.; Shao, B.; Li, H.; Chen, H. Determination of Alternaria Mycotoxins in Fresh Sweet Cherries and Cherry-Based Products: Method Validation and Occurrence. *J. Agric. Food Chem.* **2018**, *66*, 11846–11853. [CrossRef]
21. Aichinger, G.; Kruger, F.; Puntscher, H.; Preindl, K.; Warth, B.; Marko, D. Naturally occurring mixtures of Alternaria toxins: Anti-estrogenic and genotoxic effects in vitro. *Arch. Toxicol.* **2019**, *93*, 3021–3031. [CrossRef] [PubMed]
22. Prelle, A.; Spadaro, D.; Garibaldi, A.; Gullino, M.L. A new method for detection of five alternaria toxins in food matrices based on LC-APCI-MS. *Food Chem.* **2013**, *140*, 161–167. [CrossRef] [PubMed]
23. Tolgyesi, A.; Stroka, J.; Tamosiunas, V.; Zwickel, T. Simultaneous analysis of Alternaria toxins and citrinin in tomato: An optimised method using liquid chromatography-tandem mass spectrometry. *Food Addit. Contam. Part A* **2015**, *32*, 1512–1522. [CrossRef]
24. Dong, M.; Nie, D.; Tang, H.; Rao, Q.; Qu, M.; Wang, W.; Han, L.; Song, W.; Han, Z. Analysis of amicarbazone and its two metabolites in grains and soybeans by liquid chromatography with tandem mass spectrometry. *J. Sep. Sci.* **2015**, *38*, 2245–2252. [CrossRef] [PubMed]
25. Gonçalves, C.; Tölgyesi, Á.; Bouten, K.; Cordeiro, F.; Stroka, J. Determination of Alternaria Toxins in Food by SPE and LC-IDMS: Development and In-House Validation of a Candidate Method for Standardisation. *Separations* **2022**, *9*, 70. [CrossRef]
26. Wu, L.-h.; Zhou, C.; Long, G.-y.; Yang, X.-b.; Wei, Z.-y.; Liao, Y.-j.; Yang, H.; Hu, C.-x. Fitness of fall armyworm, Spodoptera frugiperda to three solanaceous vegetables. *J. Integr. Agric.* **2021**, *20*, 755–763. [CrossRef]
27. Zhang, X.; Zhu, J.; Sun, L.; Yuan, Q.; Cheng, G.; Argyropoulos, D.S. Extraction and characterization of lignin from corncob residue after acid-catalyzed steam explosion pretreatment. *Ind. Crops Prod.* **2019**, *133*, 241–249. [CrossRef]
28. Xiao, L.; Sha, W.; Tao, C.; Hou, C.; Xiao, G.; Ren, J. Effect on purine releasement of Lentinus edodes by different food processing techniques. *Food Chem. X* **2022**, *13*, 100260. [CrossRef]
29. He, Q.; Liang, J.; Zhao, Y.; Yuan, Y.; Wang, Z.; Gao, Z.; Wei, J.; Yue, T. Enzymatic degradation of mycotoxin patulin by an extracellular lipase from Ralstonia and its application in apple juice. *Food Control* **2022**, *136*, 108870. [CrossRef]
30. Asam, S.; Lichtenegger, M.; Muzik, K.; Liu, Y.; Frank, O.; Hofmann, T.; Rychlik, M. Development of analytical methods for the determination of tenuazonic acid analogues in food commodities. *J. Chromatogr. A* **2013**, *1289*, 27–36. [CrossRef]
31. European Commission. Analytical Quality Control and Method Validation Procedures for Pesticide Residues Analysis in Food and Feed (SANTE/12682/2019). Available online: https://www.eurl-pesticides.eu/userfiles/file/EurlALL/AqcGuidance_SANTE_2019_12682.pdf (accessed on 1 March 2022).
32. Maldonado Haro, M.L.; Cabrera, G.; Fernández Pinto, V.; Patriarca, A. Alternaria toxins in tomato products from the Argentinean market. *Food Control* **2023**, *147*, 109607. [CrossRef]
33. López, P.; Venema, D.; Mol, H.; Spanjer, M.; de Stoppelaar, J.; Pfeiffer, E.; de Nijs, M. Alternaria toxins and conjugates in selected foods in the Netherlands. *Food Control* **2016**, *69*, 153–159. [CrossRef]
34. Ji, X.; Xiao, Y.; Jin, C.; Wang, W.; Lyu, W.; Tang, B.; Yang, H. Alternaria mycotoxins in food commodities marketed through e-commerce stores in China: Occurrence and risk assessment. *Food Control* **2022**, *140*, 109125. [CrossRef]

35. Kecojević, I.; Đekić, S.; Lazović, M.; Mrkajić, D.; Baošić, R.; Lolić, A. Evaluation of LC-MS/MS methodology for determination of 179 multi-class pesticides in cabbage and rice by modified QuEChERS extraction. *Food Control* **2021**, *123*, 107693. [CrossRef]
36. Tang, H.; Ma, L.; Huang, J.; Li, Y.; Liu, Z.; Meng, D.; Wen, G.; Dong, M.; Wang, W.; Zhao, L. Residue behavior and dietary risk assessment of six pesticides in pak choi using QuEChERS method coupled with UPLC-MS/MS. *Ecotoxicol. Environ. Saf.* **2021**, *213*, 112022. [CrossRef] [PubMed]

toxins

MDPI

Article

Effect of Abiotic Conditions on Growth, Mycotoxin Production, and Gene Expression by *Fusarium fujikuroi* Species Complex Strains from Maize

Ting Dong [1], Shouning Qiao [1], Jianhong Xu [2,3,4,5], Jianrong Shi [1,2,3,4,5], Jianbo Qiu [1,2,3,4,5,*] and Guizhen Ma [1,*]

1 School of Environmental and Chemical Engineering, Jiangsu Ocean University, Lianyungang 222005, China
2 Jiangsu Key Laboratory for Food Quality and Safety-State Key Laboratory Cultivation Base, Ministry of Science and Technology, Nanjing 210014, China
3 Key Laboratory for Control Technology and Standard for Agro-Product Safety and Quality, Ministry of Agriculture and Rural Affairs, Nanjing 210014, China
4 Collaborative Innovation Center for Modern Grain Circulation and Safety, Nanjing 210014, China
5 Institute of Food Safety and Nutrition, Jiangsu Academy of Agricultural Sciences, Nanjing 210014, China
* Correspondence: qiujianbo@jaas.ac.cn (J.Q.); 2005000036@jou.edu.cn (G.M.)

Abstract: *Fusarium fujikuroi* species complex (FFSC) strains are a major concern for food quantity and quality due to their strong ability to synthesize mycotoxins. The effects of interacting conditions of water activity, temperature, and incubation time on the growth rate, toxin production, and expression level of biosynthetic genes were examined. High temperature and water availability increased fungal growth. Higher water activity was in favor of toxin accumulation. The maximum amounts of fusaric acid (FA) and fumonisin B1 (FB1) were usually observed at 20–25 °C. *F. andiyazi* could produce a higher content of moniliformin (MON) in the cool environment than *F. fujikuroi*. The expression profile of biosynthetic genes under environmental conditions varied wildly; it was suggested that these genes might be expressed in a strain-dependent manner. FB1 concentration was positively related to the expression of *FUM1*, while a similar correlation of *FUB8* and *FUB12* with FA production could be observed in *F. andiyazi*, *F. fujikuroi*, and *F. subglutinans*. This study provides useful information in the monitoring and prevention of such toxins entering the maize production chain.

Keywords: *Fusarium fujikuroi* species complex; water activity; temperature; mycotoxin; gene expression

Key Contribution: Water activity (a_w), temperature, incubation time, and their interactions influenced the growth, toxin concentration, and expression level of biosynthetic genes within *Fusarium fujikuroi* species complex strains.

1. Introduction

Maize is one of the most important cereal crops in the world; in addition to supplying necessary nutrients for humans and animals, it is also an important industrial raw material for the synthesis of oil, starch, bioenergy, and even medicine [1]. According to the National Bureau of Statistics in China, maize has been the largest cereal food crop in China since 2013 [2]. The national maize planting area has reached 43 million hectares, and the yield was approximately 280 million tons at the end of 2022, which accounted for more than 40% of the cereal yield in China [3].

In the process of cultivation, maize is always susceptible to fungal diseases, among which ear rot is the most important in all maize-growing regions worldwide due to its effect on production. Moreover, diseased maize grains are often contaminated with various mycotoxins produced by pathogens, resulting in a reduction in quality, which precludes them from being used as food or feed [4]. The amount and type of fungi in the kernels depend on the climatic conditions, presence of insects, and variety. The *Fusarium fujikuroi* species complex (FFSC) has been proven to be associated with maize in recent studies [5–8].

This complex includes more than 60 phylogenetically distinct species that comprise three biogeographically structured clades according to phylogenetic studies [9]. *F. proliferatum* and *F. verticillioides* have traditionally been recognized as ear rot pathogens, and recent combined studies on the etiology of the disease and phylogenetic analyses of the fungi have revealed that *F. andiyazi* [10], *F. fujikuroi* [11], *F. temperatum* [12], and *F. subglutinans* [13] are also very pathogenic to maize kernels and can cause typical symptoms.

FFSC members can produce a series of secondary metabolites in the process of infecting the host, such as fusaric acid (FA), moniliformin (MON), beauvericin (BEA), fusarin C (FUS C), fusaproliferin (FUS), and fumonisins (FBs) [14]. FBs are the notorious contaminants in maize due to their high occurrence and potential carcinogenicity [15], and they have been the focus of many research studies. Many countries have set up strict limits for FBs due to the health risk associated with the consumption of contaminated cereals. MON, a sodium or potassium salt of 1-hydroxy-cyclobut-1-ene-3,4-dione, is a highly toxic metabolite with acute toxicity to plants and extreme toxicity to various animal species. It inhibits oxidation of the tricarboxylic acid intermediate α-ketoglutarate and pyruvate dehydrogenase by interrupting pyruvate incorporation into the tricarboxylic acid cycle [16]. MON is known to be a natural contaminant in cereals worldwide [17]. In addition, the cooccurrence of mycotoxins in food and feed represents a natural trend, and the effects of mixtures of mycotoxins in feed on farm animals are yet to be fully understood. Several studies on the combined effects of MON and FB1 concluded additive or less than additive toxicity, focusing on the relative weight of specific organs, mortality, or kidney lesions [18,19]. Fusaric acid is also widely distributed in nature, as it can be produced by many *Fusarium* species. It is well recognized that FA can distort the mitochondrial membrane and inhibit adenosine triphosphate synthesis [20], yet it seems to be mildly toxic to some animals [21]. The main concern is the synergistic interactions of FA with other cooccurring mycotoxins, such as FB1 [22]. In order to protect public health and avoid trade barriers, more and more countries have set maximum levels for the most often regulated mycotoxins, including aflatoxins, trichothecenes, ochratoxin A, zearalenone, and fumonisins [23]. However, there is no related legislation on fumonisins in China at present, due to the complicated distribution of FBs [24,25].

Mycotoxin biosynthesis is a complex process with various external environmental factors forming a regulatory network, including pathway-specific and global regulators that are often responsive to carbon and nitrogen sources, pH, ambient light, and oxidative stress [26]. Water stress and temperature are the most relevant environmental factors that influence fungal growth and mycotoxin production; therefore, they are essential to understand the overall process and to predict and prevent plant diseases and mycotoxin production. The type of mycotoxin is genetically determined, highlighting the presence or absence of necessary biosynthetic genes. With the development of genomic and transcriptomic approaches, an increasing number of secondary metabolite biosynthesis pathways have been deciphered. Mycotoxin biosynthetic genes are usually clustered together. Fumonisins and fusaric acid are the products of the *FUM* (consisting of at least 17 genes) and *FUB* (consisting of at least 12 genes) clusters, respectively [27,28]. Gene expression is also influenced by abiotic factors, and *FUM* gene expression has been used to evaluate these effects. However, contradictory relationships between *FUM* transcript levels and FBs concentrations have been reported [29–31]. Medina et al. [32] suggested that only some elements in the pathway could be directly correlated with toxin synthesis. Information about the dynamic expression of the *FUB* cluster under various conditions is rare, and it is meaningful to determine the consistency of *FUM* and *FUB* expression profiles under the same stress.

In a situation of changing climatic conditions, information about the behavior of different fungal species related to growth patterns and toxin production in environmental stress conditions might become critical to improve the prediction and control of mycotoxin risk. Previous results were usually based on one or two species in the complex; systematic research on the common species could provide an explanation for the population characteristics of maize pathogens. The objective of this study was to compare the impact of strain, temperature, water activity, and incubation time on the fungal growth, mycotoxin production, and gene expression of FFSC strains isolated from maize samples in China.

2. Results

2.1. Effect of Environmental Conditions on Growth

The effect of water activity and temperature on the growth rate and biomass production of FFSC strains on potato dextrose agar (PDA) is shown in Figure 1. All strains exhibited similar behavior under all studied factors. Both water activity and temperature parameters significantly affected the growth rate and dry weight of mycelium of all species. The optimal conditions for growth rate and biomass accumulation for all strains were 0.99 a_w and 30 °C, which were the highest conditions evaluated. Both variables of all strains significantly decreased with the water activity and temperature, and all strains were able to grow at the lowest a_w or temperature. A multifactorial analysis of variance (ANOVA) showed significant effects on the growth and biomass accumulation for water and temperature, as well as their interaction, in all species. (Tables 1 and 2).

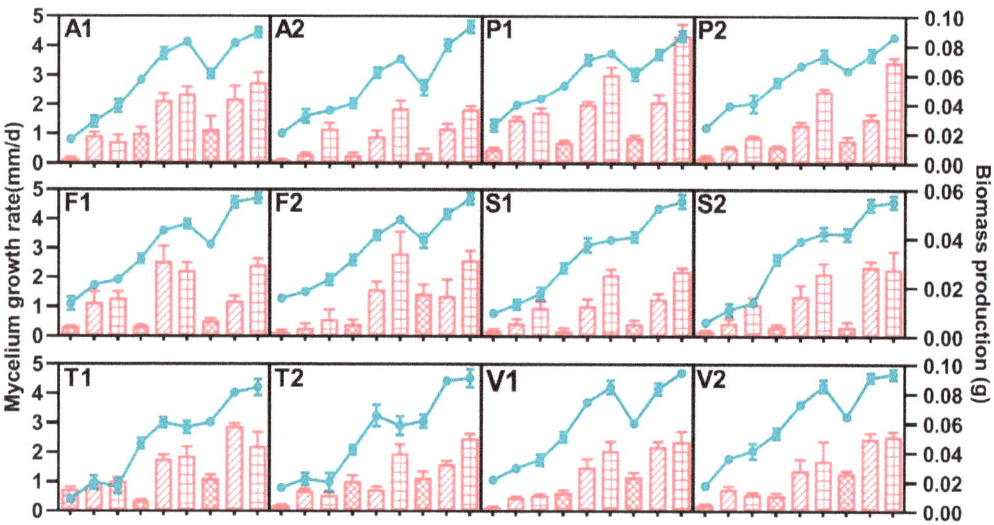

Figure 1. Effect of a_W and temperature on growth rate (line) and biomass production (column) of *Fusarium andiyazi* (A1: C19239; A2: C20180), *F. fujikuroi* (F1: C19232; F2: C20076), *F. proliferatum* (P1: C19215; P2: C20281), *F. subglutinans* (S1: C19068; S2: C20215), *F. temperatum* (T1: C20277; T2: C20278), and *F. verticillioides* (V1: C19216; V2: C20334) strains. The first, middle, and last three points and columns represent the results at 20 °C, 25 °C, and 30 °C, respectively. The columns filled with grids, oblique lines, and squares represent the results under 0.95, 0.98, and 0.99 a_W, respectively. Error bars represent the standard error measured between independent replicates.

Table 1. Analysis of variance on the effects of different strains (S), water activity (a_w), temperature (T), and their interactions on growth rate of *Fusarium fujikuroi* species complex strains on PDA plate.

Source of Variation	*F. andiyazi*			*F. fujikuroi*			*F. proliferatum*			*F. subglutinans*			*F. temperatum*			*F. verticillioides*		
	df	F	MS	df	F	MS	df	F	MS	df	F	MS	df	F	MS	df	F	MS
S	1	84.487 *	1.327	1	1.002	0.025	1	0.579	0.009	1	0.019	0.001	1	6.564 *	0.415	1	61.880 *	1.015
a_w	2	3584.760 *	56.315	2	2369.382 *	58.365	2	2485.337 *	37.368	2	4643.667 *	95.257	2	1333.440 *	84.275	2	4064.686 *	66.647
T	2	1144.040 *	17.972	2	606.962 *	14.951	2	695.150 *	10.452	2	407.928 *	8.368	2	129.193 *	8.165	2	1224.686 *	20.081
S × a_w	2	92.893	1.459	2	2.228	0.055	2	1.751	0.026	2	28.363 *	0.582	2	1.793	0.113	2	14.637 *	0.240
S × T	2	8.156 *	0.128	2	11.987 *	0.295	2	10.279 *	0.155	2	0.175	0.004	2	0.482	0.030	2	10.490 *	0.172
a_w × T	4	39.299 *	0.617	4	26.796 *	0.660	4	12.725 *	0.191	4	23.990 *	0.492	4	13.069 *	0.826	4	41.187 *	0.675
S × a_w × T	4	14.184 *	0.223	4	1.170	0.029	4	1.191	0.018	4	1.454	0.030	4	1.514	0.096	4	5.030 *	0.082

* Significant $p < 0.01$; df: degrees of freedom; F: Snedecor-F; MS: mean square.

Table 2. Analysis of variance on the effects of different strains (S), water activity (a_w), temperature (T), and their interactions on biomass accumulation of *Fusarium fujikuroi* species complex strains on PDA plate.

Source of Variation	*F. andiyazi*			*F. fujikuroi*			*F. proliferatum*			*F. subglutinans*			*F. temperatum*			*F. verticillioides*		
	df	F	MS	df	F	MS	df	F	MS	df	F	MS	df	F	MS	df	F	MS
S	1	177.033 *	0.003	1	0.392	0.001	1	72.807 *	0.003	1	6.794 *	0.001	1	14.267 *	0.001	1	0.526	0.001
a_w	2	211.398 *	0.004	2	30.937 *	0.001	2	139.288 *	0.005	2	52.366 *	0.001	2	75.640 *	0.004	2	129.654 *	0.008
T	2	271.129 *	0.005	2	16.586 *	0.001	2	333.793 *	0.013	2	323.153 *	0.002	2	44.037 *	0.002	2	50.965 *	0.003
S × a_w	2	35.106 *	0.001	2	4.338 *	0.001	2	0.883	0.001	2	3.964 *	0.001	2	2.208	0.001	2	1.827	0.001
S × T	2	16.171 *	0.001	2	2.794	0.001	2	10.701 *	0.000	2	10.620 *	0.001	2	11.134 *	0.001	2	0.440	0.001
a_w × T	4	9.384 *	0.001	4	3.435 *	0.001	4	35.363 *	0.001	4	26.373 *	0.001	4	8.796 *	0.001	4	5.466 *	0.001
S × a_w × T	4	3.596 *	0.001	4	3.460	0.001	4	0.676	0.001	4	8.383 *	0.001	4	8.960 *	0.001	4	0.087	0.001

* Significant $p < 0.01$; df: degrees of freedom; F: Snedecor-F; MS: mean square.

2.2. *Effect of Environmental Conditions on Mycotoxin Production*

Fusaric acid was the main metabolite within FFSC evaluated strains, and it was only absent in *F. verticillioides*. All these strains generally responded to water activity similarly, and higher water stress significantly decreased the FA concentration compared to a_w at 0.98 or 0.99. The FA peak value of *F. andiyazi* appeared just after 1 week, which was earlier than that of the other producers. The optimal conditions were 0.98 a_w/25 °C and 0.99 a_w/30 °C, respectively (Figure 2). Both *F. fujikuroi* strains showed dramatic differences under variable abiotic factors. FA synthesis was induced by water stress, and the maximal concentration was reached at 20 °C/0.95 a_w, while the other produced the largest amounts of FA at 25 °C/0.99 a_w (Figure 3). Temperature had a comparable effect on FA accumulation in *F. proliferatum* (Figure 4), *F. subglutinans* (Figure 5), and *F. temperatum* (Figure 6). The maximum FA levels of the three species were produced at 20 °C after 14 days of incubation when the water activity was 0.98 or 0.99. ANOVA showed statistically significant effects for all factors considered, except when considering the strain and interactions between strain and other factors (Table 3).

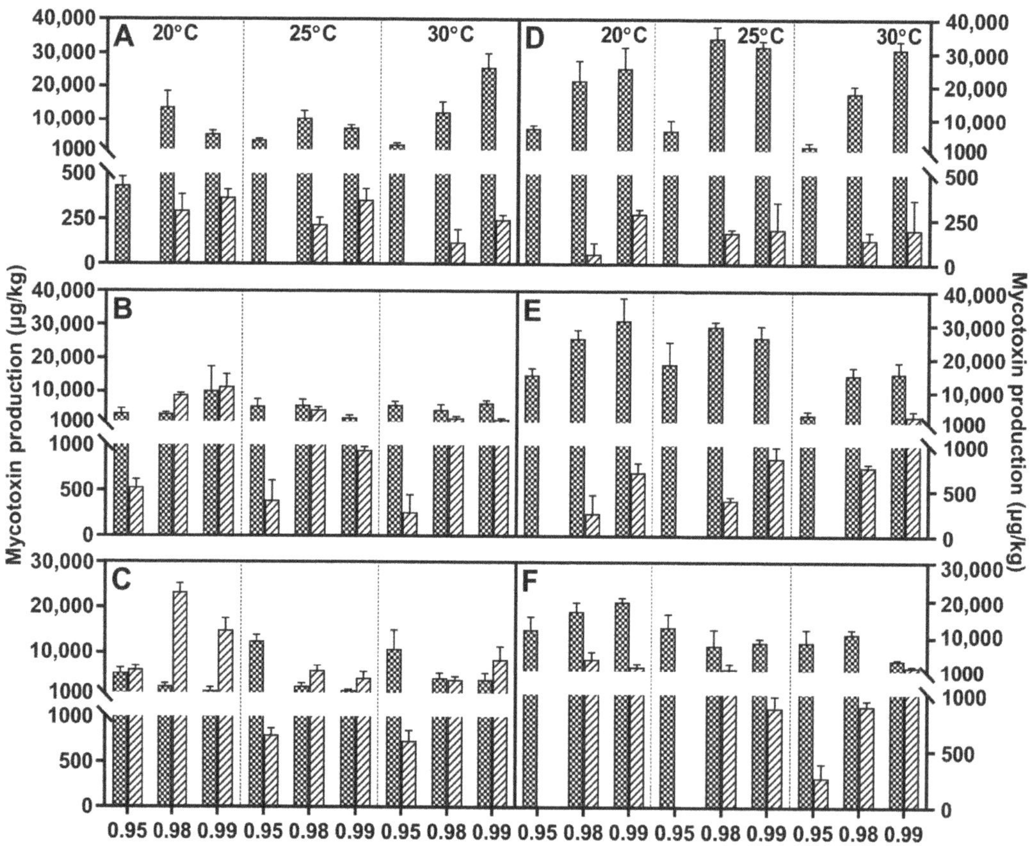

Figure 2. Fusaric acid (grid) and moniliformin (oblique line) levels produced by *Fusarium andiyazi* strain A1 (**A–C**) and A2 (**D–F**) growing on PDA adjusted to different water activity, temperature, and incubation time: 7 days (**A**,**D**), 14 days (**B**,**E**), and 21 days (**C**,**F**). Error bars represent the standard error measured between independent replicates.

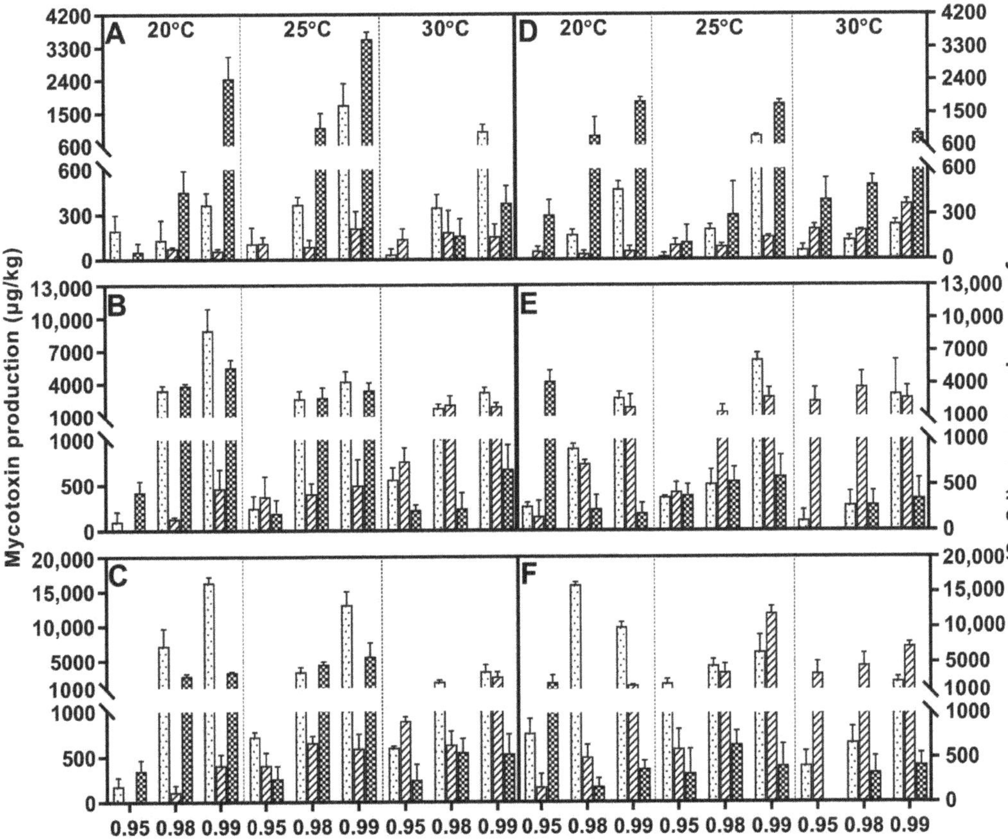

Figure 3. Fusaric acid (grid), fumonisin B1 (black dot) and moniliformin (oblique line) levels produced by *Fusarium fujikuroi* strain F1 (**A**–**C**) and F2 (**D**–**F**) growing on PDA adjusted to different water activity, temperature, and incubation time: 7 days (**A**,**D**), 14 days (**B**,**E**), and 21 days (**C**,**F**). Error bars represent the standard error measured between independent replicates.

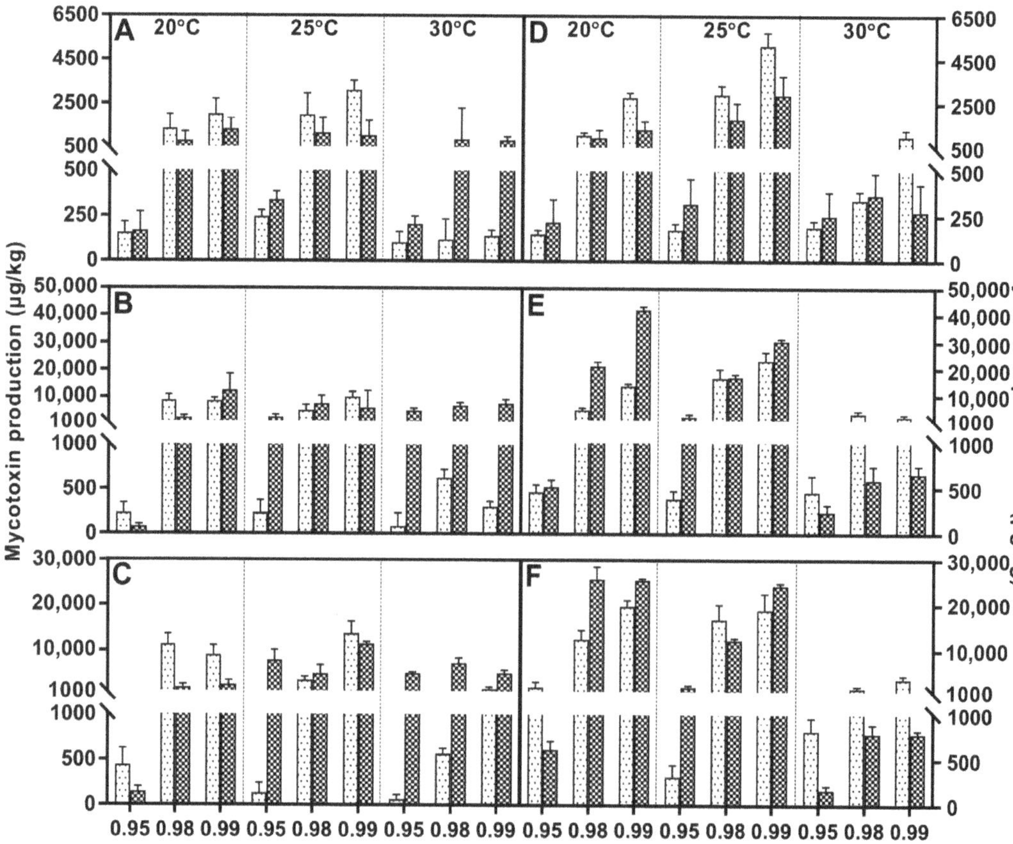

Figure 4. Fusaric acid (grid) and fumonisin B1 (black dot) levels produced by *Fusarium proliferatum* strain P1 (**A**–**C**) and P2 (**D**–**F**) growing on PDA adjusted to different water activity, temperature, and incubation time: 7 days (**A**,**D**), 14 days (**B**,**E**), and 21 days (**C**,**F**). Error bars represent the standard error measured between independent replicates.

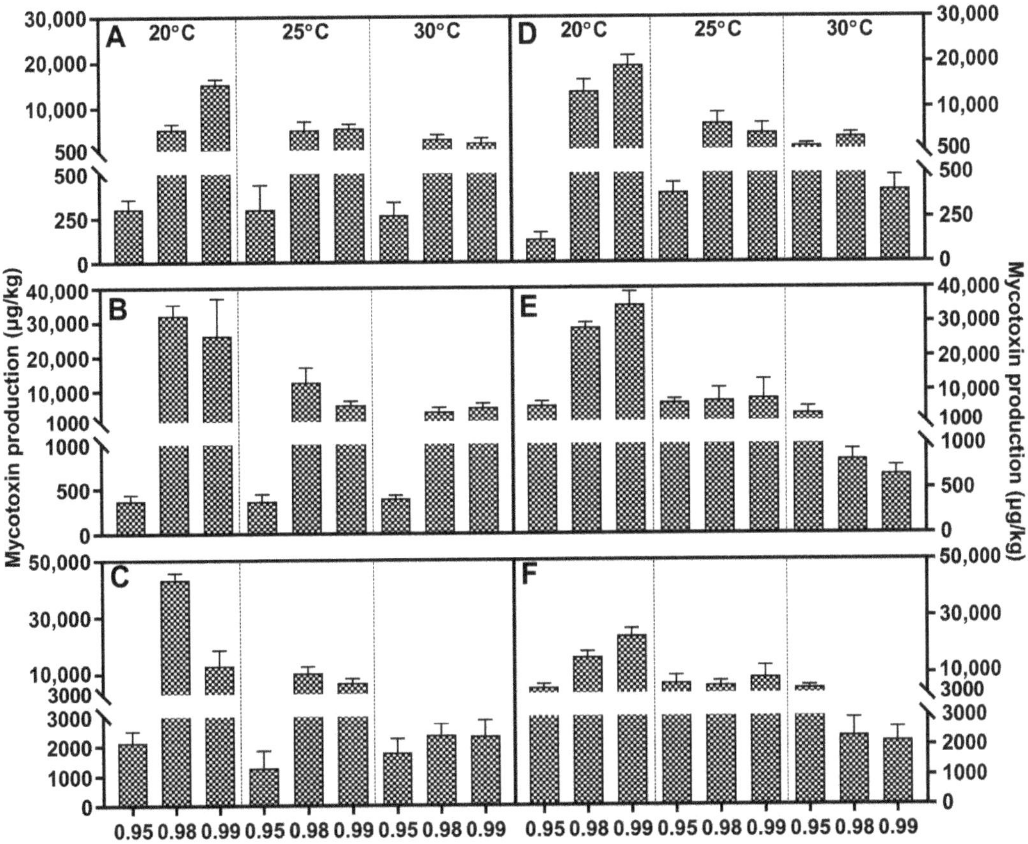

Figure 5. Fusaric acid levels produced by *Fusarium subglutinans* strain S1 (**A–C**) and S2 (**D–F**) growing on PDA adjusted to different water activity, temperature, and incubation time: 7 days (**A**,**D**), 14 days (**B**,**E**), and 21 days (**C**,**F**). Error bars represent the standard error measured between independent replicates.

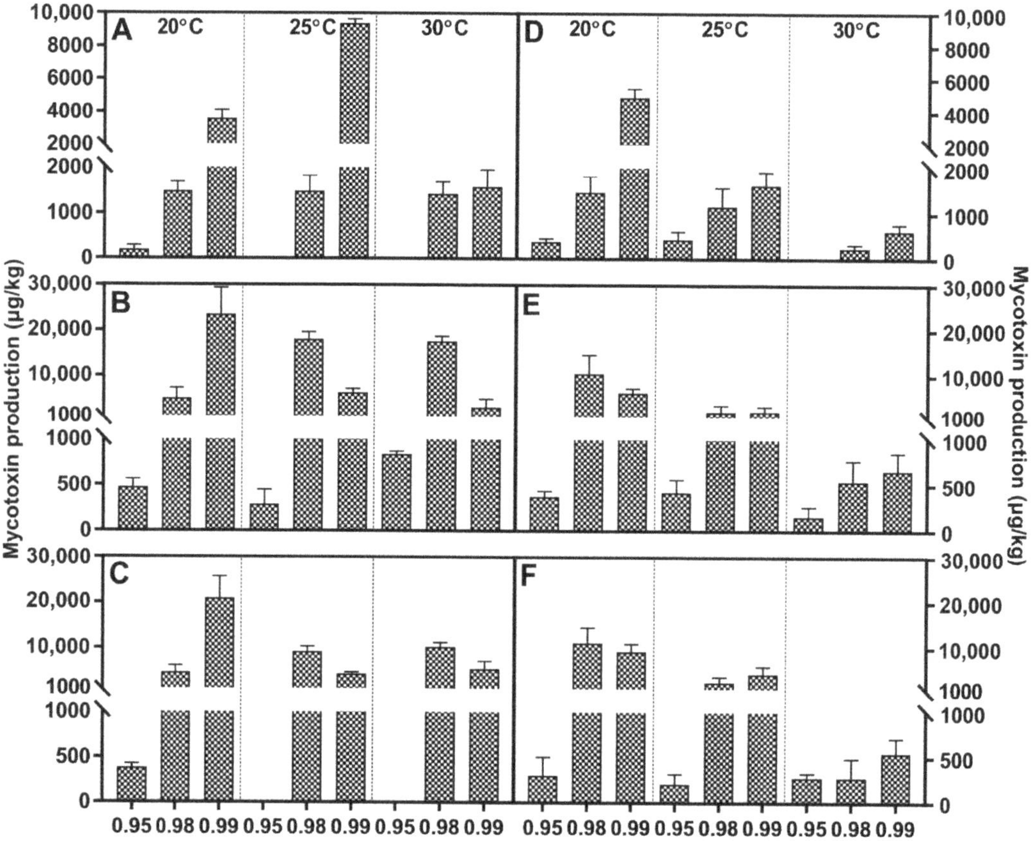

Figure 6. Fusaric acid levels produced by *Fusarium temperatum* strain T1 (**A**–**C**) and T2 (**D**–**F**) growing on PDA adjusted to different water activity, temperature, and incubation time: 7 days (**A**,**D**), 14 days (**B**,**E**), and 21 days (**C**,**F**). Error bars represent the standard error measured between independent replicates.

Table 3. Analysis of variance on the effects of strain (S), water activity (a_w), temperature (T), incubation time (I), and their interactions on fusaric acid production by *Fusarium fujikuroi* species complex strains on PDA plate.

Source of Variation	*F. andiyazi*			*F. fujikuroi*			*F. proliferatum*			*F. subglutinans*			*F. temperatum*		
	df	F	MS	df	F	MS	df	F	MS	df	F	MS	df	F	MS
S	1	283.253 *	4.498×10^9	1	64.840 *	3.497×10^7	1	244.883 *	6.994×10^8	1	0.491	7.803×10^6	1	61.901 *	4.077×10^8
a_w	2	46.065 *	7.316×10^8	2	40.200 *	2.168×10^7	2	325.476 *	9.296×10^8	2	80.839 *	1.286×10^9	2	83.057 *	5.471×10^8
T	2	5.482 *	8.706×10^7	2	49.426 *	2.665×10^7	2	176.286 *	5.035×10^8	2	167.441 *	2.664×10^9	2	24.695 *	1.627×10^8
I	2	36.490 *	5.795×10^8	2	7.334 *	3.955×10^6	2	376.419 *	1.075×10^9	2	26.050 *	4.144×10^8	2	32.805 *	2.161×10^8
S × a_w	2	28.696 *	4.557×10^8	2	50.772 *	2.738×10^7	2	135.728 *	3.877×10^8	2	11.628 *	1.850×10^8	2	16.350 *	1.077×10^8
S × T	2	31.014 *	4.925×10^8	2	21.302 *	1.149×10^7	2	258.669 *	7.388×10^8	2	0.274	4.356×10^6	2	3.477 *	2.290×10^7
S × I	2	15.001 *	2.382×10^8	2	12.886 *	6.949×10^6	2	59.652 *	1.704×10^8	2	1.333	2.120×10^7	2	12.038 *	7.930×10^7
a_w × T	4	3.664 *	5.818×10^7	4	9.254 *	4.990×10^6	4	73.850 *	2.109×10^8	4	41.383 *	6.583×10^8	4	20.105 *	1.324×10^8
a_w × I	4	37.044 *	5.883×10^8	4	2.873 *	1.549×10^6	4	76.443 *	2.183×10^8	4	4.342 *	6.908×10^7	4	10.866 *	7.158×10^7
T × I	4	7.559 *	1.200×10^8	4	7.911 *	4.266×10^6	4	41.091 *	1.174×10^8	4	10.955 *	1.743×10^8	4	4.342 *	2.860×10^7
S × a_w × T	4	1.654	2.627×10^7	4	14.865 *	8.016×10^6	4	44.084 *	1.259×10^8	4	5.210 *	8.289×10^7	4	20.633 *	1.359×10^8
S × a_w × I	4	1.173	1.863×10^7	4	7.362 *	3.970×10^6	4	35.331 *	1.009×10^8	4	7.537 *	1.199×10^8	4	3.127 *	2.060×10^7
S × T × I	4	4.912 *	7.801×10^7	4	2.588 *	1.396×10^6	4	65.870 *	1.881×10^8	4	3.064 *	4.874×10^7	4	0.937	6.173×10^6
a_w × T × I	8	5.666 *	8.999×10^7	8	2.671 *	6.831×10^6	8	29.330 *	8.377×10^7	8	5.800 *	9.226×10^7	8	4.584 *	3.019×10^7
S × a_w × T × I	8	1.364	2.165×10^7	8	5.004 *	2.699×10^6	8	11.696 *	3.340×10^7	8	3.498 *	5.565×10^7	8	9.041 *	5.956×10^7

* Significant $p < 0.01$; df: degrees of freedom; F: Snedecor-F; MS: mean square.

Fumonisin was detected in *F. fujikuroi*, *F. proliferatum*, and *F. verticillioides* strains, and higher levels of FB1 were present at 0.98/0.99 a_w. With regard to *F. fujikuroi*, FB1 contents increased with increasing incubation time, and the highest values were reached at 21 days at 20 °C, depending on the water activity (Figure 3). Regarding *F. proliferatum*, 30 °C reduced FB1 production at each water activity. The temporal variation feature was identical to that of *F. verticillioides*, and the most advantageous conditions for FB1 synthesis were 25 °C/0.99 a_w, with 14 and 21 days of incubation (Figure 4). At the early stage, the variable temperature had the same effect on *F. verticillioides*. When the inoculation time was extended to 14 and 21 days, FB1 increased gradually until it tended to be stable. Low temperature favored FB1 accumulation, and maximum production was obtained at 20 °C/0.99 a_w after 14 days of incubation for the two *F. verticillioides* strains (Figure 7). Individual environmental factors, i.e., water activity, temperature, and incubation time, and their interactions had a significant effect on FB1 levels in fumonisin producers according to ANOVA, and both *F. verticillioides* strains showed similar patterns of responses to the treatments (Table 4).

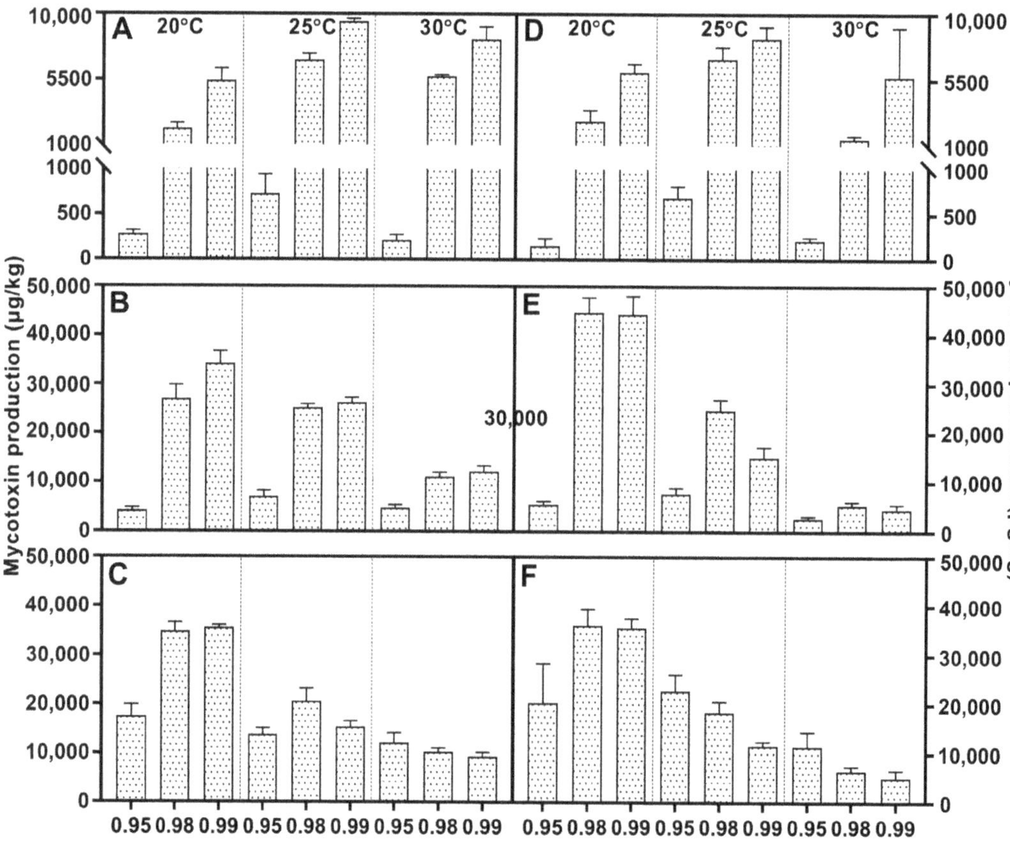

Figure 7. Fumonisin B1 levels produced by *Fusarium verticillioides* strain V1 (**A–C**) and V2 (**D–F**) growing on PDA adjusted to different water activity, temperature, and incubation time: 7 days (**A**,**D**), 14 days (**B**,**E**), and 21 days (**C**,**F**). Error bars represent the standard error measured between independent replicates.

Table 4. Analysis of variance on the effects of strain (S), water activity (a_w), temperature (T), incubation time (I), and their interactions on fumonisin production by *Fusarium fujikuroi*, *F. proliferatum*, and *F. verticillioides* strains on PDA plate.

Source of Variation	*F. fujikuroi*			*F. proliferatum*			*F. verticillioides*		
	df	F	MS	df	F	MS	df	F	MS
S	1	8.009 *	1.609×10^7	1	114.647 *	2.952×10^8	1	0.232	1.700×10^6
a_w	2	126.308 *	2.537×10^8	2	294.806 *	7.592×10^8	2	193.393 *	1.417×10^7
T	2	48.340 *	9.709×10^7	2	197.072 *	5.075×10^8	2	325.243 *	2.383×10^7
I	2	146.829 *	2.949×10^8	2	161.533 *	4.160×10^8	2	481.660 *	3.529×10^7
S × a_w	2	13.125 *	2.636×10^7	2	25.346 *	6.527×10^7	2	5.666 *	4.151×10^7
S × T	2	0.001	7.669×10^2	2	24.966 *	6.429×10^7	2	24.826	1.819×10^8
S × I	2	1.754	3.523×10^6	2	19.119 *	4.923×10^7	2	0.519	3.800×10^6
a_w × T	4	15.294 *	3.072×10^7	4	56.099 *	1.445×10^8	4	57.532 *	4.215×10^8
a_w × I	4	31.036 *	6.234×10^7	4	37.923 *	9.766×10^7	4	51.646 *	3.784×10^8
T × I	4	30.765 *	6.179×10^7	4	30.665 *	7.896×10^7	4	96.912 *	7.100×10^8
S × a_w × T	4	6.606 *	1.327×10^7	4	15.778 *	4.063×10^7	4	6.327 *	4.635×10^7
S × a_w × I	4	11.726 *	2.355×10^7	4	4.182 *	1.077×10^7	4	4.257 *	3.118×10^7
S × T × I	4	4.315 *	8.666×10^6	4	8.442 *	2.174×10^7	4	10.196 *	7.470×10^7
a_w × T × I	8	9.219 *	1.852×10^7	8	8.389 *	2.160×10^7	8	19.139 *	1.402×10^8
S × a_w × T × I	8	5.032 *	1.011×10^7	8	5.470 *	1.408×10^7	8	2.347 *	1.719×10^7

* Significant $p < 0.01$; df: degrees of freedom; F: Snedecor-F; MS: mean square.

F. andiyazi and *F. fujikuroi* strains produced MON. *F. andiyazi* strains produced larger amounts of MON in cooler conditions compared with *F. fujikuroi*, and maximum production always occurred after 21 incubation days at 0.98 a_w/20 °C, although there were great variations in the yield between the two evaluated strains (Figure 2). The concentrations were extremely low in the first week and continued to increase rapidly at higher water activity levels. The highest MON contents were observed at 0.99 a_w after 21 days of incubation by the two *F. fujikuroi* strains at 25 °C or 30 °C, respectively (Figure 3). ANOVA showed that the MON concentration was significantly influenced by all individual factors and their interactions (Table 5).

Table 5. Analysis of variance on the effects of strain (S), water activity (a_w), temperature (T), incubation time (I), and their interactions on moniliformin production by *Fusarium andiyazi* and *F. fujikuroi* on PDA plate.

Source of Variation	*F. andiyazi*			*F. fujikuroi*		
	df	F	MS	df	F	MS
S	1	165.819 *	3.793×10^8	1	201.331 *	7.644×10^7
a_w	2	53.713 *	1.229×10^8	2	74.975 *	2.847×10^7
T	2	62.539 *	1.431×10^8	2	86.057 *	3.267×10^7
I	2	112.895 *	2.583×10^8	2	155.341 *	5.898×10^7
S × a_w	2	20.811 *	4.761×10^7	2	34.034 *	1.292×10^7
S × T	2	52.985 *	1.212×10^8	2	23.524 *	8.932×10^6
S × I	2	57.470 *	1.315×10^8	2	81.936 *	3.111×10^7
a_w × T	4	10.817 *	2.474×10^7	4	9.711 *	3.687×10^6
a_w × I	4	14.864 *	3.400×10^7	4	35.043 *	1.331×10^7
T × I	4	23.853 *	5.457×10^7	4	29.757 *	1.130×10^7
S × a_w × T	4	5.738 *	1.313×10^7	4	16.148 *	6.131×10^6
S × a_w × I	4	5.619 *	1.286×10^7	4	20.626 *	7.832×10^6
S × T × I	4	15.636 *	3.577×10^7	4	16.677 *	6.332×10^6
a_w × T × I	8	6.476 *	1.481×10^7	8	8.784 *	3.335×10^6
S × a_w × T × I	8	3.991 *	9.129×10^6	8	10.463 *	3.973×10^6

* Significant $p < 0.01$; df: degrees of freedom; F: Snedecor-F; MS: mean square.

In general, *F. andiyazi* and *F. subglutinans* strains were strong producers of FA, whereas *F. verticillioides* and *F. proliferatum* strains exhibited more enhanced FB1 synthetic ability than *F. fujikuroi*. The effect of water activity and temperature on FA and FB1 production was similar as there is a significant positive relationship between the two toxin levels in *F. fujikuroi* and *F. proliferatum* (Figure 8).

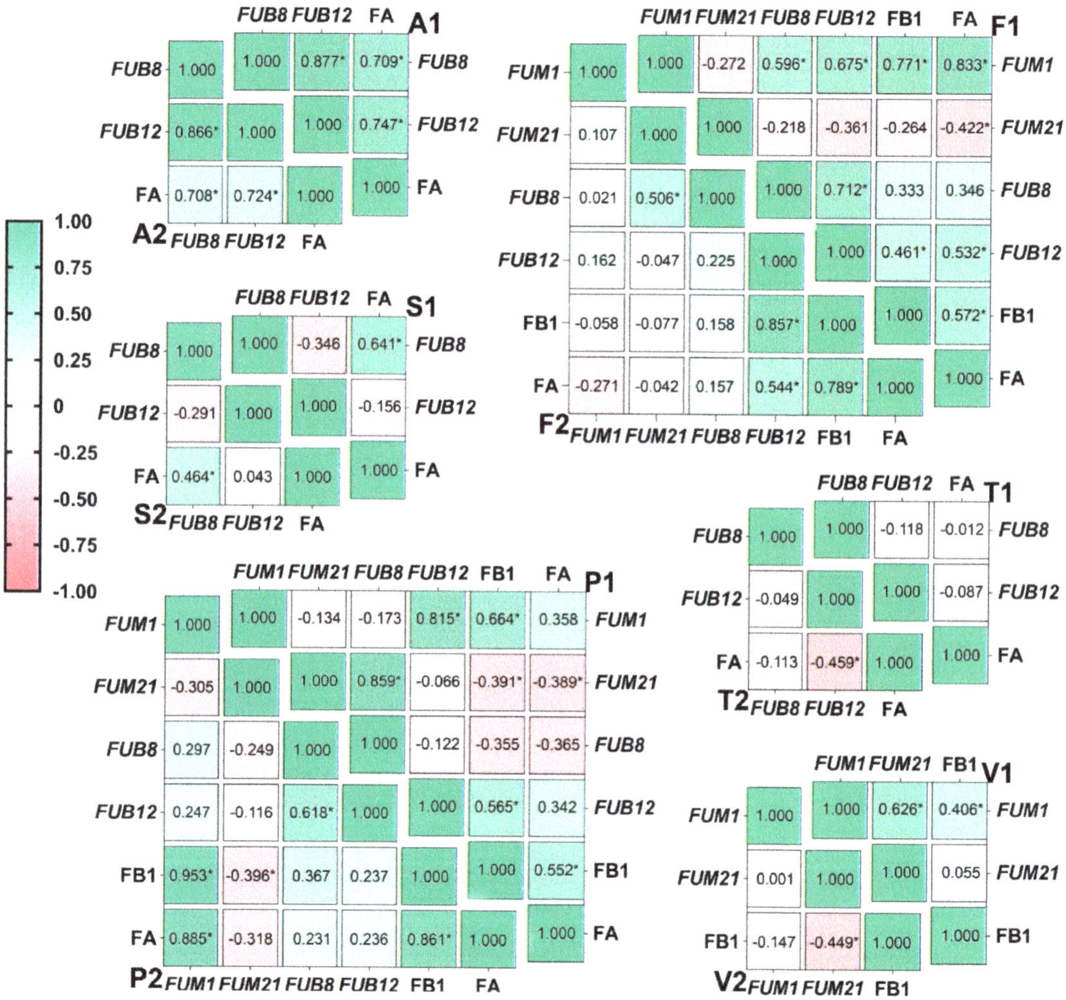

Figure 8. The heatmap of the pairwise Pearson's correlation between variables of *Fusarium* strains incubated for 7 days on PDA plate. The strains included *Fusarium andiyazi* (A1: C19239; A2: C20180), *F. fujikuroi* (F1: C19232; F2: C20076), *F. proliferatum* (P1: C19215; P2: C20281), *F. subglutinans* (S1: C19068; S2: C20215), *F. temperatum* (T1: C20277; T2: C20278), and *F. verticillioides* (V1: C19216; V2: C20334). *FUM1* and *FUM21* coded the polyketide synthase and Zn(II)2Cys6DNA-binding transcription factor in fumonisin synthetic cluster, respectively. *FUB8* and *FUB12* code the non-ribosomal peptide synthetase-like carboxylic acid reductase and Zn(II)2Cys6DNA-binding transcription factor in the fusaric acid synthetic cluster, respectively. FB1 and FA denote fumonisin B1 and fusaric acid, respectively. * $p < 0.05$.

2.3. *Effect of Environmental Conditions on Gene Expression*

Figure 9 shows the expression of *FUM1* and *FUM21* in fumonisin producers on PDA for 7 days in response to temperature and water activity, reflecting the difference between species. Compared with the values resolved at 25 °C and 1 a_w, the expression of both genes of *F. fujikuroi* seemed to be slightly repressed under most conditions, and *FUM1* expression reached the highest value was at 25 °C. For *F. proliferatum*, two targets displayed distinct rules. *FUM21* was intensively induced at the lowest water activity, which was in agreement with FB1 accumulation, while the decrease in a_w resulted in a gradual reduction in the expression of *FUM1*, especially at 20 and 25 °C. *FUM1* and *FUM21* were more highly expressed in *F. verticillioides* strain C19216, and they were induced at 25 °C. However, there was no obvious variation in all environment factors for the other strain (C20334). The nonparametric test showed that all factors had a significant influence on the expression of *FUM21* in *F. proliferatum* and *FUM21* in *F. verticillioides* (Table 6).

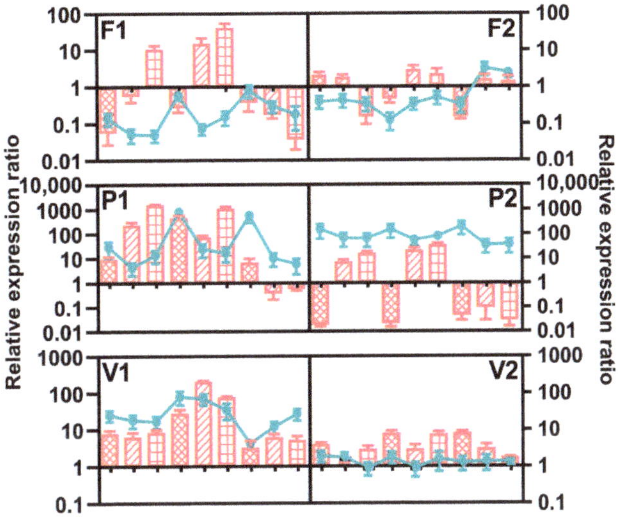

Figure 9. Relative expression of *FUM1* (line) and *FUM21* (column) of *Fusarium fujikuroi* (F1: C19232; F2: C20076), *F. proliferatum* (P1: C19215; P2: C20281), and *F. verticillioides* (V1: C19216; V2: C20334) strains incubated for 7 days under various conditions. The measured quantity of cDNA was normalized using the Cq values obtained for *Act* cDNA amplifications. The values represent the expression level of the target gene in the different culture conditions relative to the control culture (25 °C, $a_W = 1$). The data represent the means of three independent repetitions. Error bars represent the standard error measured between independent replicates. The first, middle, and last three points and columns represent the results at 20 °C, 25 °C, and 30 °C, respectively. The columns filled with grids, oblique lines, and squares represent the results under 0.95, 0.98, and 0.99 a_w, respectively. Error bars represent the standard error measured between independent replicates.

Table 6. The result of nonparametric analysis of *FUM1* and *FUM21* expression under different conditions.

FUM1	N	Mean ± SD	a_w		Temperature		Strain	
			H	*P*	*H*	*P*	*Z*	*P*
F. fujikuroi	54	2.43 ± 3.03	5.836	0.054	3.125	0.210	−2.829	0.005
F. proliferatum	54	247.78 ± 427.05	4.728	0.296	25.129	0.000	−2.638	0.008
F. verticillioides	54	16.53 ± 31.12	15.569	0.000	13.151	0.001	−3.555	0.000
FUM21	**N**	**Mean ± SD**	a_w		**Temperature**		**Strain**	
			H	*P*	*H*	*P*	*Z*	*P*
F. fujikuroi	54	2.91 ± 2.96	14.875	0.001	8.255	0.016	−1.055	0.291
F. proliferatum	54	118.17 ± 204.60	9.078	0.002	6.735	0.034	−3.694	0.000
F. verticillioides	54	17.86 ± 25.57	2.056	0.358	0.842	0.656	−6.306	0.000

SD: standard deviation; H: Kruskal–Wallis test; Z: Mann–Whitney test.

FUB8 and *FUB12* were selected to evaluate the effect of abiotic factors on FFSC strains, and the results are presented in Figure 10. *F. fujikuroi*, *F. proliferatum*, and *F. temperatum* strains had similar behavior regarding FUB expression under different conditions, indicating that *FUB8* and *FUB12* transcript levels were mainly influenced by strain and temperature, respectively. The suppression of *FUB12* expression was obvious at lower temperatures by the two *F. fujikuroi* strains, while *FUB8* expression showed the opposite trend. The expression of *FUB8* in *F. proliferatum* strain C19215 improved upon decreasing the a_w from 0.99 to 0.95, but the effect of the conditions in the other strain was irregular. Both target genes were more highly expressed in *F. temperatum* strain C20278, and two strains had the highest expression levels of *FUB8* and *FUB12* at 20 °C/0.99 a_w and 25 °C/0.98 a_w, respectively. Temperature was the main influencing factor for *F. andiyazi* and *F. subglutinans*. *FUB8* and *FUB12* was upregulated at 30 °C in C19239, while the highest expression of the target genes in the other *F. andiyazi* strain occurred at 25 °C. Low temperature contributed to the enhanced *FUB12* mRNA synthesis in both *F. subglutinans* strains. Despite the trace amounts of FA in *F. verticillioides*, the highest temperature and water activity strongly induced *FUB8* expression in strain C19216, and the response of C20334 to the factors was different, but no rules were followed. The nonparametric test indicated that individual factors had no remarkable effect on *FUB8* and *FUB12* expression in *F. andiyazi* and *F. verticillioides*, whereas strain significantly influenced both genes in the other species (Table 7).

The significant correlation between expression levels of both *FUM* genes appeared only in the *F. verticillioides* strain C19216, while *FUB8* expression was significantly related to *FUB12* expression in *F. proliferatum* strain C20281, *F. fujikuroi* strain C19232, and both *F. andiyazi* strains. In addition, a significant positive correlation was observed between *FUM1* expression and FB1 amounts in four out of six FB1-producing strains, but *FUM21* expression was negatively correlated with this toxin. The FA concentration in *F. andiyazi* exhibited an extremely significant and positive correlation with two *FUB* genes. A similar relationship of *FUB8* and *FUB12* with FA content occurred in *F. subglutinans* and *F. fujikuroi* (Figure 8).

In summary, increased expression of mycotoxin synthetic genes could be found in most treatments with various factors that indicated the induction of water stress compared with the control (a_w = 1). Considering the influence of temperature and species, expression of biosynthetic genes might be strain-dependent.

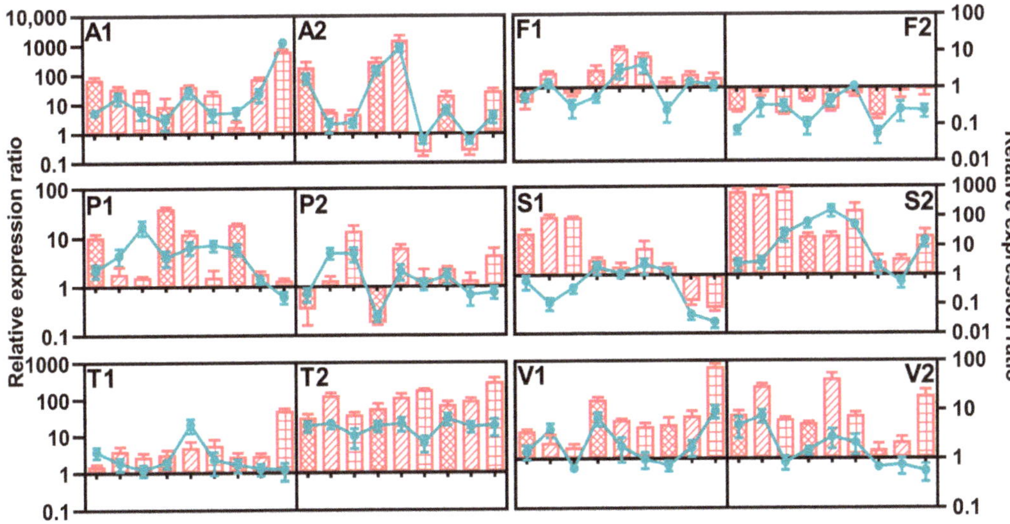

Figure 10. Relative expression of *FUB8* (line) and *FUB12* (column) of *Fusarium andiyazi* (A1: C19239; A2: C20180), *F. fujikuroi* (F1: C19232; F2: C20076), *F. proliferatum* (P1: C19215; P2: C20281), *F. subglutinans* (S1: C19068; S2: C20215), *F. temperatum* (T1: C20277; T2: C20278), and *F. verticillioides* (V1: C19216; V2: C20334) strains incubated for 7 days under various conditions. The measured quantity of cDNA was normalized using the Cq values obtained for *Act* cDNA amplifications. The values represent the expression level of the target gene in the different culture conditions relative to the control culture (25 °C, a_W = 1). The data represent the means of three independent repetitions. Error bars represent the standard error measured between independent replicates. The first, middle, and last three points and columns represent the result at 20 °C, 25 °C, and 30 °C, respectively. The columns filled with grids, oblique lines, and squares represent the results under 0.95, 0.98, and 0.99 a_W, respectively.

Table 7. The result of nonparametric analysis of *FUB8* and *FUB12* expression under different conditions.

FUB8	N	Mean ± SD	a_W		Temperature		Strain	
			H	*P*	*H*	*P*	*Z*	*P*
F. andiyazi	54	189.99 ± 389.46	1.37	0.504	2.737	0.254	−1.047	0.295
F. fujikuroi	54	2.02 ± 2.97	5.347	0.069	6.528	0.038	−5.216	0.000
F. proliferatum	54	6.80 ± 9.77	1.009	0.411	0.269	0.874	−2.673	0.008
F. subglutinans	54	353.82 ± 1109.92	1.448	0.485	35.219	0.000	−3.99	0.001
F. temperatum	54	65.16 ± 84.90	6.034	0.049	3.658	0.161	−5.821	0.000
F. verticillioides	54	9.74 ± 9.90	2.667	0.264	4.376	0.112	−1.791	0.073
FUB12	**N**	**Mean ± SD**	**a_W**		**Temperature**		**Strain**	
			H	*P*	*H*	*P*	*Z*	*P*
F. andiyazi	54	80.76 ± 229.14	2.376	0.305	4.224	0.121	−0.268	0.789
F. fujikuroi	54	0.78 ± 0.86	16.753	0.000	3.723	0.155	−4.593	0.000
F. proliferatum	54	3.50 ± 4.25	0.374	1.000	4.987	0.083	−3.746	0.000
F. subglutinans	54	16.17 ± 31.82	2.67	0.263	14.821	0.001	−4.922	0.000
F. temperatum	54	8.87 ± 7.99	4.079	0.130	0.429	0.807	−6.116	0.000
F. verticillioides	54	2.55 ± 2.58	4.712	0.095	4.01	0.135	−1.237	0.216

SD: standard deviation; H: Kruskal–Wallis test; Z: Mann–Whitney test.

3. Discussion

In this study, the impact of water activity and temperature in different culture periods on growth and mycotoxin production by various FFSC strains was systematically investigated. The detailed results generated for all strains exhibited similar patterns. All species could adapt to a wide range of conditions, particularly at higher temperatures and water potentials, and mycelial growth and biomass accumulation decreased as the water availability of the media and incubation temperature were reduced. These phenomena support previous studies [31–37]. At present, research on the response of *F. fujikuroi* to variable conditions is lacking. It can be speculated that more species in this complex should have similar characteristics according to the published literature. Moreover, *F. verticillioides* did not show significantly better performance in favorable or unfavorable environments compared with other species. This may indicate that the dominance of *F. verticillioides* in the pathogens causing maize ear rot does not stem directly from its superior ecological adaptability. The ecosystem, even in an entire plant, is a highly complex entity comprising the crop and various biotic and abiotic factors. The dominant species niche should be the result of competition with other microorganisms in the plant-associated microbiota and long-term adaptation to the host and environmental conditions.

There has been extensive work on the combined effect of water activity and temperature on mycotoxin production, and the research described here provides the first detailed study on toxic secondary metabolite accumulation on PDA medium. Fumonisins are the most important contaminants in maize due to their strong carcinogenicity and high incidence, and we found that the optimal conditions for FB1 are lower temperatures (20–25 °C) and higher water activity (0.98–0.99 a_W). This pattern is consistent with previous studies of mycotoxin production profiles of *F. proliferatum* [38] and *F. verticillioides* [32]. Two emerging toxins, fusaric acid and moniliformin, are often ignored in current food safety monitoring due to the lack of limit standards, as well as their low occurrence and ambiguous toxic effects. It is still meaningful and valuable to examine toxin characteristics under distinct environmental factors. Although their toxicity to humans can be mild or unknown, they could enhance the effects of the other mycotoxins and induce combined toxicity once the cereal grains are contaminated with multiple mycotoxins [22,39]; furthermore, they are usually synthesized by nondominant maize pathogens, such as *F. andiyazi* and *F. temperature*, but more species could be pandemic as the field population is always in dynamic status. The dynamic change in FA accumulation was similar to that of FB1, while MON production was favored by wet and hot conditions. Our results are in agreement with the few papers published by Pena et al. [37] and Fumero et al. [36,40], who observed the maximum FA/MON production at the highest water activity and temperature.

The results generated from different studies are sometimes nonuniform or even contradictory because of diversities in the culture conditions, such as shaking or stationary cultivation, nutrients from natural cereal grains or synthetic medium, and the origin of the strain evaluated, e.g., the soil or crop, a hot or cold region. Otherwise, with the same strain, the production of different types of fumonisin could be favored by different temperatures, despite sharing the same biosynthetic pathway [32]. We cannot give a definitive conclusion about the optimal conditions for growth and/or mycotoxin production, but it is suggested that the effect of environmental factors on the phenotypic parameters could be similar or at least overlap. It is possible to restrict the pathogen itself and its metabolites at the same time.

Fumonisins are synthesized by a polyketide pathway and involve at least 16 clustered genes that encode biosynthetic enzymes and regulatory and transport proteins in *Fusarium* [28]. In the current work, we focused on the expression profile of *FUM* genes in response to temperature and moisture stress. The expression pattern of the fumonisins-producing strains studied showed strain-dependent variation. The transcriptional regulation was reported by Jurado et al. [33] and Marín et al. [34], who revealed the significant inductive effect of water and temperature stress on FB-related gene transcripts in *F. verticillioides* and *F. proliferatum*, respectively. However, unlike the previous results, the great expression of *FUM1* was usually detected at 25 °C, and the *FUM21* transcript was induced at the lowest a_w. Contradictory results were also found in similar studies [30,35]. The differences may have stemmed from the target evaluated, because only some key genes have a direct connection with toxin biosynthesis. The location of *FUM1*, the key enzyme of fumonisin biosynthesis, can be sensed, and it is more susceptible to changes in water activity, while the other elements in the cluster remain stable. The functional diversity of *FUM* genes may lead to various intrinsic expression patterns and responses to the changing environment. Medina et al. [32] developed a predictive model of nine *FUM* genes in relation to environmental factors with a microarray analysis and showed that the relative expression of *FUM1*, *FUM11*, *FUM13*, *FUM14*, and *FUM19* is influenced by external environments. Lazzaro et al. [41] also hypothesized that *FUM21*, encoding a transcription factor, primarily plays its activation role before the other *FUM* cluster genes, while *FUM2* remains highly active for longer. In *Aspergillus flavus*, the transcript levels of the regulatory genes did not change significantly under varying conditions, and the structural genes appeared to be highly expressed only in correspondence with the greatest AFB1 production [42,43]. Therefore, it is important to determine the appropriate test time and genes to examine, allowing some preferable results to be obtained. As a reaction to abiotic stress, the activation of *Fusarium* mycotoxin biosynthetic genes has been described in several studies and is considered part of a more general phenomenon [44–46]. Gene clusters of some secondary metabolites, such as fujikurins, beauvericin, and trichosetin, have been proven to be silent under all conditions and can be activated only after the application of molecular techniques [47–49]. In the present work, there was no significant variation of *FUM* or *FUB* expression levels in *F. fujikuroi*, which could indicate that the ecological conditions used here are less important, or that the other metabolites contribute more to adaption to stress.

Fusaric acid is the product of the *FUB* cluster containing 12 necessary genes that have been deciphered in *Fusarium* [27,50]. To our knowledge, this is the first attempt to unveil the relationship between FA accumulation and *FUB* gene expression in FFSC strains under various abiotic factors. The similar effects of temperature and water activity on *FUB* and *FUM* gene expression indicate the shared regulatory pathways controlling the production of these two metabolites. It has been stated that the expression of secondary metabolite synthetic clusters in response to multiple environmental signals is modulated by a complicated network of global regulators, including components of the VELVET complex, GATA transcription factor *Csm1* [51,52]. For example, under inducing or repressing conditions, transcript levels of gibberellin, fusarins, fumonisin, fusarubin, and even the silent genes significantly increased in the lae1-overexpressing mutant [53]. Although the expression of the *FUM* and *FUB* clusters show some correlations, strain-specific differences in the production of the corresponding toxins are apparent, possibly due to the different genetic backgrounds and other regulatory elements. In *F. fujikuroi*, the transcription factor *AreB* is necessary for the expression of most mycotoxin synthetic clusters, including *FUB* and *FUM*, in high or low environments, while another transcription factor *AreA* was reported to be effective only in fumonisin and gibberellin synthesis [54,55]. Moreover, water stress achieved by the addition of glycerol to the medium is a form of osmotic pressure, and fungi have evolved a high-osmolarity glycerol mitogen-activated protein kinase pathway (HOG) to recognize various environmental stresses. The regulatory effect of HOG kinase on gene expression and toxin production needs to be taken into account.

It is also notable that there was some inconsistency between the effect of abiotic conditions on gene expression and mycotoxin production, which means that these results did not show significant correlation. Although López-Errasquín et al. [29] and Lazzaro et al. [41] identified a correlation between FB concentration and *FUM* gene transcripts, the correlation was usually found under a certain factor or specific parameters of the factor. Gene expression could mirror toxin production profiles under all conditions except temperature [35], which implied the different effects of varied conditions. In addition, Medina et al. [31] showed that the synthesis of fumonisins increased at higher transcript levels of *FUM19*, while higher expression of *FUM14* was correlated with decreased concentration of FBs.

FUM11 positively or negatively regulated the production of FB1 and FB2, respectively. The correlation of gene expression and production of the toxin seems to be highly variable because the expression of only a few genes is usually directly correlated with the toxin biosynthesis. The lack of correlation could be explained by a post-transcriptional regulatory mechanism. While transcriptional to translational correlations are often not very strong, it is essential to determine the relationship of gene expression to protein translation in fumonisin production to better understand the mechanism of biosynthesis.

In conclusion, fungal growth and mycotoxin accumulation are influenced by environmental factors. Mycotoxin contamination is the result of multiple factors in the field, including fungicide, pH, and light, and a better understanding of the optimal conditions is useful in the design of prevention and control strategies.

4. Materials and Methods

4.1. Fungal Strains

Two strains each of six species in the *Fusarium fujikuroi* species complex were isolated from maize samples in various regions of China between 2019 and 2020. These strains were identified by morphological characteristics, toxin profile, and phylogenetic analysis of partial elongation factor 1α. Single-spore cultures were stored as spore suspensions in 20% glycerol at −80 °C. Detailed information on the strains used in this study is listed in Table 8.

Table 8. Origin of the *Fusarium fujikuroi* species complex strains used in this study.

Strain	Code Name	Species	Host	Location	Year
C19239	A1	*Fusarium andiyazi*	Maize	Guizhou Province, southwestern China	2019
C20180	A2	*Fusarium andiyazi*	Maize	Heliongjiang Province, northeastern China	2020
C19232	F1	*Fusarium fujikuroi*	Maize	Guizhou Province, southwestern China	2019
C20076	F2	*Fusarium fujikuroi*	Maize	Hubei Province, central China	2020
C19215	P1	*Fusarium proliferatum*	Maize	Guizhou Province, southwestern China	2019
C20281	P2	*Fusarium proliferatum*	Maize	Heliongjiang Province, northeastern China	2020
C19068	S1	*Fusarium subglutinans*	Maize	Heliongjiang Province, northeastern China	2019
C20215	S2	*Fusarium subglutinans*	Maize	Shaanxi Province, northwestern China	2020
C20277	T1	*Fusarium temperatum*	Maize	Yunnan Province, southwestern China	2020
C20278	T2	*Fusarium temperatum*	Maize	Yunnan Province, southwestern China	2020
C19216	V1	*Fusarium verticillioides*	Maize	Henan Province, central China	2019
C20334	V2	*Fusarium verticillioides*	Maize	Jiangsu Province, eastern China	2020

4.2. Inoculation, Incubation, and Growth Assessment

Subsequent studies were carried out with potato dextrose agar medium. Water activity was modified to 0.95, 0.98, and 0.99 by the addition of 245, 198, and 108 mL of glycerol per litre of culture medium. The accuracy of water activity was confirmed with an AquaLab Series 4TE (Decagon Devices, Inc., Pullman, WA, USA).

A 5 mm diameter agar disc from the margin of a fresh colony was transferred and inoculated onto the center of each plate. Inoculated plates were incubated at 20, 25, and 30 °C for 7, 14, and 21 days. Three biological replicates were conducted for each experimental condition. Two diameters of the colonies were measured at right angles every day until the colony reached the edge of the plate. The radii of the colonies were plotted against time, and linear regression was used to obtain the growth rate (mm/day). After 7 days of incubation, the whole mycelium was removed, frozen at −80 °C, lyophilized, and weighed as the biomass.

4.3. Mycotoxin Analysis

After 7, 14, and 21 days, agar plates were weighed, ground, and shaken in 20 mL of acetonitrile/water (4:1) at 180 rpm for 2 h. Then, the extract was filtered through Whatman No. 4 filter paper, and an aliquot (4 mL) of the supernatant was transferred to a glass tube until evaporation to dryness under N_2 flow. The dried samples were resuspended in 1 mL of acetonitrile, filtered through a 0.22 μm Millipore membrane, and analyzed using an LC-20ADXR liquid chromatograph (Shimadzu, Kyoto, Japan) coupled to an AB SCIEX Triple-Quad mass spectrometer (Applied Biosystems, Foster City, CA, USA).

The analytical column was a Kinetex 100A C18 column (100 × 2.3 mm, 2.6 μm) from Phenomenex (Torrance, CA, USA), and the column temperature was maintained at 40 °C. The flow rate was 0.5 mL/min, and the injection volume was 2 μL. The mass spectrometric analyses were performed with the following operation parameters: gas temperature, 500 °C; gas flow rate, 10 L/min; nebulizer gas pressure, 50 psi; capillary voltage, 5500 V. Nitrogen was used as the ion source and in the collision cell. Mycotoxins were analyzed via multiple reaction monitoring (MRM).

The mobile phase consisted of 5 mM ammonium acetate/acetic acid (99.9/0.1, v/v) (A) and methanol (B). The stepwise high-performance liquid chromatography conditions were as follows: 0–0.01 min, solvent A linearly increased to 10%; 1.6–2 min, solvent A linearly increased from 10% to 35%; 2–4.44 min, solvent A linearly increased to 55%; 4.44–6.5 min, solvent A linearly increased from 55% to 90%; 6.5–12 min, solvent A remained constant at 90%; 12–15 min, solvent A remained constant at 10%. The values of the limit of quantitation and limit of detection were 10 and 5 μg/kg, respectively.

4.4. Gene Expression Analysis

Total RNA was extracted from 7 day old mycelia of *Fusarium* strains grown on PDA with or without glycerol using an Axyprep multisource total RNA miniprep kit (Axygen Scientific, Inc., Union City, CA, USA). The amount and quality of RNA were estimated by Nanodrop (Thermo Fisher Scientific, Waltham, MA, USA). First-strand cDNA was obtained by the Prime ScriptTM RT Master Mix Kit (Takara Co., Ltd., Dalian, China) according to the manufacturer's instructions and stored at −80 °C until use in PCR.

Universal primers were designed with Primer Premier 5 software (PREMIER Biosoft International, Palo Alto, California, USA) on exon–exon junctions in the target mRNA to avoid the amplification of genomic DNA. Primers QFFFUB8F (5′–CTSTTCCAYGTTGCTGGWATMTG–3′)/QFFFUB8R (5′–CTTCMGAGTARGGACGCATCTC–3′) and QFFFub12F (5′–GCMTCTTTCTTYTCAAAGGC–3′)/QFFFub12R (5′–CTGTTATGYAAAACCCACCA–3′) were used to amplify *FUB8* and *FUB12* in every species studied. Primers QFFFUM1F1 (5′–TACCACTCWCATCACATGCAAG–3′)/QFFFUM1R1 (5′–TACRGGRCTCTCCAGATTTTG–3′) and QFFFUM21F1 (5′–ATTRCCGYTGATCCACGACGA–3′)/QFFFUM21R1 (5′–AGCTYGCGCTMTRCTSAGASG–3′) were applied in the amplification of *FUM1* and *FUM21* in *F. verticillioides*, *F. fujikuroi*, and *F. proliferatum*, respectively. The actin gene was selected as the internal control, and the primers QFFActF (5′–CCACCAGACATGACAATGTT–3′)/QFFActR (5′–CGTGATCTTACCGACTACCTC–3′) were used for all stains. Real-time PCR was conducted by TB Green Premix Ex Taq (Takara Co., Ltd., Dalian, China) according to the manufacturer's protocol. Real-time PCR reactions were performed in a LightCycler 96 system (Roche, Basel, Switzerland), and the data were analyzed with sequence detector software. Gene expression profiles were normalized to actin expression, and relative changes in gene expression levels were calculated using the $2^{-\Delta\Delta Ct}$ method.

4.5. Statistical Analysis

The statistical effects of the individual factor and their interactions on growth rate, biomass accumulation, and mycotoxin production were evaluated by ANOVA using GraphPad Prism version 8.3.0 (GraphPad Software, San Diego, CA, USA). Relative expression levels of mycotoxin synthetic genes under different conditions were analyzed by nonparametric tests for multiple comparisons using SPSS Statistics 26.0 (SPSS Inc., Chicago, IL, USA).

Author Contributions: Conceptualization, J.X., J.S. and G.M.; methodology, T.D. and S.Q.; validation, T.D. and J.Q.; formal analysis, J.X.; investigation, J.S.; data curation, J.X. and J.Q.; writing—original draft preparation, T.D. and J.Q.; writing—review and editing, G.M.; visualization, T.D.; supervision, G.M.; project administration, J.S. and G.M.; funding acquisition, J.X. All authors have read and agreed to the published version of the manuscript.

Funding: This work was supported by the National Key R&D Program of China (2022YFD1400101), the National Special Project for Agro-Product Safety Risk Evaluation of China (GJFP20220105, GJFP20220102), the Jiangsu Province Science and Technology Support Program (BA2022034, BE2022377), and the Jiangsu Agriculture Science and Technology (CX(21)1005).

Institutional Review Board Statement: Not applicable.

Informed Consent Statement: Not applicable.

Data Availability Statement: Not applicable.

Conflicts of Interest: The authors declare no conflict of interest.

References

1. Nuss, E.T.; Tanumihardjo, S.A. Maize: A paramount staple crop in the context of global nutrition. *Compr. Rev. Food Sci. Food Saf.* **2010**, *9*, 417–436. [CrossRef] [PubMed]
2. National Bureau of Statistics of China. *China Statistical Year Book*; China Statistics Press: Beijing, China, 2014.
3. National Bureau of Statistics of China. *Statistical Bulletin on National Economic and Social Development*; National Bureau of Statistics of China: Beijing, China, 2023.
4. Munkvold, G.P.; Desjardins, A.E. Fumonisins in maize: Can we reduce their occurrence? *Plant Dis.* **1997**, *81*, 556–565. [CrossRef] [PubMed]
5. Choi, J.H.; Lee, S.; Nah, J.Y.; Kim, H.K.; Paek, J.S.; Lee, S.; Ham, H.; Hong, S.K.; Yun, S.H.; Lee, T. Species composition of and fumonisin production by the *Fusarium fujikuroi* species complex isolated from Korean cereals. *Int. J. Food Microbiol.* **2018**, *267*, 62–69. [CrossRef] [PubMed]
6. Jabłońska, E.; Piątek, K.; Wit, M.; Mirzwa-Mróz, E.; Wakuliński, W. Molecular diversity of the *Fusarium fujikuroi* species complex from maize. *Eur. J. Plant Pathol.* **2020**, *158*, 859–877. [CrossRef]
7. Qin, P.W.; Xu, J.; Jiang, Y.; Hu, L.; Van Der Lee, T.; Waalwijk, C.; Zhang, W.M.; Xu, X.D. Survey for toxigenic *Fusarium* species on maize kernels in China. *World Mycotoxin J.* **2020**, *13*, 213–224. [CrossRef]
8. Qiu, J.; Lu, Y.; He, D.; Lee, Y.W.; Ji, F.; Xu, J.; Shi, J. *Fusarium fujikuroi* species complex associated with rice, maize, and soybean from Jiangsu Province, China: Phylogenetic, pathogenic, and toxigenic analysis. *Plant Dis.* **2020**, *104*, 2193–2201. [CrossRef]
9. Yilmaz, N.; Sandoval-Denis, M.; Lombard, L.; Visagie, C.M.; Wingfield, B.D.; Crous, P.W. Redefining species limits in the *Fusarium fujikuroi* species complex. *Persoonia* **2021**, *46*, 129–162. [CrossRef]
10. Zhang, H.; Luo, W.; Pan, Y.; Xu, J.; Xu, J.S.; Chen, W.Q.; Feng, J. First report of *Fusarium* ear rot of maize caused by *Fusarium andiyazi* in China. *Plant Dis.* **2014**, *98*, 1428. [CrossRef]
11. Duan, C.X.; Wang, B.B.; Sun, F.F.; Yang, Z.H.; Zhu, Z.D.; Wang, X.M. Occurrence of maize ear rot caused by *Fusarium fujikuroi* in China. *Plant Dis.* **2020**, *104*, 587. [CrossRef]
12. Czembor, E.; Stępień, Ł.; Waśkiewicz, A. *Fusarium temperatum* as a new species causing ear rot on maize in Poland. *Plant Dis.* **2014**, *98*, 1001. [CrossRef]
13. Munkvold, G.P. Epidemiology of *Fusarium* diseases and their mycotoxins in maize ears. *Eur. J. Plant Pathol.* **2003**, *109*, 705–713. [CrossRef]
14. O'Donnell, K.; McCormick, S.P.; Busman, M.; Proctor, R.H.; Ward, T.J.; Doehring, G.; Geiser, D.M.; Alberts, J.F.; Rheeder, J.P. Marasas et al. 1984 "Toxigenic *Fusarium* species: Identity and mycotoxicology" revisited. *Mycologia* **2018**, *110*, 1058–1080. [CrossRef] [PubMed]
15. Marasas, W.F.O.; Riley, R.T.; Hendricks, K.A.; Stevens, V.L.; Sadler, T.W.; Gelineau-van Waes, J.; Missmer, S.A.; Cabrera, J.; Torres, O.; Gelderblom, W.C.A.; et al. Fumonisins disrupt sphingolipid metabolism, folate transport, and neural tube development in embryo culture and in vivo: A potential risk factor for human neural tube defects among populations consuming fumonisin-contaminated maize. *J. Nutr.* **2004**, *134*, 711–716. [CrossRef] [PubMed]
16. Fremy, J.M.; Alassane-Kpembi, I.; Oswald, I.P.; Cottrill, B.; Van Egmond, H.P. A review on combined effects of moniliformin and co-occurring *Fusarium* toxins in farm animals. *World Mycotoxin J.* **2019**, *12*, 281–291. [CrossRef]
17. Radić, B.Đ.; Kos, J.J.; Kocić-Tanackov, S.D.; Janić-Hajnal, E.; Mandić, A.I. Occurrence of moniliformin in cereals. *Food Feed Res.* **2019**, *46*, 149–159. [CrossRef]
18. Ledoux, D.R.; Broomhead, J.N.; Bermudez, A.J.; Rottinghaus, G.E. Individual and combined effects of the *Fusarium* mycotoxins fumonisin B1 and moniliformin in broiler chicks. *Avian Dis.* **2003**, *47*, 1368–1375. [CrossRef]

19. Sharma, D.; Asrani, R.K.; Ledoux, D.R.; Jindal, N.; Rottinghaus, G.E.; Gupta, V.K. Individual and combined effects of fumonisin B1 and moniliformin on clinicopathological and cell-mediated immune response in Japanese quail. *Poultry Sci.* **2008**, *87*, 1039–1051. [CrossRef] [PubMed]
20. Pavlovkin, J.; Mistrik, I.; Prokop, M. Some aspects of the phytotoxic action of fusaric acid on primary Ricinus roots. *Plant Soil Environ.* **2004**, *50*, 397–401. [CrossRef]
21. Porter, J.K.; Bacon, C.W.; Wray, E.M.; Hagler Jr, W.M. Fusaric acid in *Fusarium moniliforme* cultures, maize, and feeds toxic to livestock and the neurochemical effects in the brain and pineal gland of rats. *Nat. Toxins* **1995**, *3*, 91–100. [CrossRef]
22. Bacon, C.W.; Porter, J.K.; Norred, W.P.; Leslie, J. Production of fusaric acid by *Fusarium* species. *App. Environ. Microb.* **1996**, *62*, 4039–4043. [CrossRef]
23. Eskola, M.; Kos, G.; Elliott, C.T.; Hajšlová, J.; Mayar, S.; Krska, R. Worldwide contamination of food-crops with mycotoxins: Validity of the widely cited 'FAO estimate'of 25%. *Crit. Rev. Food Sci.* **2020**, *60*, 2773–2789. [CrossRef] [PubMed]
24. Chen, J.; Jia, Z.; Song, J.; Yuan, Y.; Zhang, L. Fumonisins in China: Update on occurrence, epidemiology, exposure and regulation. *Qual. Assur. Saf. Crops Foods* **2015**, *7*, 63–72. [CrossRef]
25. Shang, Y.; Yang, W. Variation analysis of cereals mycotoxin limit standards of CAC, EU, USA, and China. *J. Food Sci. Technol.* **2019**, *37*, 10–15.
26. Yu, J.H.; Keller, N. Regulation of secondary metabolism in filamentous fungi. *Annu. Rev. Phytopathol.* **2005**, *43*, 437–458. [CrossRef] [PubMed]
27. Studt, L.; Janevska, S.; Niehaus, E.M.; Burkhardt, I.; Arndt, B.; Sieber, C.M.; Humpf, H.U.; Dickschat, J.S.; Tudzynski, B. Two separate key enzymes and two pathway-specific transcription factors are involved in fusaric acid biosynthesis in *Fusarium fujikuroi*. *Environ. Microbiol.* **2016**, *18*, 936–956.
28. Li, T.; Su, X.; Qu, H.; Duan, X.; Jiang, Y. Biosynthesis, regulation, and biological significance of fumonisins in fungi: Current status and prospects. *Crit. Rev. Microbiol.* **2022**, *48*, 450–462. [CrossRef] [PubMed]
29. López-Errasquín, E.; Vázquez, C.; Jiménez, M.; González-Jaén, M.T. Real-time RT-PCR assay to quantify the expression of *FUM1* and *FUM19* genes from the fumonisin-producing *Fusarium verticillioides*. *J. Microbiol. Meth.* **2007**, *68*, 312–317. [CrossRef] [PubMed]
30. Lazzaro, I.; Busman, M.; Battilani, P.; Butchko, R.A.E. *FUM* and *BIK* gene expression contribute to describe fumonisin and bikaverin synthesis in *Fusarium verticillioides*. *Int. J. Food Microbiol.* **2012**, *160*, 94–98. [CrossRef]
31. Cendoya, E.; Pinson-Gadais, L.; Farnochi, M.C.; Ramirez, M.L.; Chéreau, S.; Marcheguay, G.; Ducos, C.; Barreau, C.; Richard-Forget, F. Abiotic conditions leading to *FUM* gene expression and fumonisin accumulation by *Fusarium proliferatum* strains grown on a wheat-based substrate. *Int. J. Food Microbiol.* **2017**, *253*, 12–19. [CrossRef]
32. Medina, A.; Schmidt-Heydt, M.; Cárdenas-Chávez, D.L.; Parra, R.; Geisen, R.; Magan, N. Integrating toxin gene expression, growth and fumonisin B1 and B2 production by a strain of *Fusarium verticillioides* under different environmental factors. *J. R. Soc. Interface* **2013**, *10*, 20130320. [CrossRef]
33. Jurado, M.; Marín, P.; Magan, N.; González-Jaén, M.T. Relationship between solute and matric potential stress, temperature, growth, and *FUM1* gene expression in two *Fusarium verticillioides* strains from Spain. *App. Environ. Microb.* **2008**, *74*, 2032–2036. [CrossRef] [PubMed]
34. Marín, P.; Magan, N.; Vázquez, C.; González-Jaén, M.T. Differential effect of environmental conditions on the growth and regulation of the fumonisin biosynthetic gene *FUM1* in the maize pathogens and fumonisin producers *Fusarium verticillioides* and *Fusarium proliferatum*. *FEMS Microbiol. Ecol.* **2010**, *73*, 303–311. [CrossRef] [PubMed]
35. Fanelli, F.; Iversen, A.; Logrieco, A.F.; Mulè, G. Relationship between fumonisin production and *FUM* gene expression in *Fusarium verticillioides* under different environmental conditions. *Food Addit. Contam. A* **2013**, *30*, 365–371. [CrossRef]
36. Fumero, M.V.; Sulyok, M.; Ramirez, M.L.; Leslie, J.F.; Chulze, S.N. Effects of water activity and temperature on fusaric and fusarinolic acid production by *Fusarium temperatum*. *Food Control* **2020**, *114*, 107263. [CrossRef]
37. Pena, G.A.; Sulyok, M.; Chulze, S.N. Effect of interacting conditions of water activity, temperature and incubation time on *Fusarium thapsinum* and *Fusarium andiyazi* growth and toxin production on sorghum grains. *Int. J. Food Microbiol.* **2020**, *318*, 108468. [CrossRef] [PubMed]
38. Cendoya, E.; Farnochi, M.C.; Chulze, S.N.; Ramirez, M.L. Two-dimensional environmental profiles of growth and fumonisin production by *Fusarium proliferatum* on a wheat-based substrate. *Int. J. Food Microbiol.* **2014**, *182*, 9–17. [CrossRef]
39. Mamur, S.; Ünal, F.; Yılmaz, S.; Erikel, E.; Yüzbaşıoğlu, D. Evaluation of the cytotoxic and genotoxic effects of mycotoxin fusaric acid. *Drug Chem. Toxicol.* **2020**, *43*, 149–157. [CrossRef]
40. Fumero, M.V.; Sulyok, M.; Chulze, S. Ecophysiology of *Fusarium temperatum* isolated from maize in Argentina. *Food Addit. Contam. A* **2016**, *33*, 147–156.
41. Lazzaro, I.; Susca, A.; Mulè, G.; Ritieni, A.; Ferracane, R.; Marocco, A.; Battilani, P. Effects of temperature and water activity on *FUM2* and *FUM21* gene expression and fumonisin B production in *Fusarium verticillioides*. *Eur. J. Plant Pathol.* **2012**, *134*, 685–695. [CrossRef]
42. O'Brian, G.R.; Georgianna, D.R.; Wilkinson, J.R.; Yu, J.; Abbas, H.K.; Cleveland, D.; Bhatnagar, T.E.; Nierman, W.; Payne, G.A. The effect of elevated temperature on gene transcription and aflatoxin biosynthesis. *Mycologia* **2007**, *99*, 232–239. [CrossRef]
43. Gallo, A.; Solfrizzo, M.; Epifan, F.; Panzarini, G.; Perrone, G. Effect of temperature and water activity on gene expression and aflatoxin biosynthesis in *Aspergillus flavus* on almond medium. *Int. J. Food Microbiol.* **2016**, *217*, 162–169. [CrossRef]

44. Ochiai, N.; Tokai, T.; Nishiuchi, T.; Takahashi-Ando, N.; Fujimura, M.; Kimura, M. Involvement of the osmosensor histidine kinase and osmotic stress-activated protein kinases in the regulation of secondary metabolism in *Fusarium graminearum*. *Biochem. Biophys. Res. Commun.* **2007**, *363*, 639–644. [CrossRef]
45. Schmidt-Heydt, M.; Magan, N.; Geisen, R. Stress induction of mycotoxin biosynthesis genes by abiotic factors. *FEMS Microbiol. Lett.* **2008**, *284*, 142–149. [CrossRef] [PubMed]
46. Kohut, G.; Ádám, A.L.; Fazekas, B.; Hornok, L. N-starvation stress induced *FUM* gene expression and fumonisin production is mediated via the HOG-type MAPK pathway in *Fusarium proliferatum*. *Int. J. Food Microbiol.* **2009**, *130*, 65–69. [CrossRef] [PubMed]
47. von Bargen, K.W.; Niehaus, E.M.; Krug, I.; Bergander, K.; Würthwein, E.U.; Tudzynski, B.; Humpf, H.U. Isolation and structure elucidation of fujikurins A-D: Products of the PKS19 gene cluster in *Fusarium fujikuroi*. *J. Nat. Prod.* **2015**, *78*, 1809–1815. [CrossRef] [PubMed]
48. Niehaus, E.M.; Studt, L.; von Bargen, K.W.; Kummer, W.; Humpf, H.U.; Reuter, G.; Tudzynski, B. Sound of silence: The beauvericin cluster in *Fusarium fujikuroi* is controlled by cluster-specific and global regulators mediated by H3K27 modification. *Environ. Microbiol.* **2016**, *18*, 4282–4302. [CrossRef]
49. Janevska, S.; Arndt, B.; Baumann, L.; Apken, L.H.; Mauriz Marques, L.M.; Humpf, H.U.; Tudzynski, B. Establishment of the inducible Tet-on system for the activation of the silent trichosetin gene cluster in *Fusarium fujikuroi*. *Toxins* **2017**, *9*, 126. [CrossRef] [PubMed]
50. Brown, D.W.; Lee, S.H.; Kim, L.H.; Ryu, J.G.; Lee, S.; Seo, Y.; Kim, H.Y.; Busman, M.; Yun, S.H.; Proctor, R.H.; et al. Identification of a 12-gene fusaric acid biosynthetic gene cluster in *Fusarium* species through comparative and functional genomics. *Mol. Plant-Microbe Interact.* **2015**, *28*, 319–332. [CrossRef] [PubMed]
51. Janevska, S.; Tudzynski, B. Secondary metabolism in *Fusarium fujikuroi*: Strategies to unravel the function of biosynthetic pathways. *Appl. Microbiol. Biotechnol.* **2018**, *102*, 615–630. [CrossRef]
52. Atanasoff-Kardjalieff, A.K.; Studt, L. Secondary Metabolite Gene Regulation in Mycotoxigenic *Fusarium* Species: A Focus on Chromatin. *Toxins* **2022**, *14*, 96. [CrossRef]
53. Niehaus, E.M.; Rindermann, L.; Janevska, S.; Münsterkötter, M.; Güldener, U.; Tudzynski, B. Analysis of the global regulator *Lae1* uncovers a connection between *Lae1* and the histone acetyltransferase *HAT1* in *Fusarium fujikuroi*. *Appl. Microbiol. Biotechnol.* **2018**, *102*, 279–295. [CrossRef] [PubMed]
54. Pfannmüller, A.; Leufken, J.; Studt, L.; Michielse, C.B.; Sieber, C.M.; Güldener, U.; Hawat, S.; Hippler, M.; Fufezan, C.; Tudzynski, B. Comparative transcriptome and proteome analysis reveals a global impact of the nitrogen regulators *AreA* and *AreB* on secondary metabolism in *Fusarium fujikuroi*. *PLoS ONE* **2017**, *12*, e0176194. [CrossRef] [PubMed]
55. Wiemann, P.; Sieber, C.M.; von Bargen, K.W.; Studt, L.; Niehaus, E.M.; Espino, J.J.; Huß, K.; Michielse, B.C.; Albermann, S.; Wagner, D.; et al. Deciphering the cryptic genome: Genome-wide analyses of the rice pathogen *Fusarium fujikuroi* reveal complex regulation of secondary metabolism and novel metabolites. *PLoS Pathog.* **2013**, *9*, e1003475. [CrossRef]

toxins MDPI

Article

MicotoXilico: An Interactive Database to Predict Mutagenicity, Genotoxicity, and Carcinogenicity of Mycotoxins

Josefa Tolosa [1,*], Eva Serrano Candelas [2], José Luis Vallés Pardo [2], Addel Goya [2], Salvador Moncho [2], Rafael Gozalbes [2,3] and Martina Palomino Schätzlein [2,*]

1 Laboratory of Food Chemistry and Toxicology, Faculty of Pharmacy, University of Valencia, Av. Vicent Andrés Estellés, Burjasot, 46100 Valencia, Spain
2 ProtoQSAR S.L., CEEI-Technology Park of Valencia, Av. Benjamín Franklin, 12, 46980 Paterna, Spain; eserrano@protoqsar.com (E.S.C.); jlvalles@protoqsar.com (J.L.V.P.); agoya@protoqsar.com (A.G.); smoncho@protoqsar.com (S.M.); rgozalbes@protoqsar.com (R.G.)
3 Moldrug AI Systems S.L., Olimpia Arozena Torres, 45, 46018 Valencia, Spain
* Correspondence: josefa.tolosa@uv.es (J.T.); mpalomino@protoqsar.com (M.P.S.)

Abstract: Mycotoxins are secondary metabolites produced by certain filamentous fungi. They are common contaminants found in a wide variety of food matrices, thus representing a threat to public health, as they can be carcinogenic, mutagenic, or teratogenic, among other toxic effects. Several hundreds of mycotoxins have been reported, but only a few of them are regulated, due to the lack of data regarding their toxicity and mechanisms of action. Thus, a more comprehensive evaluation of the toxicity of mycotoxins found in foodstuffs is required. In silico toxicology approaches, such as Quantitative Structure–Activity Relationship (QSAR) models, can be used to rapidly assess chemical hazards by predicting different toxicological endpoints. In this work, for the first time, a comprehensive database containing 4360 mycotoxins classified in 170 categories was constructed. Then, specific robust QSAR models for the prediction of mutagenicity, genotoxicity, and carcinogenicity were generated, showing good accuracy, precision, sensitivity, and specificity. It must be highlighted that the developed QSAR models are compliant with the OECD regulatory criteria, and they can be used for regulatory purposes. Finally, all data were integrated into a web server that allows the exploration of the mycotoxin database and toxicity prediction. In conclusion, the developed tool is a valuable resource for scientists, industry, and regulatory agencies to screen the mutagenicity, genotoxicity, and carcinogenicity of non-regulated mycotoxins.

Keywords: mycotoxins; QSAR; mutagenicity; genotoxicity; carcinogenicity

Key Contribution: For the first time; a database containing almost 4400 mycotoxins classified in 170 different categories attending to their chemical structure has been constructed. Diverse QSAR models have been developed for the prediction of mycotoxin mutagenicity, genotoxicity, and carcinogenicity. The developed models have shown good performance and can be applied for regulatory purposes.

1. Introduction

Mycotoxins are common contaminants present in several human food products and animal feed. They are produced during the secondary metabolism of different filamentous fungi (molds), being the most common mycotoxin-producing fungal species *Aspergillus*, *Fusarium*, *Penicillium*, *Claviceps*, and *Alternaria* [1]. These genera are responsible for the production of an important variety of mycotoxins, which also include the main mycotoxins reported, such as aflatoxins, trichothecenes, fumonisins, ochratoxins, patulin, and zearalenone [2]. Some differences can be observed between strains: while *Fusarium* species usually infect growing crops in the field, *Aspergillus* and *Penicillium* species frequently grow on foods and feeds during the storage stage [3]. As a result, there is a wide range of foodstuffs susceptible to contamination by mycotoxins, including cereals, nuts, pasta, fruits, coffee, and by-products of animal origin [4]. Thus, mycotoxins can enter the food chain directly from plant-based food components contaminated with mycotoxins or due to the consumption of animal-derived products from animals fed contaminated feedstuffs, due to the carry-over of mycotoxins to animal-derived products such as milk, meat, and eggs [3]. Even when excellent agronomic, storage, and processing techniques are used, mycotoxin contamination of food and feed is still an unavoidable and unpredictable hazard, creating a challenging risk for food safety. Therefore, mycotoxin occurrence in foodstuffs is an actual problem; indeed, mycotoxins are the main hazard reported in European border rejection notifications by the Rapid Alert System for Food and Feed (RASFF) [3].

Moreover, it is worth highlighting that the pattern of mycotoxin production by several fungi in different geographical distribution is being affected by climate change, generating increasing concern [5]. For instance, the rises in temperature and rainfall in some geographical regions may result in more favorable environmental conditions for *Fusarium*, as in Europe. On the other hand, longer and more frequent droughts may encourage *Aspergillus flavus* to produce aflatoxins under both pre-harvest and post-harvest settings. In addition, recent investigations have demonstrated that the growth of mycotoxin-producing fungi can be stimulated even by a slight elevation in CO_2 levels [6,7]. Thus, the changing climatic conditions we are facing nowadays could change mycotoxin production and distribution worldwide.

Consumption of mycotoxin-contaminated food or feed can induce acute or chronic toxicity in humans and animals, with chronic effects being the most prevalent, due to prolonged exposure to lower concentrations. As a result, regulations concerning mycotoxins have been implemented in many countries to safeguard consumers from the harmful effects of these compounds [3,8]. Nevertheless, as regulations are primarily based on known toxic effects, maximal allowed limits, or tolerable daily intakes (TDI) were determined only for a few mycotoxins, as there are many mycotoxins for which no experimental data exist [9].

Therefore, additional risk assessment surveys on non-regulated mycotoxins are urgently required. However, the performance of traditional in vivo assays on thousands of different compounds would be extremely expensive and unethical. In this context, the application of alternative methods, such as in silico strategies could be extraordinarily useful. Indeed, several guidance documents have been drafted to improve standardization, harmonization, and uptake of in silico methods by regulatory authorities including the EFSA (European Food Safety Authority) and the ECHA (European Chemicals Agency) [10,11]. In this sense, the Commission Regulation No. 1907/2006 called REACH (Registration, Evaluation and Authorisation of Chemicals) (http://ecb.jrc.it/reach/reachlegislation/) (accessed on 23 January 2023) foresees the use of in silico methods such as (Quantitative) Structure–Activity Relationship ([Q]SAR) models when the same level of information can be obtained by means other than in vivo testing [10–12]. More concretely, information relating to the genotoxic potential of chemicals by using in silico prediction approaches has become an important source, as recommended by the REACH regulation as well as the ICH M7 guideline for the assessment and control of DNA reactive or mutagenic impurities [12–15].

Among the toxic effects that can be caused by mycotoxins, the induction of genetic alterations is an important matter of concern [14,16–18], as several mycotoxins and some of their metabolites have been described as genotoxic compounds, including aflatoxins, ochratoxins, citrinin, and HT-2 and T-2 toxins [19]. However, scarce information has been reported regarding the capability of other mycotoxins to cause these adverse effects.

A genotoxic compound can induce mutations (mutagenicity) and/or cause the generation of tumors (carcinogenicity). To characterize these properties, an in silico toxicology (IST) protocol template [20] as well as a protocol for genetic toxicology (the GIST protocol) [14] have been designed and developed by an international consortium comprising several industry, academia, and government agencies. Therefore, in the present study, in vitro tests recommended for the mutagenicity and genotoxicity endpoints by the GIST have been taken into account to search experimental data for QSAR model building. To define the mutagenicity, the in vitro bacterial reverse mutation assay (commonly referred to as the Ames test) provides robust and high-quality data, which have been previously used to develop QSAR models with a good performance in predicting mutagenic activity. On the other hand, for genotoxicity assessment, data on in vitro micronucleus (MN) assay have been widely used, while carcinogenicity is evaluated by detecting tumor generation in in vivo models.

The aim of the present study was the development of specific QSAR models to predict genotoxicity, mutagenicity, and carcinogenicity of a wide range of mycotoxins. To this end, we build, for the first time, a comprehensive database including almost 4400 different mycotoxins, clustered in different categories according to their chemical structure. We then overlapped this list with different databases of genotoxicity, mutagenicity, and carcinogenicity to obtain experimental data based on the Ames test for mutagenicity, in vitro and in vivo MN assay for genotoxicity, and data from in vivo models for carcinogenicity. These data were then applied for the building and validation of scientifically valid and robust QSAR models that predicted the endpoints on a test set of mycotoxins with a high accuracy, sensitivity, and selectivity. Finally, the mycotoxins database together with the predicted toxicity values was integrated in a new, open access web server that can be explored interactively.

2. Results

2.1. Mycotoxin Database and Clustering

Our search resulted in a data set of 4360 mycotoxins identified by their name, isomeric SMILES (Simplified Molecular Input Line Entry System) and PubChem CID. In addition, we grouped the mycotoxins according to their chemical structure in 170 different families. To our knowledge, this is the first time that such a comprehensive database of mycotoxins has been published, as previous works only made reference to several hundred mycotoxins [21–25]. To provide an easy access to the database, we have created a web application, MicotoXilico, that allows easy exploration of the data (https://chemopredictionsuite.com/MicotoXilico, accessed on 20 May 2023).

In order to better visualize the structural diversity of the mycotoxins, a clustering based on k-nearest neighbor approach from structural fingerprints was performed (https://tmap.gdb.tools/ accessed on 20 May 2023). The resulting graphs can be explored interactively at the MicotoXilico web. In Figure 1, a clustering plot indicating the major categories of mycotoxins is depicted. In the graph, the number of linkages between compounds is proportional to their chemical similarity, meaning that similar mycotoxins will be connected through short pathways.

As we see, myctoxins have a very high structural variability, representing almost the full range of natural products. They include alkaloids, terpenoids, peptids, fatty acids, lactones, nucleotids, phenols, and anthraquinones, among others. Until now, even if there were some known categories such as zerealones, ergot alkaloids, or trichothecenes, there exists no general systematic classification of mycotoxins. In our classification, we have maintained the groups based on specific structural motives, including between 5 and 50 different compounds. An overview of the categories and the number of compound in each category can be found in the Appendix A (Table A1). In some cases, when there was a very high similarity between structures, we have created unified groups, such enniatins and beauvericins, emodins and rubrofusarins, asperlins and asperlactones, fusarins and fusariens, or usnic acids and ustins. In other cases, the structural similarity was not big enough to unify categories, but it was still remarkable, for instance, between sphingofungins and fumonisins, alternariolides and enniatins, roquefortines and ergot alkaloids, or brefeldins and zearealones. These similarities can be appreciated in the graphical clustering visualization.

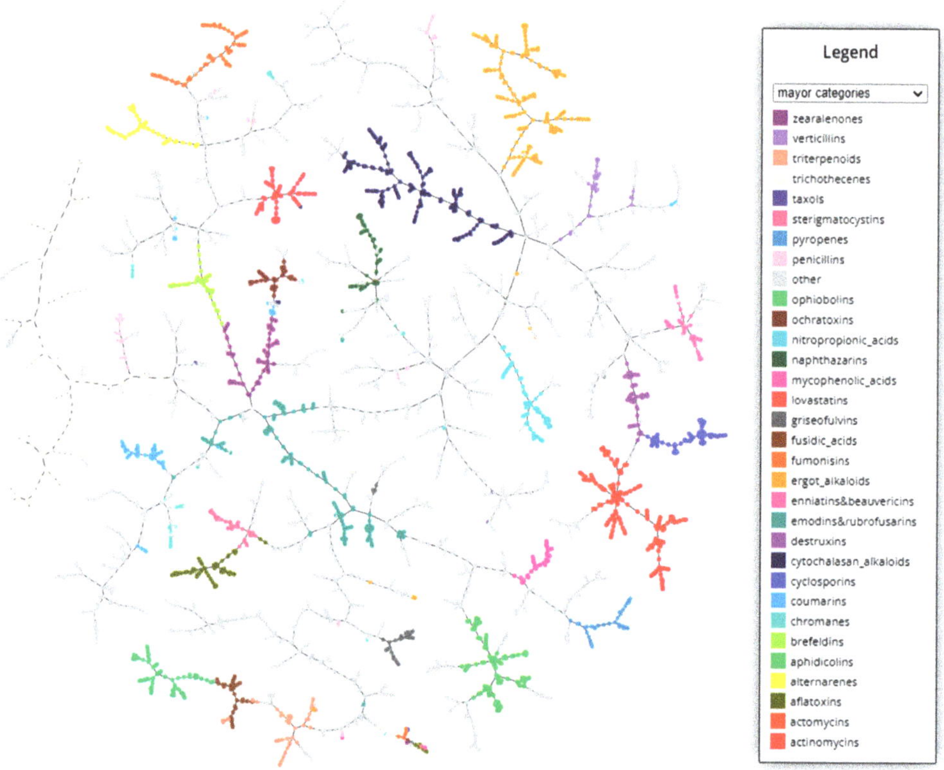

Figure 1. TMAP clustering graph of the database containing 4360 mycotoxins. The most important categories are labelled in different colors. For an interactive exploration of all categories, please visit MicotoXilico (https://chemopredictionsuite.com/MicotoXilico, accessed on 20 May 2023).

Some of the larger, more traditional categories have a higher structural variability because they include several subgroups of compounds. For instance, cytochalasan alcaloids can be subclassified into chaetoglobosins, daldinins, and chalasins. In particular, trichothecenes, the most abundant category, with more than 350 compounds, has a high number of subcategories, as can be explored in the web server (https://chemopredictionsuite.com/MicotoXilico, accessed on 20 May 2023).

Some compounds could not be associated with a specific structural category, and they have therefore been classified into more general groups comprising alkaloids, terpenoids, amino acid derivatives, peptides, clyclic peptides, nucleotides, anthraquinones, benzoquinones, naphthalenes, phenols, phtalates, furans, lactones, thiazolidines, and fatty acid-like compounds. Furthermore, 79 compounds could not been associated with any category and were labelled as "not classified".

2.2. QSAR Models Building

Once we had generated a comprehensive database of mycotoxins, we wanted to optimize several machine learning models to predict the genotoxicity, mutagenicity, and carcinogenicity of these compounds. The tests chosen for the characterization of each type of toxicity follow the recommended endpoint protocols of the OECD [26]. For data searching, we overlapped our ensemble of mycotoxins with different databases of mutagenicity, genotoxicity, and carcinogenicity to obtain experimental data based on the Ames test for mutagenicity, in vitro and in vivo MN assay for genotoxicity, and data from in vivo models for carcinogenicity. As expected, experimental data could only be found for a relatively low number of compounds (350–100 compounds, depending on the endpoint).

These data were then used for the building and validation of robust QSAR models to predict the four endpoints. For model building, we followed the protocols described in the materials and methods section that meet the requirement of the five principles of the OECD for QSAR model building in a regulatory context [27].

A summary of the characteristics of the final classification models can be found in Figure 2, including the number of compounds in the training and test set, the descriptor selection method, the number of descriptors, the compound/descriptor ratio, and the model algorithm. In the Appendix A, a list with the molecular descriptors selected for each model can be found (Tables A2–A6). It is worth mentioning that the descriptors were selected from a panel of more than 4000, including 2D and 3D descriptors, in order to obtain the best chemical description of the complex mycotoxin structure.

For mutagenicity, models based only on mycotoxin data could be build, as data for up to 365 compounds could be retrieved, with a balanced proportion between mutagenic and non-mutagenic compounds. Two mutagenicity QSAR models (model A and model B) were generated applying two different data selection criteria (for details, see Section 5.3). In both cases, all the parameters included in the metrics were higher than 0.8, thus showing a good performance. Both models were able to correctly predict the mutagenicity of almost 90% of compounds in the internal cross validation.

For genotoxicity, data retrieval was much more complicated, as very few data from non-genotoxic mycotoxins could be detected. Therefore, we decided to build mixed models including mycotoxins and other organic compounds (for details, see Section 5.4). Regarding the in vitro genotoxicity QSAR model, information on the in vitro MN assay from 455 compounds was considered (Figure 2). For a more comprehensive analysis of genotoxicity, we decided to also include an in vivo genotoxicity model. In this case, the ProtoPRED model (https://protopred.protoqsar.com/, accessed on 20 May 2023) based on the in vivo MN assay was applied, built from a training set that included mycotoxins. Regarding the performance of these models, most parameters were higher than 0.8 (Figure 3d). Only the specificity was moderate, as some non-genotoxic compounds were predicted as genotoxic in the internal validation. This result could relate to the fact that only a few data from non-genotoxic mycotoxins could be found.

For carcinogenicity, again we applied the in vivo carcinogenicity QSAR model from ProtoPRED (https://protopred.protoqsar.com/, accessed on 20 May 2023), containing mycotoxins in the training set, as not enough carcinogenicity data were retrieved to build a specific model. The metrics obtained for the carcinogenicity QSAR model showed a good performance on the training set, with all parameters being higher than 0.9. Parameters on the test set showed lower values, especially for precision and sensitivity, but are still close to 0.7, the value recommended for QSAR models by ECHA [28].

When we compare the metrics of the training and test sets in Figure 2, we can see that there are almost no differences for the mutagenicity models, small differences in the in vitro genotoxicity model, and a higher difference in the carcinogenicity and the in vivo genotoxicity. This result is coherent with the fact that the mutagenicity models were built only with mycotoxins, while in the other models, only part of the training set were mycotoxins, meaning that we have more structural differences between the training and test set.

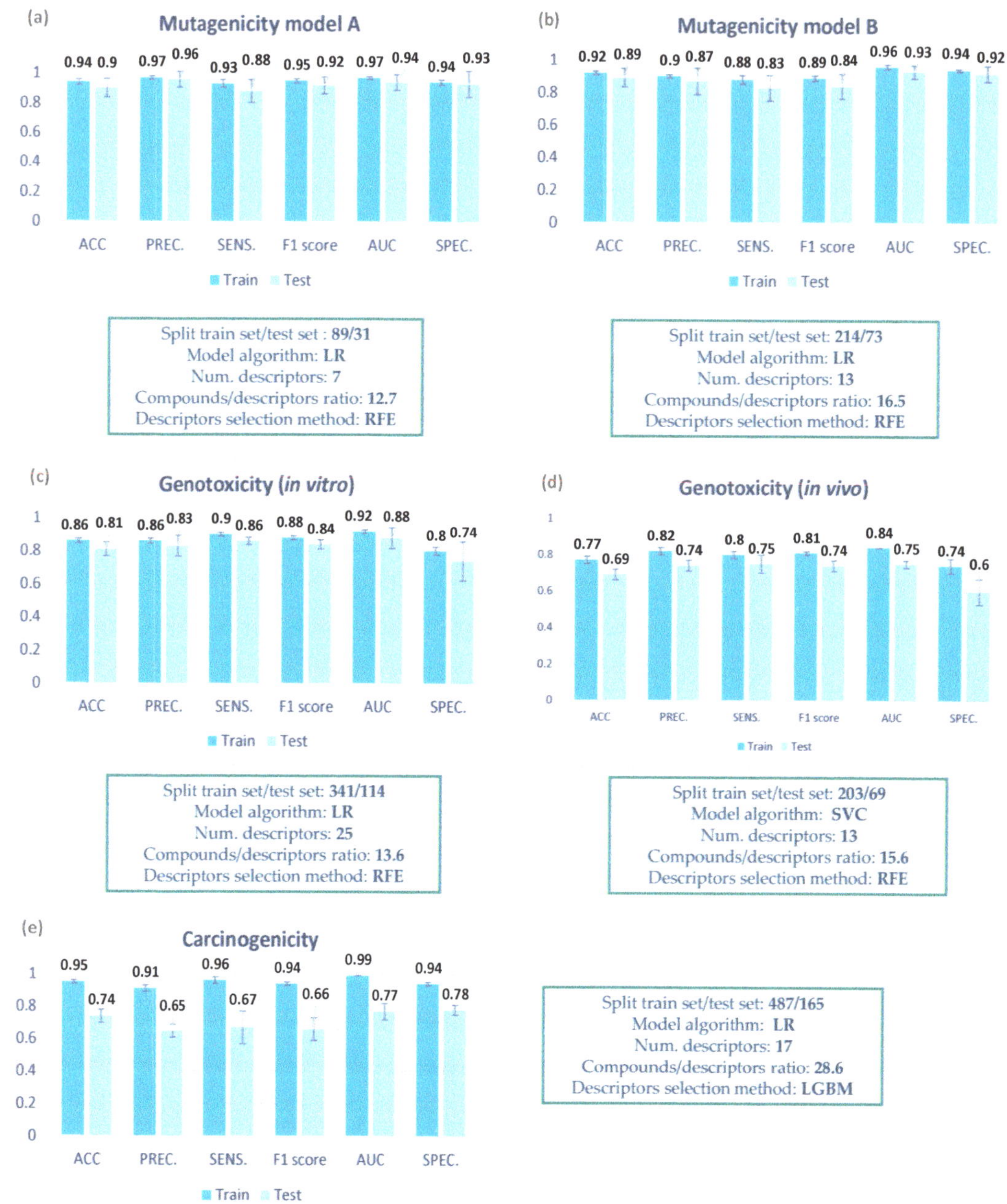

Figure 2. Metrics for QSAR models of (**a**) mutagenicity model A, (**b**) mutagenicity model B, (**c**) in vitro genotoxicity, (**d**) in vivo genotoxicity, and (**e**) carcinogenicity. ACC: accuracy; PREC: precision; SENS: sensitivity; AUC: area under the curve; SPEC: specificity; LGBM: Light Gradient Boosting Machine Classifier; LR: logistic regression; RFE: recursive feature elimination; SVC: Support Vector Machine Classifier (SVC).

(a)

Experimental values	Non-mutagen	**Mutagen**	
	Mutagenicity QSAR predictions		
Non-mutagen	9	3	0.75 (**SPEC.**)
Mutagen	1	11	0.92 (**SENS.**)
Total	0.9 (pred NEG)	0.79 (pred POS)	0.83 (**ACC**)

(b)

Experimental values	Non-genotoxic	**Genotoxic**	
	Genotoxicity QSAR predictions (*in vitro*)		
Non-genotoxic	6	1	0.86 (**SPEC.**)
Genotoxic	0	8	1 (**SENS.**)
Total	1 (pred NEG)	0.89 (pred POS)	0.93 (**ACC**)

(c)

Experimental values	Non-genotoxic	**Genotoxic**	
	Genotoxicity QSAR predictions (*in vivo*)		
Non-genotoxic	6	7	0.46 (**SPEC.**)
Genotoxic	7	52	0.88 (**SENS.**)
Total	0.46 (pred NEG)	0.88 (pred POS)	0.81 (**ACC**)

(d)

Experimental values	Non-carcinogen	**Carcinogen**	
	Carcinogenicity QSAR predictions		
Non-carcinogen	25	3	0.89 (**SPEC.**)
Carcinogen	12	35	0.74 (**SENS.**)
Total	0.67 (pred NEG)	0.92 (pred POS)	0.8 (**ACC**)

Figure 3. Matrix confusion for the external validation of (**a**) the mutagenicity QSAR model A applied to 24 mycotoxins, (**b**) the in vitro genotoxicity QSAR model applied to 15 mycotoxins, (**c**) the in vivo genotoxicity QSAR model applied to 72 mycotoxins, and (**d**) the carcinogenicity QSAR model applied to 75 mycotoxins. SPEC: specificity; SENS: sensitivity; ACC: accuracy; pred NEG: predicted negatives; pred POS: predicted positives.

2.3. QSAR Model Application to an External Validation Set

After model building and internal validation, we decided to further confirm the performance of the models by performing an external validation with an independent set of mycotoxins with known experimental data. This allows us to evaluate if the model is truly predictive or if the model has been overfit to the data used for model building.

Since only the mutagenicity model A is considered valid from a regulatory point of view, we decided to subject only this model to external validation. The list containing the different mycotoxins used for the external validation for each model can be found in Tables A7–A10 of the Appendix A. Figure 3 shows the resulting confusion matrix for all model validations, proving that the experimental data were in general well predicted.

For mutagenicity (Figure 3a), the confusion matrix showed that the model was capable of correctly classifying more than 0.8 of the compounds (0.83 accuracy). The in vitro genotoxicity QSAR model showed the highest accuracy (0.93) within all developed models. All genotoxic compounds were predicted as positive (1.00 of sensitivity), while only one compound out of 7 non-genotoxic mycotoxins was predicted as genotoxic (0.86 of specificity) (Figure 3b), thus improving the specificity of the internal validation of the in vitro model. For the in vivo genotoxicity model, however, although obtaining good values for accuracy (0.81) and sensitivity (0.88), the specificity was again low (0.46), proving that this model was not accurate for the prediction of non-genotoxic compounds.

The carcinogenicity model was applied to a validation set of 75 mycotoxins, showing an accuracy of 0.81. The 89% of non-carcinogenic mycotoxins were predicted as inactive compounds, while 77% of carcinogenic mycotoxins were predicted as active compounds (Figure 3d).

Thus, even if the models were built with a small data set of experimental data from mycotoxins, they seem to have a good predictive power on the external validation set of mycotoxins.

In order to see how well our models were adapted to mycotoxins, in comparison with other, more general toxicological QSAR models, we also performed a prediction with the validation data set applying three other reference QSAR tools: VEGA [29], Leadscope, and Case Ultra (the last two being integrated into QSARToolbox) [30]. We predicted the same four endpoints, obtained from the same or a very similar protocol. The metrics of these predictions can be found on Table A11.

We can observe that, in general, our models provide a better prediction for the selected test mycotoxins. Only for mutagenicity do we obtain a better prediction with Case Ultra; however, six compounds could not be predicted by this model because they were outside the applicability domain. Also, for the other models, the prediction of several mycotoxins could not be performed because they were not in the applicability domain. We have also found that, in some cases, VEGA could not correctly read and normalize structures with aromatic rings, probably due to the aromaticity model that the software uses. These results show that our models are better adapted to complex structures with several aromatic rings, which are typical mycotoxin structures.

2.4. Mutagenicity, Genotoxicity, and Carcinogenicity Prediction of the General Mycotoxin Database

After model building, we wanted to perform a prediction of the genotoxicity, mutagenicity, and carcinogenicity of the whole mycotoxin database described in Section 2.1. A general overview of the results is presented in Figure 4, and detailed predictions can be explored in the MicotoXilico web application (https://chemopredictionsuite.com/MicotoXilico, accessed on 20 May 2023).

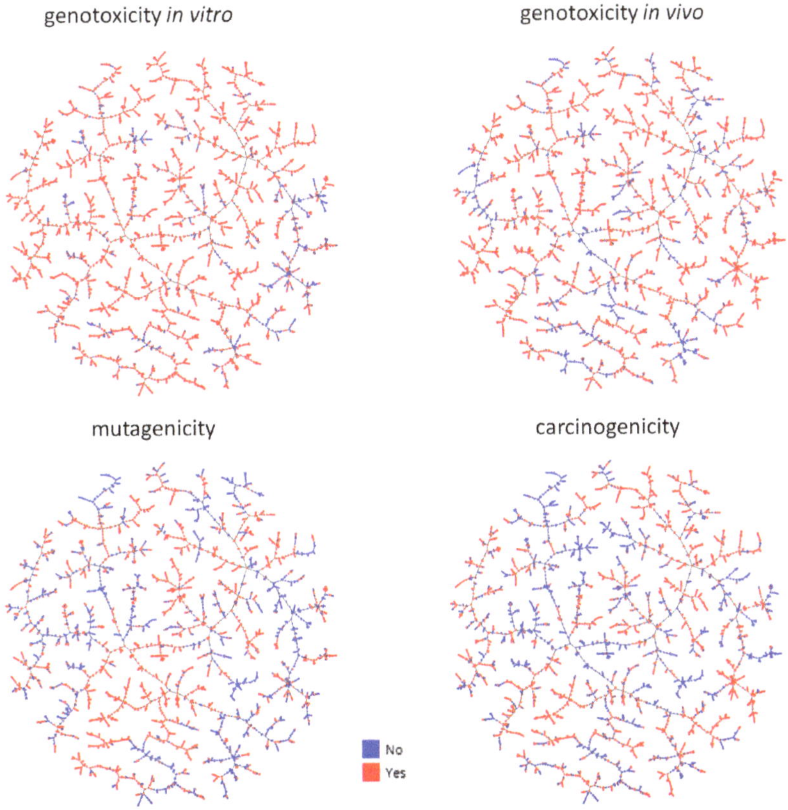

Figure 4. Global overview of genotoxicity, mutagenicity, and carcinogenicity predictions of the whole database of mycotoxins. Clustering was performed with the TMAP package. Blue = non-toxic; red = toxic. For more details, please visit MicotoXilico (https://chemopredictionsuite.com/MicotoXilico accessed on 20 May 2023).

From the figure, we can appreciate that a very high percentage of mycotoxins are predicted to be genotoxic, mutagenic, and/or carcinogenic. This result is not surprising, as the definition of mycotoxin already assumes that we are dealing with toxic compounds. In particular, genotoxicity seems to be a property of most mycotoxins, as 80–90% of these compounds are predicted as genotoxic. However, further studies are required to confirm these results, as the models were built with only a few data from non-genotoxic mycotoxins, and the genotoxicity models only have a moderate specificity.

We also used the Benigni and Bossa rules for mutagenicity [31], implemented in Proto-ICH software (https://protopred.protoqsar.com/, accessed on 20 May 2023) to identify the presence of structural mutagenicity alerts in our database. The number of molecules detected as positive based on structural alerts were significantly less in comparison with the molecules detected as positive based on QSAR models (https://chemopredictionsuite.com/MicotoXilico, accessed on 20 May 2023). On the contrary, when we performed the same comparison with a set of over 6000 general organic compounds, most of the compounds predicted as mutagenic had a mutagenicity alert (data not shown).

In order to compare the toxicity between mycotoxin categories, we represented the percentage of genotoxic, mutagenic, and carcinogenic compounds for the 30 major mycotoxin categories in a heatmap (Figure 5).

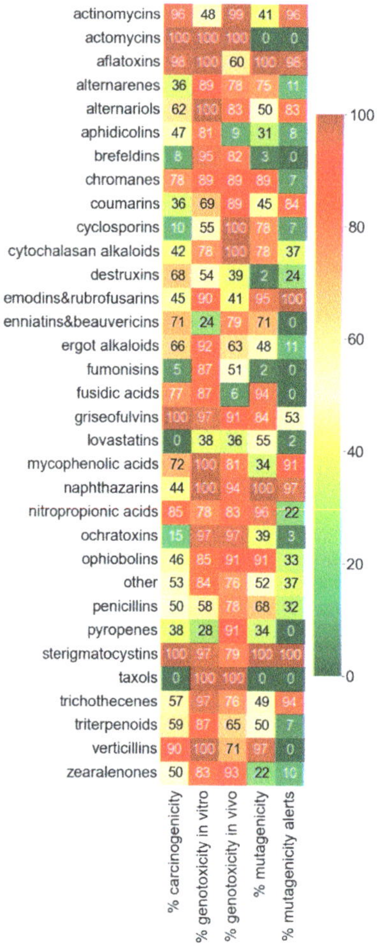

Figure 5. Graphical representation of the percentage of genotoxic, carcinogenic, and mutagenic mycotoxins from the major categories obtained after prediction with the corresponding QSAR models.

As we can see, there are several categories that are specially concerning, as they have a very high toxicity prediction in mutagenicity, genotoxicity, and carcinogenicity (Figure 5), such as aflatoxins, chromanes, fusidic acids, griseofulvins, nitropropionic acids, sterigmatocystins, and verticillins. Some categories have non-positive predictions in some of the toxicity endpoints (taxols and lovastatins), but there is no type of compound that is negative in all three toxicities.

In many categories, the mutagenicity alerts correspond to a very high mutagenicity prediction (<80%), such as in aflatoxins, emodins, napththazarins, and sterigmatocystins. However, it is worth highlighting that several categories without alerts showed a high predicted mutagenicity index, such as alternarenes, chromanes, enniatins and beauvericins, fusidic acids, nitropropionic acids, and verticillins.

3. Discussion

In this work, we explored the application of in silico QSAR models to obtain toxicity data of mycotoxin, which are urgently needed for public health regulations. In order to obtain a general overview of the type and number of mycotoxins that exist, we performed a comprehensive search, generating for the first time a database containing almost 4400 compounds. Our research revealed that the number and diversity of the identified mycotoxins is much higher than generally assumed: most publications only indicate the existence of several hundreds of mycotoxins, while we found more than 4000 known mycotoxins. It is worth mentioning the high structural diversity of these compounds, which generally have relatively high molecular weights and many structures including sugar, aromatic, or peptide rings.

The constructed database allowed us to perform a search of mutagenicity (Ames test), genotoxicity (in vivo and in vitro MN assay), and carcinogenicity (in vivo models) data by overlapping with different experimental databases. The search confirmed a low availability of experimental data, covering only a small percentage of the total existing mycotoxins. Nevertheless, we could obtain and validate robust QSAR models that predicted the four endpoints on an external validation set of mycotoxins with a relatively high accuracy, sensitivity, and specificity. These good results could be achieved by taking into account the specific characteristics of mycotoxins during the model building process, creating an appropriate applicability domain by the inclusion of mycotoxin and mycotoxin-like structure in the training set, and using a broad panel of 2D and 3D chemical descriptors. This was further confirmed by a comparison of the prediction of the mycotoxin's validation set with three other, more general, QSAR reference tools (Table A11), which provided a worse prediction and only included part of the molecules in their applicability domain. In the case of the genotoxicity, the specificity of the predictions was moderate to low, probably due the fact that only a few data from non-genotoxic mycotoxins were retrieved and incorporated in the models. Thus, further experiments are required to confirm the existence of non-genotoxic categories of mycotoxins. However, following the caution principle, a model with lower specificity (predicting false positives) is preferred over a model with lower sensitivity (predicting false negatives).

When applying the prediction models to the database constructed containing almost 4400 mycotoxins, we obtained a very high proportion of mutagenic, carcinogenic, and especially genotoxic compounds. This result is not unexpected, as mycotoxins are defined *per se* as toxic compounds, and compounds of many categories proved to induce acute toxicity. However, differences between categories can be observed, mainly due to differences in the chemical structure. Concerning genotoxicity, all major categories included compounds with a positive genotoxicity prediction. Among them, some categories are well known to be genotoxic, such as aflatoxins, ochratoxins, or sterigmatocystins. Indeed, 97% of mycotoxins from the ochratoxin family have been predicted as genotoxic compounds. This result agrees with the scientific opinion published by the EFSA in 2020 [32] indicating the genotoxicity of ochratoxin A and thus eliminating the previously established TDI and establishing instead an MOE (Margin Of Exposure), as no threshold can be allowed for genotoxic compounds. In the case of sterigmatocystin, a mycotoxin structurally related to aflatoxin B1, it has been demonstrated to induce tumors in diverse animal species, and thus, it is a known carcinogen mycotoxin [33], which agrees with the prediction performed with our QSAR models.

However, some categories showing a high percentage of genotoxic potential, are not well studied. For instance, griseofulvins were predicted as genotoxic by both in vitro and in vivo QSAR models. In the literature, animal studies have shown evidence that they are able to cause a variety of acute and chronic toxic effects, including liver and thyroid cancer in rodents, abnormal germ cell maturation, teratogenicity, and embryotoxicity in various species [34].

Regarding enniatins and beauvericins, commonly named as emerging *Fusarium* mycotoxins, the EFSA concluded in 2014 that a risk assessment was not possible given the lack of relevant toxicity data [35]. On one hand, in vitro genotoxicity data available suggested a potential genotoxic effect for beauvericin, while in vitro genotoxicity data for enniatins were negative. These results agree with those predicted by the in vitro QSAR model developed in our study (Figure 6). On the other hand, there are no in vivo genotoxicity data for either beauvericin or enniatins and no studies on carcinogenicity of beauvericin and enniatins have been identified, and thus, the use of in silico predictions for these endpoints can provide valuable information. Thus, according to predictions on enniatins and beauvericin, 79% and 71% of compounds from this category were predicted as genotoxic (in vivo model) and carcinogenic, respectively. In addition, 71% of enniatins and beauvericin were also predicted as mutagenic, thus suggesting a careful assessment of the emerging *Fusarium* mycotoxins toxicity.

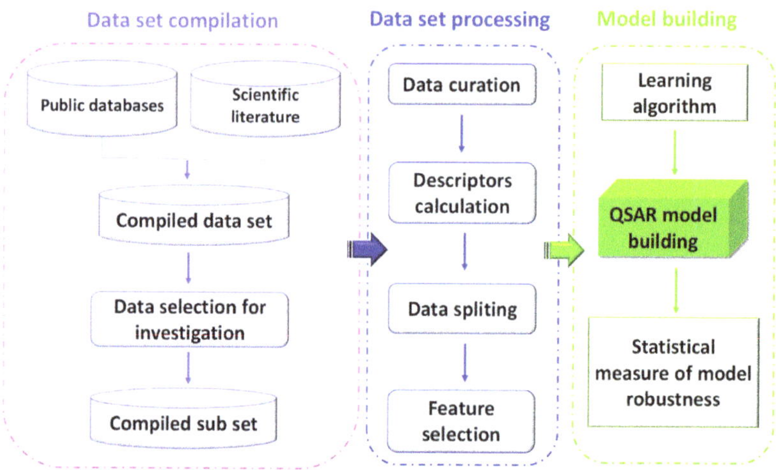

Figure 6. Workflow diagram for QSAR model building.

Compounds from several categories have been classified as carcinogenic by the IARC [36], such as aflatoxins, trichothecenes, or fumonisins. Other classes, such as actinomycins, cyclosporings, and lovastatins, have still not been classified, but they are labelled as potentially carcinogenic by the ECHA. Furthermore, in several categories, carcinogenicity predictions proved to have an impact on the DNA of cells. For instance, aphidicolin is an inhibitor of eucaryotic nuclear DNA. Brevianamide produced a slightly teratogenic effect in chick embryos [37]. Emodin is suspected to create DNA strand breaks and/or non-covalently binding to DNA and inhibiting the catalytic activity of topoisomerase II (Toxin and Toxin Target Database (T3DB)); nevertheless, a genotoxic effect could be confirmed [38]. Mycophenolic acid inhibits the de novo pathway of guanosine nucleotide synthesis without incorporation into DNA (Toxin and Toxin Target Database (T3DB)).

Alternariols and alternarenes are *Alternaria* mycotoxins that can be found in cereals around the world, but little relevance is still given to this fact. Currently, the toxicity of several altenariols is being investigated, including alternariol, alternariol monomethyl ether, altertoxins, altenuene, tenuazonic acid, and tentoxin. Among them, tenuazonic acid, alternariol, alternariol monomethyl ether, altenuene, and altertoxin I are the most important mycotoxins that can be found as contaminants in fruits and vegetables [39]. In 2011, the EFSA carried out a risk assessment on *Alternaria* toxins, as they were reported to induce genotoxicity, cytotoxicity, and reproductive and developmental toxicity, among other adverse effects [40]. Regarding their genotoxic effects, it was reported that alternariol, alternariol monomethyl ether, and altertoxins could induce gene locus mutation, DNA damage or synthesis disorder, chromosome aberration, and other effects in in vitro studies. In fact, according to the in vitro genotoxicity model developed, 100% of mycotoxins from the alternariols category was predicted as genotoxic compounds. In addition, alternariols have been related to the high incidence of esophageal cancer in Linxian, China [39], which can be related to our findings, as 62% of compounds belonging to this family were predicted as carcinogenic mycotoxins. Thus, special attention should be paid to this mycotoxin category. In this sense, maximum levels have been recently recommended by the EU in the

Commission Recommendation 2022/553 [41] for alternariol, alternariol monomethyl ether, and tenuazonic acid.

A high percentage of trichothecenes have been predicted as genotoxic by the in vitro model (97%) and the in vivo model (76%). The main trichothecene reported to occur in food commodities is deoxynivalenol. Although deoxynivalenol is not genotoxic by itself, it has recently been shown that this toxin exacerbates the genotoxicity induced by model or bacterial genotoxins. In addition, other trichothecenes, namely, T-2 toxin, diacetoxyscirpenol, nivalenol, fusarenon-X, and the newly discovered NX toxin, were also reported as compounds able to exacerbate the DNA damage inflicted by various genotoxins [42]. In addition, in the study reported by Yang et al. [43], deoxynivalenol was able to cause damage to the membrane, the chromosomes, and the DNA at all times of culture in human peripheral blood lymphocytes, thus concluding that deoxynivalenol potentially triggers genotoxicity in human lymphocytes. In other study performed on Sprague Dawley rats, deoxynivalenol increased the percentage of chromosomal aberration, DNA fragmentation, and comet score [44].

Citrinin, a mycotoxin classified in the chromanes category, has been reported to be genotoxic at high concentrations in cultured human lymphocytes, as it caused a significant concentration-dependent increase in MN frequency in human lymphocytes [45], a result according to our genotoxicity predictions for the chromanes category, where almost 90% of compounds were predicted as genotoxic by both the in vitro and the in vivo models.

The same occurs with zearalenones, which have been predicted as genotoxic by both genotoxicity models, according to some data reported showing that zearalenone and some of its metabolites increased the percentage of chromosome aberrations in mouse bone-marrow cells and in HeLa cells [46] and can increase the frequencies of polychromatic erythrocytes micronucleated and chromosomal aberrations in bone marrow cells from Balb/c female mice [47].

Regarding fumonisins, 87% of mycotoxins belonging to this family were predicted as genotoxic by the in vitro model. The most predominant, fumonisins, fumonisin B1, fumonisin B2, and fumonisin B3, are carcinogenic and genotoxic secondary metabolites found in corn-based foods worldwide and are produced by *Fusarium verticillioides* and *F. proliferatum* [48]. Fumonisin B1 is defined by IARC as a possible human carcinogen in Group 2B, and it shows genotoxic activity via oxidative stress, DNA damage, cell cycle arrest, apoptosis, inhibition of mitochondrial respiration, and deregulation of calcium homeostasis [49]. Some studies revealed that exposure to fumonisin B1 caused a significant increase in micronucleus frequency in a concentration- and time-dependent manner in rabbit kidney cells [50], and in HepG2 cells, fumonisin B1 has shown clastogenic effects [51].

Studies on the genotoxic activity of ergot alkaloids, also predicted as genotoxic by our models, are very limited. In the scientific report delivered by EFSA in 2012 [52], it was stated that genotoxicity studies on ergot alkaloids were insufficient, and more concretely, some studies evaluating the genotoxic and mutagenic effects of ergotamine revealed different results. In the literature, it has been reported that ergotamine is able to induce chromosomal abnormalities in human lymphocytes and leukocytes [53] but does not show mutagenic effects in mouse lymphoma cells [54]. Other authors have demonstrated that ergotamine and ergometry can induce sister chromatid exchange in ovarian cells [55]. Due to the scarce and different data obtained, further studies are necessary to evaluate the genotoxic and mutagenic potential of ergot alkaloids.

Furthermore, our results reveal that several mycotoxin categories are predicted as mutagenic but have no mutagenicity alert following ICH-M7 criteria. This did not happen when we performed the same comparison with a general database of organic compounds, where almost all molecules predicted as mutagenic had an alert. For some of these categories, no mutagenicity has been detected previously (alternarenes, enniatins and beauvericins, fusidic acids, and verticillins), while for others, some experimental evidence exists already (chromanes, nitropropionic acids, among others) [56,57]. This suggests that the structure of the mycotoxins could have been underestimated in the expert analysis of the mutagenicity, and that ICH-M7 criteria do not take into account specific mutagenic structural motives present in mycotoxins. Regulatory agencies should take this into account and request a revised version of these criteria to obtain a better coverage of mycotoxins, which are a danger for public health, and thus prioritize mycotoxins based on their mutagenic, genotoxic, and carcinogenic potential, as already suggested in other studies [58–60].

4. Conclusions

The web server developed in this work represents a valuable resource for scientists, industry, and regulatory agencies, including a comprehensive database of over 4000 mycotoxins divided into categories that can be easily explored by an interactive visualization. To our knowledge, this is the first database containing such a high number of mycotoxins. Furthermore, the user can directly access a prediction of the mutagenicity, genotoxicity, and carcinogenicity of the whole ensemble of mycotoxins, without the need to perform a prediction workflow. The data are based on mycotoxin specific and robust QSAR models, that were built up according to OECD principles and are adapted to REACH criteria, which means that they can be used for regulatory purposes. Thus, the developed models are a valuable tool for screening toxicity of non-regulated mycotoxins. Future perspectives of our work include the experimental validation of our models by analyzing selected mycotoxins from categories with no data in the training set, which have to be synthesized or purified, as most of them are not commercially available. Furthermore, the new QSAR models will be included in our ProtoPRED platform.

5. Materials and Methods

To achieve the development of adequate and robust QSAR models, several elements are required. First of all, a data set providing experimental values of a biological activity or property for a group of already tested chemicals is necessary; in this case, the biological properties evaluated were mutagenicity, genotoxicity, and carcinogenicity; thus, experimental results derived from the mutagenic assay *Ames test* and the in vitro and in vivo MN test, and in vivo long-term carcinogenicity assay on rodents have been collected, as described in Section 5.1. Secondly, statistical methods (often called chemometric methods) are employed to find and validate the relationship between the calculated descriptors and the toxicity properties of the mycotoxins. The exact workflow is summarized in Figure 6 and described in the following sections.

5.1. Mycotoxin Data Set Construction and Chemical Data Curation

For the construction of a general data set of mycotoxins, an initial list of mycotoxins was collected from the literature [22,61–66] and specific mycotoxin databases (https://zenodo.org/record/2648816#.ZClfI3ZBy5d; https://sciex.com/products/spectral-library/mycotoxin-libraries) (accessed on 2 March 2022).

This list was further completed by searching in PubChem for compounds of the same family, and their metabolites, resulting in a list of 4360 compounds. Only compounds that were directly produced by fungi and their metabolites were included. For each compound, the isomeric SMILES together with the CAS number and the PubChem CID code was retrieved. The resulting database was further curated by normalizing the smiles and removing counterions, salts, and mixtures. Finally, each compound was assigned to a specific family or category to their chemical structure, obtaining 170 different categories (Table A1).

5.2. Mycotoxin Clustering and Chemical Space Distribution

To study the chemical space of the generated mycotoxins data set, a fingerprinting system was generated using the LSH Forest algorithm from TMAP (https://tmap.gdb.tools/, accessed on 20 May 2023) for dimensionality reduction. Each point in the TMAP represents a fingerprint of a unique chemical transformation generated using the fully trained model. The points were colored by categories or their outcome in the toxicity predictions. The Faerun package (https://pypi.org/project/faerun/, accessed on 20 May 2023) was used to create an interactive visualization tool for the clustering scheme.

5.3. Mycotoxin Mutagenicity QSAR Model

For QSAR mutagenicity model development, the endpoint Bacterial Reverse Mutation Test (Ames test) (OECD test guideline 471) was selected (Figure 7), considered the first outcome to assess the possible mutagenicity of a substance [67,68]. We scanned several high-quality databases (Carcinogenic Potency Database CEBS, CCRIS Mutagenicity assay, Vega, QSAR Toolbox, EFSA OpenFoodTox, ECVAM, etc.) as well as the scientific literature [68] for data of the compounds of our previously generated mycotoxin database. For each compound, CAS number, isomeric SMILES, and experimental data expressed as active (1) or inactive (0) were compiled in a table.

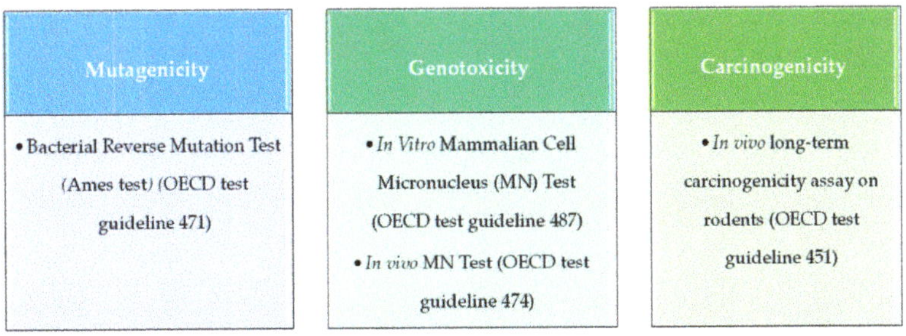

Figure 7. In vitro and in vivo assays consulted for model development and their corresponding OECD test guideline.

As a result, raw data of 356 mycotoxins were recovered, which were further refined as follows: For a first model A, data of 120 compounds (75 mutagens and 45 no mutagens) were selected that strictly fulfil the OECD guideline TG471: assayed with at least five *Salmonella* strains (TA1535, TA1537 or TA97 or TA 97a, TA98, TA100, and TA102) with and without microsomal activation by S9 fraction, since often the interaction with genetic material occurs after metabolic activation. For a second model B, data were selected less strictly, to cover a greater part of the chemical space, including data from at least two strains with or without metabolic activation. The data set of model B was composed of 287 compounds, with a ratio of mutagens to non-mutagens of 108/179. A third data set of 24 compounds, fulfilling the OECD guideline, was saved for external validation (Section 5.6).

For the development of the QSAR models, around 4000 different chemical descriptors from 15 different categories were calculated for each compound with an in-house python script. Then, descriptors were selected by recursive feature elimination (RFE) to 7 for model A and 13 for model B (Tables A2 and A3). In the next step, the data set was divided into a training set for model building, and a test set, in a proportion of 75–25%, respectively. Mutagenic and non-mutagenic compounds were distributed homogeneously in both sets. For model building, several algorithms were tested on the training set, obtaining the best metrics for Logistic Regression.

5.4. Mycotoxin Genotoxicity QSAR Models

For in vitro genotoxicity QSAR modelling, the in vitro MN assay was chosen, a robust and quantitative assay of chromosome damage with the capacity to detect not only clastogenic and aneugenic events but also some epigenetic effects and recommended by the OECD for genotoxicity evaluation in test guideline 487 (Figure 7).

Databases (Genetic Toxicology Data Bank in PubChem and eChemPortal) and the scientific literature [69–71] were scanned for data for the in vitro MN assay, fulfilling the OECD guideline, retrieving data for 91 compounds, of which 15 were saved for external validation (Section 5.6). To enable model building, the data set was further completed with 379 compounds that were not mycotoxins but close in the chemical space, generating a final data set of 455 compounds. The ratio of genotoxic to non-genotoxic compounds was 264/191, respectively. For model building, the same procedure as in Section 5.3 was employed, also obtaining model based on 25 descriptors (Table A4) based on Logistic Regression.

For in vivo genotoxicity QSAR modelling, the in vivo MN assay on mammalian erythrocyte was chosen (OECD test guideline 474), which identifies substances that cause micronuclei in erythroblasts sampled from bone marrow and/or peripheral blood cells of animals, usually rodents. As not enough data for mycotoxins could be retrieved (72 mycotoxins from the ISSMIC public database), the QSAR model from ProtoPRED (https://protopred.protoqsar.com/, accessed on 20 May 2023) including mycotoxins in the training set, was applied. The model was built with 13 descriptors (Table A5) and based on the Support Vector Machine Classifier (SVC) algorithm. The model was then externally validated with the external set of 72 mycotoxins obtaining good results. All mycotoxins were included in the applicability domain of the model.

5.5. Mycotoxin Carcinogenicity QSAR Model

For carcinogenicity assessment, the long-term carcinogenicity study on rat was employed, fulfilling the OECD test guideline 451 (Figure 7). As there was not enough data to build a specific model for mycotoxins (75 mycotoxins from different databases, such as PubChem AID_1259411, Carcinogenic Potency Database CEBS and OpenFoodTox TX22525), the QSAR model from ProtoPRED (https://protopred.protoqsar.com/ accessed on 20 May 2023) was applied, showing excellent results when applied to the 75 mycotoxins in the external validation. All mycotoxins were included in the applicability domain of the model.

The data for developing the model was extracted from Carcinogenicity I (ISSCAN) public database retrieved from QSAR Toolbox, which contains curated information on chemical compounds tested with the long-term carcinogenicity bioassay on rodents (rat and mouse). The main primary sources of data are the NTP, CPDB, CCRIS, and IARC repositories. After curation and preprocessing, the database was formed by 652 experimental results, with 251 of positive values (38.5%) and 401 negative values (61.5%). For model building, the same procedure as in Section 5.3 was employed, also obtaining the best metrics for Light Gradient Boosting Machine (LGBM) Classifier.

5.6. External Model Validation

The predictive potential of the QSAR models was evaluated with a completely independent data set of mycotoxins, in terms of accuracy, precision, sensitivity, and specificity. The metrics considered to evaluate the performance of the built classification models were calculated according to the following formulas:

$$\text{Accuracy} = \text{TP} + \text{TN}/(\text{TP} + \text{FN} + \text{FP} + \text{TN})$$

$$\text{Precision} = \text{TP}/(\text{TP} + \text{FP})$$

$$\text{Sensitivity} = \text{TP}/(\text{TP} + \text{FN})$$

$$\text{Specificity} = \text{TN}/(\text{TN} + \text{FP})$$

where TP = true positive; FP = false positive; TN = true negative; and FN = false negative.

Details about the compounds selected for each validation can be found in Tables A7–A10.

Author Contributions: Conceptualization, R.G., M.P.S. and J.T.; methodology, M.P.S., J.T., E.S.C., J.L.V.P. and S.M.; software, E.S.C., J.L.V.P., S.M. and A.G.; validation, M.P.S., J.T., E.S.C., J.L.V.P. and S.M.; formal analysis, M.P.S., J.T., E.S.C., J.L.V.P. and S.M.; investigation, M.P.S., J.T. and E.S.C.; resources, M.P.S., E.S.C., J.L.V.P. and S.M.; data curation, M.P.S., E.S.C. and J.T.; writing—original draft preparation, M.P.S. and J.T.; writing—review and editing, M.P.S., J.T., E.S.C., J.L.V.P., S.M. and R.G.; visualization, M.P.S., E.S.C., A.G. and J.T.; supervision, R.G. and E.S.C.; project administration, R.G., M.P.S. and J.T.; funding acquisition, R.G., M.P.S. and J.T. All authors have read and agreed to the published version of the manuscript.

Funding: This research was funded by the Torres Quevedo MicotoXilico PTQ2020-011477 and the Ministerio de Ciencia e Innovación of Spain PID2020-115871RB-I00 and from the Conselleria d'Innovació, Universitats, Ciència i Societat Digital (AEST/2021/077).

Institutional Review Board Statement: Not applicable.

Informed Consent Statement: Not applicable.

Data Availability Statement: The mycotoxin database and all performed predictions can be downloaded free of charge at the webserver: https://chemopredictionsuite.com/MicotoXilico.

Conflicts of Interest: The authors declare no conflict of interest.

Appendix A

Tables A1–A11.

Table A1. Number of compounds in each mycotoxin category.

Category	Num of Cpds
trichothecenes	410
ergot alkaloids	158
cytochalasan alkaloids	147
actinomycins	141
resveratrols	121
aphidicolins	90
nucleotides	88
not classified	79
alkaloids	74
emodins	67
taxols	67
fumonisins	63
zearalenones	60
peptides	59
cyclosporins	58
lovastatins	58
aflatoxins	57
coumarins	55
nitropropionic acids	54
ophiobolins	54
terpenoids	51
penicillins	50
triterpenoids	46

Table A1. *Cont.*

Category	Num of Cpds
destruxins	41
brefeldins	38
phenols	37
fumitremorgins	36
brevianamides	36
naphthazarins	36
alternarenes	36
ETPs	34
anthraquinones	34
brassicicenes	34
ochratoxins	33
sterigmatocystins	33
mycophenolic acids	32
penitrems	32
griseofulvins	32
pyropenes	32
verticillins	31
fusidic acids	31
fatty acid like	29
cyclic peptides	29
decarestrictines	28
chromanes	27
naphthalenes	25
yanuthones	25
benzene derivatives	25
furans	25
alternariolides	24
viridins	24
alternariols	24
quinazolines	24
roquefortines	23
anthracyclines	22
territrems	22
radicicols	22
asterriquinones	22
altersolanols	21
pyrones	21
xanthones	21
apicidins	20

Table A1. *Cont.*

Category	Num of Cpds
enniatins	19
chinulins	19
rugulosins	19
tyrosols	19
beauvericins	19
wortmannins	19
austalides	19
pyrenocines	19
cyclopeptins	18
chaetomugilins	18
zaragozic acids	18
chrysophanols	18
alterporriols	18
rubrofusarins	17
gibberellins	17
andrastins	16
physcions	16
aspernolides	16
abscisic acids	15
koninginins	15
roridins	14
versicolorins	14
chetomins	14
malformins	14
radicinins	14
usnic acids&ustins	13
secalonic acids	13
aspewentins	13
gregatins	13
altertoxins	13
aspulvinones	13
phomopsins	12
rubellins	12
viridicatins	12
cotylenins	12
aureobasidins	12
communesins	12
altenuenes	12
austocystins	12

Table A1. *Cont.*

Category	Num of Cpds
austins	12
arugosins	12
fusaric acids	11
patulins	11
lactones	11
fusarins&fusarielins	11
zinniols	11
cladosporols	11
altenusins	11
emestrins	11
schweinfurthins	10
fumagillins	10
aurasperones	10
astellolides	10
penicillenols	10
tryptophols	10
kojic acids	10
averufins	10
culmorins	10
kipukasins	10
xalines	10
rubratoxins	9
brevicompanines	9
penostatins	9
ustiloxins	9
melleins	9
sphingofungins	8
asperlins&asperlactones	8
thiazolidines	8
benzoquinones	8
nigerapyrones	8
andibenins	8
cordycepins	8
asterriquinones	8
chartarlactams	8
vertinols	7
tenuazonic acids	7
xanthins	7
calphostins	7

Table A1. *Cont.*

Category	Num of Cpds
erythromycins	7
marcfortines	7
bikaverins	7
solanapyrones	7
quinolones	7
cephalosporins	7
stachybocins	7
chevalones	7
glaucins	7
fusarielins	6
phtalates	6
depudecins	6
meleagrins	6
myrothecines	6
AF toxins	6
aflavinines	6
cercosporins	5
chromanones	5
aspergillimides	5
tryprostatins	5
amino acid derivative	5
flavones	4
chlamydosporols	4
amphotericins	4
aspilactonols	4
emodins&rubrofusarins	4
orienticins	4
candidusins	4
asperpyrones	3
candidins	3

Table A2. Molecular descriptors selected by recursive feature elimination (RFE) for QSAR mutagenicity model A.

Descriptor	Definition
C-030	X–CH–X atom-centered fragments
ATSC7Z	centered Moreau–Broto autocorrelation of lag 7 weighted by atomic number
ATSC8dv	centered Moreau–Broto autocorrelation of lag 8 weighted by valence electrons
MATS1c	Moran coefficient of lag 1 weighted by Gasteiger charge
GATS1c	Geary coefficient of lag 1 weighted by Gasteiger charge
GATS3d	Geary coefficient of lag 3 weighted by sigma electrons
X1solA	average solvation connectivity index of order 1

Table A3. Molecular descriptors selected by recursive feature elimination (RFE) for QSAR mutagenicity model B.

Descriptor	Definition
C-018	=CHX atom-centered fragments
C-019	=CRX atom-centered fragments
ATSC6dv	centered Moreau–Broto autocorrelation of lag 6 weighted by valence electrons
ATSC8dv	centered Moreau–Broto autocorrelation of lag 8 weighted by valence electrons
AATSC5p	averaged and centered Moreau–Broto autocorrelation of lag 5 weighted by polarizability
MATS1c	Moran coefficient of lag 1 weighted by Gasteiger charge
nR_3_False_False_False_True	number of 3–membered rings non-aromatic hetero
R1s+	R maximal autocorrelation of lag 1/weighted by I-state GETAWAY descriptors
NPR2	normalized principal moment of order 2
De	D total accessibility index, Sanderson electronegativity-weighted
Mor28v	signal 28/weighted by van der Waals volume 3D-MoRSE descriptors
Mor02s	signal 02/weighted by I-state 3D-MoRSE descriptors
Mor23s	signal 23/weighted by I-state 3D-MoRSE descriptors

Table A4. Molecular descriptors selected by recursive feature elimination (RFE) for QSAR in vitro genotoxicity model.

Descriptor	Definition
H-047	atom-centered fragments. H attached to $C^1(sp^3)/C^0(sp^2)$
O-056	alcohol. Atom-centered fragments
O-061	O–atom-centered fragments
F-084	F attached to C1(sp2). Atom-centered fragments
AATS5dv	averaged Moreau–Broto autocorrelation of lag 5 weighted by valence electrons. 2D
ATSC4v	centered Moreau–Broto autocorrelation of lag 4 weighted by van der Waals volume 2D autocorrelations
AATSC1Z	averaged and centered Moreau–Broto autocorrelation of lag 1 weighted by atomic number. 2D
AATSC5c	averaged and centered Moreau–Broto autocorrelation of lag 5 weighted by Gasteiger charge. 2D
GATS2Z	Geary coefficient of lag 2 weighted by atomic number. 2D
BELdv0	highest eigenvalue of Burden matrix weighted by valence electrons
L3i	3rd component size directional WHIM index/weighted by ionization potential WHIM descriptors
Mor09u	signal 09/unweighted 3D-MoRSE descriptors
Mor22s	signal 22/weighted by I-state 3D-MoRSE descriptors
JhetZ	Balaban-type index from Z-weighted distance matrix (Barysz matrix)
D/Dr3	distance/detour ring index of order 3
JGI9	mean topological charge index of order 9 2D autocorrelations
B05[O-O]	presence/absence of O-O at topological distance 5 2D Atom Pairs
B07[C-N]	presence/absence of C-N at topological distance 7 2D Atom Pairs
B08[C-O]	presence/absence of C-O at topological distance 8 2D Atom Pairs
B10[C-C]	presence/absence of C-C at topological distance 10 2D Atom Pairs
nAllOx	number of allylic oxidation sites excluding steroid dienone
nByciclic	number of atoms that are in two rings.
PEOE_VSA4	MOE Charge VSA Descriptor 4 ($-0.20 <= x < -0.15$). 2D
SLogP_VSA4	MOE logP VSA Descriptor 4 ($0.00 <= x < 0.10$). 2D

Table A5. Molecular descriptors selected by Recursive Feature Elimination (RFE) based on Support Vector Machine (SVM) using linear kernel (C = 1), for QSAR in vivo genotoxicity model.

Descriptor	Definition
C-016	"=CHR."
ATSC5i	centered Moreau–Broto autocorrelation of lag 5 (log function) weighted by ionization potential.
AATSC0se	averaged centered Moreau–Broto autocorrelation of lag 0 (log function) weighted by Sanderson electronegativity.
AATSC5i	averaged centered Moreau–Broto autocorrelation of lag 5 (log function) weighted by ionization potential.
MATS4p	Moran autocorrelation of lag 4 (log function) weighted by polarizability.
GATS4i	Geary autocorrelation of lag 4 (log function) weighted by ionization potential.
B02(C-C)	presence/absence of C-C at topological distance 2.
B02(C-O)	presence/absence of C-O at topological distance 2.
B04(C-O)	presence/absence of C-O at topological distance 4.
B04(O-O)	presence/absence of O-O at topological distance 4.
B08(C-N)	presence/absence of C-N at topological distance 8.
B09(O-O)	presence/absence of O-O at topological distance 9.
EState_VSA6	EState VSA descriptor 6.

Table A6. Molecular descriptors selected by Light Gradient Boosting Machine (LGBM) for QSAR carcinogenicity model.

Descriptor	Definition
AATSC1dv	averaged centered Moreau–Broto autocorrelation of lag 1 (log function) weighted by valence electrons
SMR_VSA9	MOE MR VSA descriptor 9
SIC2	structural information content index (neighborhood symmetry of 2-order)
ATSC1pe	centered Moreau–Broto autocorrelation of lag 1 (log function) weighted by pauling EN
C-032	X–CX–X
JGI7	mean topological charge index of order 7
BELc0	highest eigenvalue of Burden matrix weighted by charge
AATSC4se	averaged centred Moreau–Broto autocorrelation of lag 4 (log function) weighted by Sanderson electronegativity
GATS6se	Geary autocorrelation of lag 6 (log function) weighted by Sanderson electronegativity
NssNH	number of ssNH
D/Dr5	distance/detour ring index of order 5
GATS1s	Geary autocorrelation of lag 1 (log function) weighted by I-state
RBF	rotatable bond fraction
JGI5	mean topological charge index of order 5
PJI2	2D Petitjean shape index
GATS1d	Geary autocorrelation of lag 1 (log function) weighted by sigma electrons
AATSC3d	averaged centred Moreau–Broto autocorrelation of lag 3 (log function) weighted by sigma electrons

Table A7. Validation set for mutagenicity data.

SMILES	Name	y
CC1=C[C@H]2[C@@H](CC3=CNC4=CC=CC2=C34)N(C1)C	agroclavine	0
C[C@H]1CCCC(=O)CCC/C=C/C2=C(C(=CC(=C2)O)O)C(=O)O1	zearalenone	0
CC1=C[C@@H]2[C@](CC1)([C@]3([C@@H]([C@H]([C@H]([C@@]34CO4)O2)O)OC(=O)C)C)COC(=O)C	diacetoxyscirpenol	0
CC1=C[C@@H]2[C@]([C@@H](C1=O)O)([C@]3(C[C@H]([C@H]([C@@]34CO4)O2)O)C)CO	deoxynivalenol	0
C[C@@H]1[C@H](O1)[C@H]2[C@H](C=CC(=O)O2)OC(=O)C	asperlin	0
CC(C)[C@@]1(C(=O)N2[C@H](C(=O)N3CCC[C@H]3[C@@]2(O1)O)CC4=CC=CC=C4)NC(=O)[C@@H]5C[C@H]6[C@@H](CC7=CNC8=CC=CC6=C78)N(C5)C	dihydroergocristine	0
CCCC[C@@H](C)[C@H]([C@H](C[C@@H](C)C[C@@H](CCCC[C@H](C[C@@H]([C@H](C)N)O)O)O)OC(=O)C[C@@H](CC(=O)O)C(=O)O)OC(=O)C[C@@H](CC(=O)O)C(=O)O	fumonisin B1	0
CCCC[C@@H](C)[C@H]([C@H](C[C@@H](C)CCCCCC[C@H](C[C@@H]([C@H](C)N)O)O)OC(=O)C[C@@H](CC(=O)O)C(=O)O)OC(=O)C[C@@H](CC(=O)O)C(=O)O	fumonisin B2	0
CCCC[C@@H](C)[C@H]([C@H](C[C@@H](C)C[C@@H](CCCCCC[C@H]([C@H](C)N)O)O)OC(=O)CC(CC(=O)O)C(=O)O)OC(=O)CC(CC(=O)O)C(=O)O	fumonisn B3	0
CC1=C[C@@H]2[C@]([C@@H](C1=O)O)([C@]3([C@@H]([C@H]([C@H]([C@@]34CO4)O2)O)OC(=O)C)C)CO	fusarenon-X	0
CC1=C[C@@H]2[C@](CC1)(C3([C@@H]([C@H]([C@H](C34CO4)O2)O)OC(=O)C)C)CO	4-Acetoxyscirpenol	0
CC1=C2COC(=O)C2=C(C(=C1OC)C/C=C(\C)/CCC(=O)O)O	mycophenolic acid	0
COC1=C2C3=C(C(=O)CC3)C(=O)OC2=C4[C@@H]5CCO[C@@H]5OC4=C1	aflatoxin B2	1
C[C@H]1[C@H]([C@H](C[C@@H](O1)O[C@H]2C[C@@](CC3=C2C(=C4C(=C3O)C(=O)C5=CC=CC=C5C4=O)O)(C(=O)C)O)N)O	idarubicin	1
C[C@@H]1[C@@H](C(=O)N[C@@H](C(=O)N2CCC[C@H]2C(=O)N(CC(=O)N([C@H](C(=O)O1)C(C)C)C)C(C)C)NC(=O)C3=C4C(=C(C=C3)C)OC5=C(C(=O)C(=C(C5=N4)C(=O)N[C@H]6[C@H](OC(=O)[C@@H](N(C(=O)CN(C(=O)[C@@H]7CCCN7C(=O)[C@H](NC6=O)C(C)C)C)C)C(C)C)C)N)C	actinomycin D	1
COC1=C2C3=C(C(=O)CC3)C(=O)OC2=C4[C@@H]5[C@@H]6[C@@H](O6)O[C@@H]5OC4=C1	2,3-Epoxyaflatoxin B1	1
COC1=C2C3=C(C(=O)CC3)C(=O)OC2=C4C5CCOC5OC4=C1	aflatoxin B2 alpha	1
COC1=C2C3=C([C@@H](CC3)O)C(=O)OC2=C4[C@@H]5C=CO[C@@H]5OC4=C1	aflatoxicol B	1
COC1=C2C(=C3[C@@H]4C=CO[C@@H]4OC3=C1)OC5=CC=CC(=C5C2=O)O	sterigmatocystin	1
COC1=C2C3=C(C(=O)C[C@@H]3O)C(=O)OC2=C4[C@@H]5C=CO[C@@H]5OC4=C1	aflatoxin Q1	1
C[C@H]1[C@H]([C@H](C[C@@H](O1)O[C@H]2C[C@@](CC3=C2C(=C4C(=C3O)C(=O)C5=C(C4=O)C(=CC=C5)OC)O)(/C(=N\NC(=O)C6=CC=CC=C6)/C)O)N)O	zorubicin	1
C1=NC(=C2C(=N1)N(C=N2)[C@H]3[C@@H]([C@@H]([C@H](O3)CO)O)O)NO	inosine oxime	1
C[C@@H]1[C@@]2([C@@H](O2)[C@](O1)(C)/C=C(\C)/C=C(\C)/[C@H]3[C@](O3)(C)C4=C(C(=C(C(=O)O4)C)OC)C)C	verrucosidin	1
C1C[C@]2([C@@H]3[C@H](CC(=O)C4=C(C=CC(=C34)C5=C2C(=C(C=C5)O)C1=O)O)O)O	altertoxin I	1

Table A8. Validation set for in vitro genotoxicity data.

SMILES	Name	y
CC(C)C[C@H]1C(=O)O[C@@H](C(=O)N([C@H](C(=O)O[C@@H](C(=O)N([C@H](C(=O)O[C@@H](C(=O)N1C)C(C)C)CC(C)C)C)C(C)C)CC(C)C)C)C(C)C	enniatin C	0
CC[C@H](C)[C@@H]1C(=O)N([C@H](C(=O)O[C@@H](C(=O)N([C@H](C(=O)O[C@@H](C(=O)N([C@H](C(=O)O1)C(C)C)C)[C@@H](C)CC)C(C)C)C)C(C)C)C(C)C)C	enniatin I	0
CC[C@H](C)[C@H]1C(=O)O[C@@H](C(=O)N([C@H](C(=O)O[C@@H](C(=O)N([C@H](C(=O)O[C@@H](C(=O)N1C)C(C)C)C)C(C)C)C(C)C)C)C(C)C)C(C)C	enniatin J3	0
CC[C@H](C)[C@@H]1C(=O)N([C@H](C(=O)O[C@@H](C(=O)N([C@H](C(=O)O[C@@H](C(=O)N([C@H](C(=O)O1)C(C)C)C)C(C)C)C(C)C)C)C(C)C)C(C)C)C	enniatin H	0
CC[C@H]1C(=O)O[C@@H](C(=O)N([C@H](C(=O)O[C@@H](C(=O)N([C@H](C(=O)O[C@@H](C(=O)N1C)C(C)C)C(C)C)C)C(C)C)C(C)C)C)C(C)C	enniatin K1	0
CCC(C)[C@@H]1C(=O)N([C@H](C(=O)O[C@@H](C(=O)N([C@H](C(=O)O[C@@H](C(=O)N([C@H](C(=O)O1)C(C)C)C)C(C)C)C(C)C)C)C(C)C)C(C)C)C	cyclo[D-OVal-N(Me)Val-D-OVal-N(Me)Val-D-OxiIle-N(Me)Val]	0
O=c1cc(CO)occ1O	kojic acid	0
CC1=C[C@@H]2[C@]([C@@H](C1=O)O)([C@]3(C[C@H]([C@H]([C@@]34CO4)O2)O)C)COC(=O)C	15-acetyldeoxynivalenol	1
COC1=CC2=C(C(=C1)OC)C(=O)C3=C(C4=C(C=C3C2=O)O[C@@H]5[C@H]4CCO5)O	6,8-O-Dimethylversicolorin	1
COC1=C2C3=C(C(=O)CC3)C(=O)OC2=C4[C@@H]5C=CO[C@@H]5OC4=C1	aflatoxin B1	1
C[C@@H]1CCC[C@H](/C=C/C(=O)O[C@]23[C@@H](/C=C/C1)[C@@H](C(=C)[C@H]([C@H]2[C@@H](NC3=O)CC4=CC=CC=C4)C)O)O	cytochalasin B	1
C1CO[C@H]2[C@@H]1C3=C(O2)C=C(C4=C3OC5=CC=CC(=C5C4=O)O)O	demethyldihydrosterigmatocystin	1
CCCC[C@@H](C)[C@H]([C@H](C[C@@H](C)C[C@@H](CCCC[C@H](C[C@@H]([C@H](C)N)O)O)O)OC(=O)C[C@@H](CC(=O)O)C(=O)O)OC(=O)C[C@@H](CC(=O)O)C(=O)O	fumonisin B1	1
C1=CO[C@H]2[C@@H]1C3=C(O2)C=C4C(=C3O)C(=O)C5=C(C4=O)C=C(C=C5O)O	versicolorin A	1
C[C@H]1CCC[C@@H](CCC/C=C/C2=C(C(=CC(=C2)O)O)C(=O)O1)O	alpha-zearalenol	1

Table A9. Validation set for in vivo genotoxicity data.

SMILES	Name	y
Cc1cc(O)cc2oc(=O)c3c(O)cc(O)cc3c12	alternariol	0
CCCCCCCCCCCCCCC(C(CO)N)O	C17-Sphinganine	0
C/C=C/C[C@@H](C)[C@@H](O)[C@H]1C(=O)N[C@@H](CC)C(=O)N(C)CC(=O)N(C)[C@@H](CC(C)C)C(=O)N[C@@H](C(C)C)C(=O)N(C)[C@@H](CC(C)C)C(=O)N[C@@H](C)C(=O)N[C@H](C)C(=O)N(C)[C@@H](CC(C)C)C(=O)N(C)[C@@H](CC(C)C)C(=O)N(C)[C@@H](C(C)C)C(=O)N1C	cyclosporin A	0
CC1=C[C@H]2O[C@@H]3[C@H](O)C[C@@](C)([C@]34CO4)[C@@]2(CO)[C@H](O)C1=O	deoxynivalenol	0
C=C1C[C@]23C[C@@]1(O)CC[C@H]2[C@@]12C=C[C@H](O)[C@@](C)(C(=O)O1)[C@H]2[C@@H]3C(=O)O	gibberellic acid	0
O=c1cc(CO)occ1O	kojic acid	0
CN1C[C@@H](C=C2[C@H]1CC3=CNC4=CC=CC2=C34)C(=O)O	lysergic acid	0
CCN(CC)C(=O)[C@H]1CN([C@@H]2CC3=CNC4=CC=CC(=C34)C2=C1)C	lysergic acid diethylamine	0
CC(=O)OC[C@]12C[C@H](OC(=O)CC(C)C)C(C)=C[C@H]1O[C@@H]1[C@H](O)[C@@H](OC(C)=O)[C@@]2(C)[C@@]12CO2	T-2 toxin	0
CC(=O)OC[C@@]12C[C@@H](OC(=O)CC(C)C)C(C)=C[C@@H]1O[C@H]1[C@H](O)[C@@H](O)[C@]2(C)[C@@]12CO2	HT-2 toxin	0
CC(C)C[C@@H]1NC(=O)[C@H](C)N(C)C(=O)CNC(=O)/C(=C/c2ccccc2)N(C)C1=O	tentoxin	0

Table A9. *Cont.*

SMILES	Name	y
CC[C@H](C)[C@@H]1NC(=O)C(C(C)=O)=C1O	tenuazonic acid	0
CCC(C)C1NC(=O)C(C(C)=O)=C1O	tenuazonic acid	0
O=c1[nH]c(=O)n([C@H]2C[C@H](O)[C@@H](CO)O2)cc1Br	5-bromo-2′-deoxyuridine	1
Cc1c2oc3c(C)ccc(C(=O)NC4C(=O)NC(C(C)C)C(=O)N5CCC[C@H]5C(=O)N(C)CC(=O)N(C)C(C(C)C)C(=O)OC4C)c3nc-2c(C(=O)NC2C(=O)NC(C(C)C)C(=O)N3CCCC3C(=O)N(C)CC(=O)N(C)C(C(C)C)C(=O)OC2C)c(N)c1=O	A-2600	1
Cc1c2oc3c(C)ccc(C(=O)NC4C(=O)N[C@H](C(C)C)C(=O)N5CCC[C@H]5C(=O)N(C)CC(=O)N(C)[C@@H](C(C)C)C(=O)O[C@@H]4C)c3nc-2c(C(=O)NC2C(=O)N[C@H](C(C)C)C(=O)N3CCC[C@H]3C(=O)N(C)CC(=O)N(C)[C@@H](C(C)C)C(=O)O[C@@H]2C)c(N)c1=O	actinomycin D	1
Cc1c2oc3c(C)ccc(C(=O)NC4C(=O)NC(C(C)C)C(=O)N5CCCC5C(=O)N(C)CC(=O)N(C)C(C(C)C)C(=O)OC4C)c3nc-2c(C(=O)NC2C(=O)NC(C(C)C)C(=O)N3CCCC3C(=O)N(C)CC(=O)N(C)C(C(C)C)C(=O)OC2C)c(N)c1=O	actinomycin D	1
Cc1c2oc3c(C)ccc(C(=O)N[C@@H]4C(=O)N[C@H](C(C)C)C(=O)N5CCC[C@H]5C(=O)N(C)CC(=O)N(C)[C@@H](C(C)C)C(=O)O[C@H]4C)c3nc-2c(C(=O)N[C@@H]2C(=O)N[C@H](C(C)C)C(=O)N3CCC[C@H]3C(=O)N(C)CC(=O)N(C)[C@@H](C(C)C)C(=O)O[C@H]2C)c(N)c1=O	actinomycin D	1
Cc1c2oc3c(C)ccc(C(=O)N[C@@H]4C(=O)N[C@H](C(C)C)C(=O)N5CCC[C@H]5C(=O)N(C)CC(=O)N(C)[C@H](C(C)C)C(=O)O[C@@H]4C)c3nc-2c(C(=O)N[C@@H]2C(=O)N[C@H](C(C)C)C(=O)N3CCC[C@H]3C(=O)N(C)CC(=O)N(C)[C@@H](C(C)C)C(=O)O[C@@H]2C)c(N)c1=O	actinomycin D	1
Cc1c2oc3c(C)ccc(C(=O)N[C@@H]4C(=O)N[C@@H](C(C)C)C(=O)N5CCC[C@H]5C(=O)N(C)CC(=O)N(C)[C@@H](C(C)C)C(=O)O[C@@H]4C)c3nc-2c(C(=O)N[C@@H]2C(=O)N[C@@H](C(C)C)C(=O)N3CCC[C@H]3C(=O)N(C)CC(=O)N(C)[C@@H](C(C)C)C(=O)O[C@@H]2C)c(N)c1=O	actinomycin D	1
Cc1c2oc3c(C)ccc(C(=O)N[C@@H]4C(=O)N[C@H](C(C)C)C(=O)N5CCC[C@H]5C(=O)N(C)CC(=O)N(C)[C@@H](C(C)C)C(=O)O[C@@H]4C)c3nc-2c(C(=O)N[C@H]2C(=O)N[C@H](C(C)C)C(=O)N3CCC[C@@H]3C(=O)N(C)CC(=O)N(C)[C@@H](C(C)C)C(=O)O[C@@H]2C)c(N)c1=O	actinomycin D	1
Cc1c2oc3c(C)ccc(C(=O)N[C@@H]4C(=O)N[C@H](C(C)C)C(=O)N5CCC[C@H]5C(=O)N(C)CC(=O)N(C)C(C(C)C)C(=O)O[C@@H]4C)c3nc-2c(C(=O)N[C@@H]2C(=O)N[C@H](C(C)C)C(=O)N3CCC[C@H]3C(=O)N(C)CC(=O)N(C)[C@@H](C(C)C)C(=O)O[C@@H]2C)c(N)c1=O	actinomycin D	1
[3H]c1c([3H])c(C([3H])([3H])[3H])c2oc3c(C([3H])([3H])[3H])c(=O)c(N([3H])[3H])c(C(=O)N([3H])[C@]4([3H])C(=O)N([3H])[C@]([3H])(C([3H])(C([3H])([3H])[3H])C([3H])([3H])[3H])C(=O)N5C([3H])([3H])C([3H])([3H])C([3H])([3H])[C@]5([3H])C(=O)N(C([3H])([3H])[3H])C([3H])([3H])C(=O)N(C([3H])([3H])[3H])[C@@]([3H])(C([3H])(C([3H])([3H])[3H])C([3H])([3H])[3H])C(=O)O[C@]4([3H])C([3H])([3H])[3H])c-3nc2c1C(=O)N([3H])[C@]1([3H])C(=O)N([3H])[C@]([3H])(C([3H])(C([3H])([3H])[3H])C([3H])([3H])[3H])C(=O)N2C([3H])([3H])C([3H])([3H])C([3H])([3H])[C@]2([3H])C(=O)N(C([3H])([3H])[3H])C([3H])([3H])C(=O)N(C([3H])([3H])[3H])[C@@]([3H])(C([3H])(C([3H])([3H])[3H])C([3H])([3H])[3H])C(=O)O[C@]1([3H])C([3H])([3H])[3H]	actinomycin d-[3h(g)]	1
COc1cc2c(c3oc(=O)c4c(c13)CCC4=O)[C@@H]1C=CO[C@@H]1O2	aflatoxin B1	1
COc1cc2c(c3oc(=O)c4c(c13)CCC4=O)[C@@H]1CCO[C@@H]1O2	aflatoxin B2	1
COc1cc2c(c3oc(=O)c4c(c13)CCOC4=O)[C@H]1C=CO[C@H]1O2	aflatoxin G1	1
COc1cc2c(c3oc(=O)c4c(c13)CCOC4=O)[C@H]1C=CO[C@@H]1O2	aflatoxin G1	1
COc1cc2c(c3oc(=O)c4c(c13)CCOC4=O)[C@@H]1C=CO[C@@H]1O2	aflatoxin G1	1
COc1cc2c(c3oc(=O)c4c(c13)CCOC4=O)C1C=COC1O2	aflatoxin G1	1

Table A9. *Cont.*

SMILES	Name	y
COc1cc2c(c3oc(=O)c4c(c13)CCOC4=O)[C@@H]1CCO[C@@H]1O2	aflatoxin G2	1
COc1cc2c(c3oc(=O)c4c(c13)CCC4=O)[C@]1(O)C=CO[C@@H]1O2	aflatoxin M1	1
COc1cc2c(c3oc(=O)c4c(c13)[C@@H](O)CC4=O)[C@@H]1C=CO[C@@H]1O2	aflatoxin Q1	1
COc1cc2c(c3oc(=O)c4c(c13)CCC4O)[C@@H]1C=CO[C@@H]1O2	aflatoxicol	1
COc1cc2c(c3oc(=O)c4c(c13)CC[C@H]4O)[C@@H]1C=CO[C@H]1O2	aflatoxicol	1
COc1cc2c(c3oc(=O)c4c(c13)CC[C@@H]4O)[C@@H]1C=CO[C@@H]1O2	aflatoxicol	1
C[C@@H]1CCC[C@H](O)CCCCCc2cc(O)cc(O)c2C(=O)O1	alpha-Zearalanol	1
C[C@H]1CCCC(O)CCC/C=C/c2cc(O)cc(O)c2C(=O)O1	alpha-Zearalenol	1
COc1cc(O)c2c(=O)oc3cc(O)cc(C)c3c2c1	alternariol monomethyl ether	1
CC(C)C1OC(=O)[C@H](Cc2ccccc2)N(C)C(=O)C(C(C)C)OC(=O)[C@H](Cc2ccccc2) N(C)C(=O)C(C(C)C)OC(=O)[C@H](Cc2ccccc2)N(C)C1=O	beauvericin	1
CC(C)[C@H]1OC(=O)[C@H](Cc2ccccc2)N(C)C(=O)[C@@H](C(C)C)OC(=O)[C@H] (Cc2ccccc2)N(C)C(=O)[C@@H](C(C)C)OC(=O)[C@H](Cc2ccccc2)N(C)C1=O	beauvericin	1
CC(C)[C@@H]1OC(=O)[C@@H](Cc2ccccc2)N(C)C(=O)[C@H](C(C)C)OC(=O)[C@@H] (Cc2ccccc2)N(C)C(=O)[C@H](C(C)C)OC(=O)[C@@H](Cc2ccccc2)N(C)C1=O	beauvericin	1
CC(C)C1OC(=O)C(Cc2ccccc2)N(C)C(=O)C(C(C)C)OC(=O)C(Cc2ccccc2)N(C)C(=O) C(C(C)C)OC(=O)C(Cc2ccccc2)N(C)C1=O	beauvericin	1
C[C@H]1CCC[C@@H](O)CCCCCc2cc(O)cc(O)c2C(=O)O1	beta-Zearalanol	1
C[C@H]1CCC[C@@H](O)CCC/C=C/c2cc(O)cc(O)c2C(=O)O1	beta-Zearalenol	1
CC1=C2C(=CO[C@H](C)[C@H]2C)C(=O)C(C(=O)O)=C1O	citrinin	1
CC1=C2C(=CO[C@@H](C)[C@@H]2C)C(=O)C(C(=O)O)=C1O	citrinin	1
Cc1c2oc3c(C)ccc(C(=O)N[C@@H]4C(=O)N[C@H](C(C)C)C(=O)N5CCC[C@H]5C(=O) N(C)CC(=O)N(C)[C@@H](C(C)C)C(=O)O[C@@H]4C)c3nc-2c(C(=O)N[C@@H]2C(=O)N[C@H](C(C)C)C(=O)N3CCC[C@H]3C(=O)N(C)CC(=O)N(C) [C@@H](C(C)C)C(=O)O[C@@H]2C)c(N)c1=O	dactinomycin	1
COc1cccc2c1C(=O)c1c(O)c3c(c(O)c1C2=O)C[C@@](O)(C(=O)CO)C[C@@H]3O[C@H] 1C[C@H](N)[C@H](O)[C@H](C)O1	doxorubicin	1
Cc1cc(O)c2c(c1)C(=O)c1cc(O)cc(O)c1C2=O	emodin	1
CC[C@H](C)[C@H]1C(=O)OC(C(C)C)C(=O)N(C)[C@@H]([C@H](C)CC)C(=O)O[C@H] (C(C)C)C(=O)N(C)[C@@H]([C@@H](C)CC)C(=O)O[C@H](C(C)C)C(=O)N1C	enniatins A	1
CC[C@H](C)[C@H]1C(=O)O[C@H](C(C)C)C(=O)N(C)[C@@H]([C@@H](C)CC)C(=O) O[C@H](C(C)C)C(=O)N(C)[C@@H](C(C)C)C(=O)O[C@H](C(C)C)C(=O)N1C	enniatin A1	1
CC(C)[C@H]1C(=O)O[C@H](C(C)C)C(=O)N(C)[C@@H](C(C)C)C(=O)O[C@H](C(C) C)C(=O)N(C)[C@@H](C(C)C)C(=O)O[C@H](C(C)C)C(=O)N1C	enniatin B	1
CC[C@H](C)[C@H]1C(=O)O[C@H](C(C)C)C(=O)N(C)[C@@H](C(C)C)C(=O)O[C@H] (C(C)C)C(=O)N(C)[C@@H](C(C)C)C(=O)O[C@H](C(C)C)C(=O)N1C	enniatin B1	1
O=c1[nH]c(=O)n([C@H]2C[C@H](O)[C@@H](CO)O2)cc1F	floxuridine	1
CCCC[C@@H](C)[C@@H](OC(=O)C[C@@H](CC(=O)O)C(=O)O)[C@H](C[C@@H] (C)CCCCCC[C@@H](O)C[C@H](O)[C@H](C)N)OC(=O)C[C@@H](CC(=O)O)C(=O)O	fumonisin B2	1
CCCCC(C)C(OC(=O)CC(CC(=O)O)C(=O)O)C(CC(C)CC(O)CCCCCCC(O)C(C)N) OC(=O)CC(CC(=O)O)C(=O)O	fumonisin B3	1
CCCCC(C)C(OC(=O)CC(CC(=O)O)C(=O)O)C(CC(C)CC(O)CCCCC(O)CC(O)C(C)N)OC (=O)CC(CC(=O)O)C(=O)O	fumonisin B	1
CO[C@@]12[C@H](COC(N)=O)C3=C(C(=O)C(C)=C(N)C3=O)N1C[C@@H]1N[C@@H]12	mitomycin C	1

Table A9. *Cont.*

SMILES	Name	y
O=c1cc(O)c1=O	monilformin	1
C[C@@H]1CC2=C(C=C(C(=C2C(=O)O1)O)C(=O)N[C@@H](CC3=CC=CC=C3)C(=O)O)Cl	ochratoxin A	1
O=C1C=C2C(=CCOC2O)O1	patulin	1
C=C(C)[C@@H]1NC(=O)[C@@H](NC)[C@@H](O)c2cc(Cl)c(O)c(c2)O[C@](C)(CC)[C@@H](C(=O)N2CC=C[C@H]2C(=O)N/C(C(=O)N/C(=C/C(=O)O)C(=O)O)=C(\C)CC)NC1=O	phomopsin A	1
COC(=O)C12Oc3ccc(-c4ccc5c(c4O)C(O)=C4C(=O)CC(C)C(O)C4(C(=O)OC)O5)c(O)c3C(O)=C1C(=O)CC(C)C2O	secalonic acid D	1
COc1c(Cl)cc2c(c1OC)N(C)[C@@H]1N3C(=O)[C@@]4(C)SS[C@@]3(C(=O)N4C)[C@H](O)[C@]21O	sporidesmin	1
COc1cc2c(c3oc4cccc(O)c4c(=O)c13)[C@@H]1C=CO[C@@H]1O2	sterigmatocystin	1
CC1=CC2OC3[C@H](O)C[C@@](C)(C34CO4)[C@@]2(CO)[C@H](O)C1=O	vomitoxin	1
C[C@H]1CCCC(=O)CCCCCc2cc(O)cc(O)c2C(=O)O1	zearalanone	1
C[C@H]1CCCC(=O)CCC/C=C/c2cc(O)cc(O)c2C(=O)O1	zearalenone	1
C[C@H]1CCCC(O)CCC/C=C/c2cc(O)cc(O)c2C(=O)O1	alpha-zearalenol	1
C[C@H]1CCC[C@H](O)CCC/C=C/c2cc(O)cc(O)c2C(=O)O1	alpha-zearalenol	1
C[C@H]1CCC[C@@H](O)CCC/C=C/c2cc(O)cc(O)c2C(=O)O1	beta-zearalenol	1
C[C@H]1CCC[C@H](O)CCCCCc2cc(O)cc(O)c2C(=O)O1	zeranol	1

Table A10. Validation set for carcinogenicity data.

SMILES	Name	y
CCCCOC(=O)c1ccc(O)cc1	butylparaben	0
O=C(NC(CO)C(O)c1ccc([N+](=O)[O-])cc1)C(Cl)Cl	bhloramphenicol	0
O=C(N[C@H](CO)[C@H](O)c1ccc([N+](=O)[O-])cc1)C(Cl)Cl	chloramphenicol	0
O=C(N[C@@H](CO)[C@@H](O)c1ccc([N+](=O)[O-])cc1)C(Cl)Cl	chloramphenicol	0
O=C(N[C@@H](CO)[C@H](O)c1ccc([N+](=O)[O-])cc1)C(Cl)Cl	chloramphenicol	0
O=C(N[C@H](CO)[C@@H](O)c1ccc([N+](=O)[O-])cc1)C(Cl)Cl	chloramphenicol	0
O=C(N[C@H](CO)[C@H](O)[13c]1[13cH][13cH][13c]([N+](=O)[O-])[13cH][13cH]1)C(Cl)Cl	chloramphenicol	0
[2H]c1c([2H])c([C@@]([2H])(O)[C@@H](CO)NC(=O)C(Cl)Cl)c([2H])c([2H])c1[N+](=O)[O-]	chloramphenicol	0
O=[13C](N[13C@H]([13CH2]O)[13C@H](O)[13c]1ccc([N+](=O)[O-])cc1)[13CH](Cl)Cl	chloramphenicol	0
[2H]C(Cl)(Cl)C(=O)N[C@]([2H])(C([2H])([2H])O)[C@]([2H])(O)c1ccc([N+](=O)[O-])cc1	chloramphenicol	0
[2H]c1c([2H])c(C([2H])(O)C(CO)NC(=O)C(Cl)Cl)c([2H])c([2H])c1[N+](=O)[O-]	chloramphenicol	0
[3H]c1cc([C@@H](O)[C@@H](CO)NC(=O)C(Cl)Cl)cc([3H])c1[N+](=O)[O-]	chloramphenicol	0
[2H]c1c([2H])c([C@@]([2H])(O)[C@H](CO)NC(=O)C(Cl)Cl)c([2H])c([2H])c1[N+](=O)[O-]	chloramphenicol	0
O=[13C](N[13C@H]([13CH2]O)[13C@H](O)[13c]1[13cH][13cH][13c]([N+](=O)[O-])[13cH][13cH]1)[13CH](Cl)Cl	chloramphenicol	0
[2H]OC[C@H]([C@H](O[2H])c1ccc([N+](=O)[O-])cc1)N([2H])C(=O)C(Cl)Cl	chloramphenicol	0
C/C=C/C[C@@H](C)[C@@H](O)[C@H]1C(=O)N[C@@H](CC)C(=O)N(C)CC(=O)N(C)[C@@H](CC(C)C)C(=O)N[C@@H](C(C)C)C(=O)N(C)[C@@H](CC(C)C)C(=O)N[C@@H](C)C(=O)N[C@H](C)C(=O)N(C)[C@@H](CC(C)C)C(=O)N(C)[C@@H](CC(C)C)C(=O)N(C)[C@@H](C(C)C)C(=O)N1C	cyclosporin A	0

Table A10. *Cont.*

SMILES	Name	y
CC1=C[C@H]2O[C@@H]3[C@H](O)C[C@@](C)([C@]34CO4)[C@@]2(CO)[C@H](O)C1=O	deoxynivalenol	0
[2H]c1c(C([2H])([2H])[2H])cc(O)c2c1C(=O)c1cc(O)cc(O)c1C2=O	emodin	0
CC(=O)O[C@@H]1[C@@H](O)[C@H]2O[C@@H]3C=C(C)C(=O)[C@@H](O)[C@]3(CO)[C@]1(C)[C@]21CO1	fusarenon X	0
CC(C)COC(=O)c1ccc(O)cc1	isobutyl 4-hydroxybenzoate	0
CO[C@@H]1[C@@H](O[C@@H]2O[C@H](C)[C@@H](O[C@H]3C[C@@](C)(O)[C@@H](OC(=O)CC(C)C)[C@H](C)O3)[C@H](N(C)C)[C@H]2O)[C@@H](CC=O)C[C@@H](C)[C@@H](O)/C=C/C=C/C[C@@H](C)OC(=O)C[C@H]1OC(C)=O	josamycin	0
COC(C(=O)O)c1ccccc1	methoxyphenylacetic acid	0
CC1=C[C@H]2O[C@@H]3[C@H](O)[C@@H](O)[C@@](C)([C@]34CO4)[C@@]2(CO)[C@H](O)C1=O	nivalenol	0
CN(C)[C@@H]1C(=O)C(C(N)=O)=C(O)[C@@]2(O)C(=O)C3=C(O)c4c(O)cccc4[C@@](C)(O)[C@H]3[C@H](O)[C@@H]12.Cl	oxytetracycline	0
O=[13C]1[13CH]=[13C]2[13C](=[13CH][13CH2]O[13CH]2O)O1	patulin	0
CN(C)[C@@H]1C(=O)C(C(N)=O)=C(O)[C@@]2(O)C(=O)C3=C(O)c4c(O)cccc4[C@@](C)(O)[C@H]3C[C@@H]12.Cl	tetracycline	0
Oc1cc(Cl)ccc1Oc1ccc(Cl)cc1Cl	triclosan	0
CC1=C[C@@H]2[C@]([C@@H](C1=O)O)([C@]3(C[C@H]([C@H]([C@]34CO4)O2)O)C)CO	vomitoxin	0
[2H]C([2H])([2H])Oc1cc2c(c3oc(=O)c4c(c13)CCC4=O)[C@@H]1CCO[C@@H]1O2	1217830-52-8	1
C[C@H]1Cc2c(Cl)cc(C(=O)N[C@@H](Cc3ccccc3)C(=O)O)c(O)c2C(=O)O1	3-epi-Ochratoxin A	1
C[C@H]1Cc2c(Cl)cc(C(=O)N[C@H](Cc3ccccc3)C(=O)O)c(O)c2C(=O)O1	3S14R-Ochratoxin A	1
O=c1[nH]c(=O)n([C@H]2C[C@H](O)[C@@H](CO)O2)cc1Br	5-BROMO-2′-DEOXYURIDINE	1
Cc1c2oc3c(C)ccc(C(=O)NC4C(=O)NC(C(C)C)C(=O)N5CCC[C@H]5C(=O)N(C)CC(=O)N(C)C(C(C)C)C(=O)OC4C)c3nc-2c(C(=O)NC2C(=O)NC(C(C)C)C(=O)N3CCCC3C(=O)N(C)CC(=O)N(C)C(C(C)C)C(=O)OC2C)c(N)c1=O	A-2600	1
Cc1c2oc3c(C)ccc(C(=O)NC4C(=O)N[C@H](C(C)C)C(=O)N5CCC[C@H]5C(=O)N(C)CC(=O)N(C)[C@@H](C(C)C)C(=O)O[C@@H]4C)c3nc-2c(C(=O)NC2C(=O)N[C@H](C(C)C)C(=O)N3CCC[C@H]3C(=O)N(C)CC(=O)N(C)[C@@H](C(C)C)C(=O)O[C@@H]2C)c(N)c1=O	actinomycin D	1
Cc1c2oc3c(C)ccc(C(=O)NC4C(=O)NC(C(C)C)C(=O)N5CCCC5C(=O)N(C)CC(=O)N(C)C(C(C)C)C(=O)OC4C)c3nc-2c(C(=O)NC2C(=O)NC(C(C)C)C(=O)N3CCCC3C(=O)N(C)CC(=O)N(C)C(C(C)C)C(=O)OC2C)c(N)c1=O	actinomycin D	1
Cc1c2oc3c(C)ccc(C(=O)N[C@@H]4C(=O)N[C@H](C(C)C)C(=O)N5CCC[C@H]5C(=O)N(C)CC(=O)N(C)[C@@H](C(C)C)C(=O)O[C@H]4C)c3nc-2c(C(=O)N[C@@H]2C(=O)N[C@H](C(C)C)C(=O)N3CCC[C@H]3C(=O)N(C)CC(=O)N(C)[C@@H](C(C)C)C(=O)O[C@H]2C)c(N)c1=O	actinomycin D	1
Cc1c2oc3c(C)ccc(C(=O)N[C@@H]4C(=O)N[C@H](C(C)C)C(=O)N5CCC[C@H]5C(=O)N(C)CC(=O)N(C)[C@H](C(C)C)C(=O)O[C@@H]4C)c3nc-2c(C(=O)N[C@@H]2C(=O)N[C@H](C(C)C)C(=O)N3CCC[C@H]3C(=O)N(C)CC(=O)N(C)[C@@H](C(C)C)C(=O)O[C@@H]2C)c(N)c1=O	actinomycin D	1
Cc1c2oc3c(C)ccc(C(=O)N[C@@H]4C(=O)N[C@@H](C(C)C)C(=O)N5CCC[C@H]5C(=O)N(C)CC(=O)N(C)[C@@H](C(C)C)C(=O)O[C@@H]4C)c3nc-2c(C(=O)N[C@@H]2C(=O)N[C@@H](C(C)C)C(=O)N3CCC[C@H]3C(=O)N(C)CC(=O)N(C)[C@@H](C(C)C)C(=O)O[C@@H]2C)c(N)c1=O	actinomycin	1

Table A10. *Cont.*

SMILES	Name	y
[3H]c1c([3H])c(C([3H])([3H])[3H])c2oc3c(C([3H])([3H])[3H])c(=O)c(N([3H])[3H])c(C(=O)N([3H])[C@]4([3H])C(=O)N([3H])[C@]([3H])(C([3H])(C([3H])([3H])[3H])C([3H])([3H])[3H])C(=O)N5C([3H])([3H])C([3H])([3H])C([3H])([3H])[C@]5([3H])C(=O)N(C([3H])([3H])[3H])C([3H])([3H])C(=O)N(C([3H])([3H])[3H])[C@@]([3H])(C([3H])(C([3H])([3H])[3H])C([3H])([3H])[3H])C(=O)O[C@]4([3H])C([3H])([3H])[3H])c-3nc2c1C(=O)N([3H])[C@]1([3H])C(=O)N([3H])[C@]([3H])(C([3H])(C([3H])([3H])[3H])C([3H])([3H])[3H])C(=O)N2C([3H])([3H])C([3H])([3H])C([3H])([3H])[C@]2([3H])C(=O)N(C([3H])([3H])[3H])C([3H])([3H])C(=O)N(C([3H])([3H])[3H])[C@@]([3H])(C([3H])(C([3H])([3H])[3H])C([3H])([3H])[3H])C(=O)O[C@]1([3H])C([3H])([3H])[3H]	actinomycin D	1
Cc1c2oc3c(C)ccc(C(=O)N[C@@H]4C(=O)N[C@H](C(C)C)C(=O)N5CCC[C@H]5C(=O)N(C)CC(=O)N(C)[C@@H](C(C)C)C(=O)O[C@@H]4C)c3nc-2c(C(=O)N[C@H]2C(=O)N[C@H](C(C)C)C(=O)N3CCC[C@@H]3C(=O)N(C)CC(=O)N(C)[C@@H](C(C)C)C(=O)O[C@@H]2C)c(N)c1=O	actinomycin D	1
Cc1c2oc3c(C)ccc(C(=O)N[C@@H]4C(=O)N[C@H](C(C)C)C(=O)N5CCC[C@H]5C(=O)N(C)CC(=O)N(C)C(C(C)C)C(=O)O[C@@H]4C)c3nc-2c(C(=O)N[C@@H]2C(=O)N[C@H](C(C)C)C(=O)N3CCC[C@H]3C(=O)N(C)CC(=O)N(C)[C@@H](C(C)C)C(=O)O[C@@H]2C)c(N)c1=O	actinomycin D	1
COc1cc2c(c3oc(=O)c4c(c13)CCC4=O)C1C=COC1O2	aflatoxin BI	1
COc1cc2c(c3oc(=O)c4c(c13)CC[C@@H]4O)[C@@H]1C=CO[C@@H]1O2	aflatoxicol	1
COc1cc2c(c3oc(=O)c4c(c13)CCC4=O)[C@@H]1C=CO[C@@H]1O2	aflatoxin B1	1
[13CH3]O[13c]1[13cH][13c]2[13c]([13c]3o[13c](=O)[13c]4[13c]([13c]13)[13CH2][13CH2][13C]4=O)[13C@@H]1[13CH]=[13CH]O[13C@@H]1O2	aflatoxin B1-13C17	1
COc1cc2c(c3oc(=O)c4c(c13)CCC4=O)[C@@H]1CCO[C@@H]1O2	aflatoxin B2	1
COc1cc2c(c3oc(=O)c4c(c13)CCC4=O)C1CCOC1O2	aflatoxin B2 alpha	1
COc1cc2c(c3oc(=O)c4c(c13)CCOC4=O)C1C=COC1O2	aflatoxin G1	1
COc1cc2c(c3oc(=O)c4c(c13)CCOC4=O)[C@@H]1C=CO[C@@H]1O2	aflatoxin G1	1
COc1cc2c(c3oc(=O)c4c(c13)CCOC4=O)[C@H]1C=CO[C@@H]1O2	aflatoxin G1	1
COc1cc2c(c3oc(=O)c4c(c13)CCOC4=O)[C@H]1C=CO[C@H]1O2	aflatoxin G1	1
[13CH3]O[13c]1[13cH][13c]2[13c]([13c]3o[13c](=O)[13c]4[13c]([13c]13)[13CH2][13CH2]O[13C]4=O)[13C@@H]1[13CH]=[13CH]O[13C@@H]1O2	aflatoxin G1-13C17	1
COc1cc2c(c3oc(=O)c4c(c13)CCOC4=O)[C@@H]1CCO[C@@H]1O2	aflatoxin G2	1
[13CH3]O[13c]1[13cH][13c]2[13c]([13c]3o[13c](=O)[13c]4[13c]([13c]13)[13CH2][13CH2]O[13C]4=O)[13C@@H]1[13CH2][13CH2]O[13C@@H]1O2	aflatoxin G2-13C17	1
CC1Cc2c(Cl)cc(C(=O)NC(Cc3ccccc3)C(=O)O)c(O)c2C(=O)O1	antibiotic 9663	1
COc1cc2c(c3oc4cccc(O)c4c(=O)c13)C1C=CO[C@@H]1O2	CHEMBL1532401	1
CC1=C2C(=CO[C@H](C)[C@H]2C)C(=O)C(C(=O)O)=C1O	citrinin	1
CC[C@@H]1NC(=O)[C@@H]2[C@H](Cl)[C@H](Cl)CN2C(=O)[C@H](CO)NC(=O)C[C@H](c2ccccc2)NC(=O)[C@H](CO)NC1=O	cyclochlorotine	1
Cc1c2oc3c(C)ccc(C(=O)N[C@@H]4C(=O)N[C@H](C(C)C)C(=O)N5CCC[C@H]5C(=O)N(C)CC(=O)N(C)[C@@H](C(C)C)C(=O)O[C@@H]4C)c3nc-2c(C(=O)N[C@@H]2C(=O)N[C@H](C(C)C)C(=O)N3CCC[C@H]3C(=O)N(C)CC(=O)N(C)[C@@H](C(C)C)C(=O)O[C@@H]2C)c(N)c1=O	dactinomycin	1
COc1cccc2c1C(=O)c1c(O)c3c(c(O)c1C2=O)C[C@@](O)(C(C)=O)C[C@@H]3O[C@H]1C[C@H](N)[C@H](O)[C@H](C)O1	daunorubicin	1
COc1cc2c(c3oc(=O)c4c(c13)CCOC4=O)C1CCOC1O2	dihydroaflatoxin G1	1
COc1cccc2c1C(=O)c1c(O)c3c(c(O)c1C2=O)C[C@@](O)(C(=O)CO)C[C@@H]3O[C@H]1C[C@H](N)[C@H](O)[C@H](C)O1	doxorubicin	1
C=C1C[C@]23C[C@@]1(O)CC[C@H]2[C@@]12C=C[C@H](O)[C@@](C)(C(=O)O1)[C@H]2[C@@H]3C(=O)O	gibberellic acid	1

Table A10. *Cont.*

SMILES	Name	y
COC1=CC(=O)C[C@@H](C)[C@]12Oc1c(Cl)c(OC)cc(OC)c1C2=O	griseofulvin	1
CC[C@H](C)C(=O)O[C@H]1C[C@H](C=C2[C@H]1[C@H]([C@H](C=C2)C)CC [C@@H]3C[C@H](CC(=O)O3)O)C	lovastatin	1
Cc1cc(O)c2c(c1O)C(=O)C13C(=C2O)C(=O)C2C(O)C1C1C(O)C3C(=O)C3=C(O)c4c(O)cc(C) c(O)c4C(=O)C321	luteoskyrin	1
Cc1cc(O)c2c(c1O)C(=O)[C@]13C(=C2O)C(=O)[C@@H]2[C@H](O)[C@H]1[C@@H]1[C@@H] (O)[C@H]3C(=O)C3=C(O)c4c(O)cc(C)c(O)c4C(=O)[C@@]321	luteoskyrin	1
C=CCc1ccc(OC)c(OC)c1	methyl eugenol	1
CO[C@@]12[C@H](COC(N)=O)C3=C(C(=O)C(C)=C(N)C3=O)N1C[C@@H]1N[C@@H]12	mitomycin C	1
C[C@@H]1Cc2c(Cl)cc(C(=O)N[C@H](Cc3ccccc3)C(=O)O)c(O)c2C(=O)O1	N-[(3R)-5-chloro-8-hydroxy-3-methyl-1-oxo-3,4-dihydro-1H-2-benzopyran-7-carbonyl]-D-phenylalanine	1
C[C@@H]1Cc2c(Cl)cc(C(=O)N[C@@H](Cc3ccccc3)C(=O)O)c(O)c2C(=O)O1	ochratoxin A	1
CC1Cc2c(Cl)cc(C(=O)N[C@H](Cc3ccccc3)C(=O)O)c(O)c2C(=O)O1	ochratoxin-A	1
CC/C(=C(\C(=O)N/C(=C/C(=O)O)/C(=O)O)/NC(=O)[C@@H]1C=CCN1C(=O)[C@@H]2[C@@] (OC3=C(C(=CC(=C3)[C@@H]([C@@H](C(=O)N[C@H](C(=O)N2)C(=C)C)NC)O)Cl)O)(C)CC)/C	phomopsin A	1
COc1cc2c(c3oc4cccc(O)c4c(=O)c13)[C@@H]1C=CO[C@@H]1O2	sterigmatocystin	1
COc1cc2c(c3oc4cccc(O)c4c(=O)c13)C1C=COC1O2	sterigmatocystin	1

Table A11. Prediction metrics of the external validation set of mycotoxins with 4 different QSAR prediction tools.

Model	Accuracy	Sensitivity	Specificity	Precision	f1	TN	FP	FN	TP	Num of Cpds
Carcinogenicity ProtoPRED	**0.80**	**0.74**	**0.89**	**0.92**	**0.82**	**25**	**3**	**12**	**35**	**75**
Carcinogenicity QTB Case Ultra	0.71	0.75	0.65	0.77	0.76	15	8	9	27	59 *
Carcinogenicity QTB Leadscope	0.59	0.62	0.53	0.65	0.63	9	8	9	15	41 *
Carcinogenicity VEGA	0.70	0.56	0.81	0.67	0.61	21	5	8	10	31 **
In vitro genotoxicity mycotoxin model	**0.93**	**1.00**	**0.86**	**1.00**	**0.94**	**6**	**0**	**1**	**8**	**15**
In vitro genotoxicity VEGA	0.54	1.00	0.00	0.54	0.70	0	6	0	7	13 **
In vivo genotoxicity ProtoPRED	**0.81**	**0.88**	**0.46**	**0.88**	**0.88**	**6**	**7**	**7**	**52**	**72**
In vivo genotoxicity QTB Case Ultra	0.62	0.60	0.67	0.75	0.67	2	1	2	3	8 *
In vivo genotoxicity QTB Leadscope	0.65	0.64	0.67	0.82	0.72	4	2	5	9	20 *
In vivo genotoxicity VEGA	0.39	0.33	0.57	0.70	0.45	4	3	14	7	28 ***
Mutagenicity mycotoxin model	**0.83**	**0.92**	**0.75**	**0.92**	**0.85**	**9**	**1**	**3**	**11**	**24**
Mutagenicity QTB Case Ultra	0.94	0.90	1.00	1.00	0.95	8	0	1	9	18 *
Mutagenicity QTB Leadscope	0.88	0.86	0.90	0.86	0.86	9	1	1	6	17 *
Mutagenicity VEGA	0.67	0.83	0.50	0.62	0.71	6	6	2	10	24

QTB = QSARToolbox; FN = false negatives; TN = true positives; FP = false positives; TP = true positives. The models described in this article are labelled in bold. * Several compounds could not be predicted because they were not included in the applicability domain of the model. ** Some compounds could not be predicted because the program did not accept the input structure. *** Some compounds could not be predicted because the program did not accept the input structure or the prediction was unclear.

References

1. Greeff-Laubscher, M.R.; Beukes, I.; Marais, G.J.; Jacobs, K. Mycotoxin production by three different toxigenic fungi genera on formulated abalone feed and the effect of an aquatic environment on fumonisins. *Mycology* **2020**, *11*, 105–117. [CrossRef]
2. Carballo, D.; Tolosa, J.; Ferrer, E.; Berrada, H. Dietary exposure assessment to mycotoxins through total diet studies. A review. *Food Chem. Toxicol.* **2019**, *128*, 8–20. [CrossRef]
3. Alshannaq, A.; Yu, J.H. Occurrence, toxicity, and analysis of major mycotoxins in food. *Int. J. Environ. Res. Public Health* **2017**, *14*, 632. [CrossRef] [PubMed]
4. Tolosa, J.; Font, G.; Mañes, J.; Ferrer, E. Nuts and dried fruits: Natural occurrence of emerging *Fusarium* mycotoxins. *Food Control* **2013**, *33*, 215–220. [CrossRef]
5. Zingales, V.; Taroncher, M.; Martino, P.A.; Ruiz, M.J.; Caloni, F. Climate change and effects on molds and mycotoxins. *Toxins* **2022**, *14*, 445. [CrossRef] [PubMed]
6. Liu, C.; Van der Fels-Klerx, H.J. Quantitative modeling of climate change impacts on mycotoxins in cereals: A review. *Toxins* **2021**, *13*, 276. [CrossRef]
7. Medina, Á.; Rodríguez, A.; Magan, N. Climate change and mycotoxigenic fungi: Impacts on mycotoxin production. *Curr. Opin. Food Sci.* **2015**, *5*, 99–104. [CrossRef]
8. European Commission. Commission Regulation (EC) No. 2023/915 of 25 April 2023 on Maximum Levels for Certain Contaminants in Food and Repealing Regulation (EC) No. 1881/2006. O.J.E.U., L 119/103. Available online: https://eur-lex.europa.eu/legal-content/EN/TXT/PDF/?uri=CELEX:32023R0915 (accessed on 16 January 2023).
9. van Egmond, H.P.; Schothorst, R.C.; Jonker, M.A. Regulations relating to mycotoxins in food. *Anal. Bioanal. Chem.* **2007**, *389*, 147–157. [CrossRef]
10. Benigni, R.; Bossa, C. Predictivity and reliability of QSAR models: The case of mutagens and carcinogens. *Toxicol. Mech. Methods* **2008**, *18*, 137–147. [CrossRef]
11. Gozalbes, R.; de Julián-Ortiz, J.V. Applications of chemoinformatics in predictive toxicology for regulatory purposes, especially in the context of the EU REACH legislation. *Int. J. Quant. Struct.-Prop. Relat. IJQSPR* **2018**, *3*, 1–24. [CrossRef]
12. Commission Regulation No. 1907/2006 REACH Regulation (EC) No 1907/2006 of the European Parliament and of the Council of 18 December 2006 Concerning the Registration, Evaluation, Authorisation and Restriction of Chemicals (REACH). O.J.E.U., L. 396. Available online: https://eur-lex.europa.eu/legal-content/EN/TXT/PDF/?uri=CELEX:02006R1907-20140410&from=EN (accessed on 27 December 2022).
13. Barber, C.; Cayley, A.; Hanser, T.; Harding, A.; Heghes, C.; Vessey, J.D.; Werner, S.; Weiner, S.K.; Wichard, J.; Giddings, A.; et al. Evaluation of a statistics-based Ames mutagenicity QSAR model and interpretation of the results obtained. *Regul. Toxicol. Pharmacol.* **2016**, *76*, 7–20. [CrossRef]
14. Hasselgren, C.; Ahlberg, E.; Akahori, Y.; Amberg, A.; Anger, L.T.; Atienzar, F.; Auerbach, S.; Beilke, L.; Bellion, P.; Benigni, R.; et al. Genetic toxicology in silico protocol. *Regul. Toxicol. Pharmacol.* **2019**, *107*, 104403. [CrossRef]
15. ICH M7 Assessment and Control of DNA Reactive (Mutagenic) Impurities in Pharmaceuticals to Limit Potential Carcinogenic Risk. EMA/CHMP/ICH/83812/2013. Available online: https://www.ema.europa.eu/en/ich-m7-assessment-control-dna-reactive-mutagenic-impurities-pharmaceuticals-limit-potential (accessed on 27 December 2022).
16. Gollapudi, B.B.; Johnson, G.E.; Hernandez, L.G.; Pottenger, L.H.; Dearfield, K.L.; Jeffrey, A.M.; Julien, E.; Kim, J.H.; Lovell, D.P.; MacGregor, J.T.; et al. Quantitative approaches for assessing dose-response relationships in genetic toxicology studies. *Environ. Mol. Mutagen.* **2013**, *54*, 8–18. [CrossRef]
17. Toropova, A.P.; Toropov, A.A. CORAL software: Prediction of carcinogenicity of drugs by means of the Monte Carlo method. *Eur. J. Pharm. Sci.* **2014**, *52*, 21–25. [CrossRef]
18. Toropov, A.A.; Toropova, A.P.; Raitano, G.; Benfenati, E. CORAL: Building up QSAR models for the chromosome aberration test. *Saudi J. Biol. Sci.* **2019**, *26*, 1101–1106. [CrossRef]
19. Ülger, T.G.; Uçar, A.; Çakıroğlu, F.P.; Yilmaz, S. Genotoxic effects of mycotoxins. *Toxicon* **2020**, *185*, 104–113. [CrossRef] [PubMed]
20. Myatt, G.J.; Ahlberg, E.; Akahori, Y.; Allen, D.; Amberg, A.; Anger, L.T.; Aptula, A.; Auerbach, S.; Beilke, L.; Bellion, P.; et al. In silico toxicology protocols. *Regul. Toxicol. Pharmacol.* **2018**, *96*, 1–17. [CrossRef] [PubMed]
21. Nielsen, K.F.; Smedsgaard, J. Fungal metabolite screening: Database of 474 mycotoxins and fungal metabolites for dereplication by standardised liquid chromatography–UV–mass spectrometry methodology. *J. Chromatogr. A* **2003**, *1002*, 111–136. [CrossRef]
22. Nielsen, K.F.; Mogensen, J.M.; Johansen, M.; Larsen, T.O.; Frisvad, J.C. Review of secondary metabolites and mycotoxins from the *Aspergillus niger* group. *Anal. Bioanal. Chem.* **2009**, *395*, 1225–1242. [CrossRef] [PubMed]
23. Zhou, J.; Xu, J. Chemistry and Biodiversity of Rhizophora-Derived Endophytic Fungi. *Mangrove Ecosyst. Ecol. Funct.* **2018**, *8*, 165–186. [CrossRef]
24. Mycotoxins High Resolution MS/MS Spectral Library, Sciex. Available online: https://sciex.com/gt/products/spectral-library/mycotoxin-libraries (accessed on 18 January 2023).
25. Aalizadeh, R.; Alygizakis, N.; Schymanski, E.; Slobodnik, J.; Fischer, S.; Cirka, L. S0 | SUSDAT | Merged NORMAN Suspect List: SusDat (Version NORMAN-SLE-S0.0.3.2). 2021. Available online: https://zenodo.org/record/4558070 (accessed on 20 May 2023). [CrossRef]

26. OECD. Guidance Document on Revisions to OECD Genetic Toxicology Test Guidelines. OECD Workgroup of National Coordinators for Test 42 Guidelines (WNT). 2015. Available online: https://www.oecd.org/chemicalsafety/testing/Genetic%20Toxicology%20Guidance%20Document%20Aug%2031%202015.pdf (accessed on 2 October 2022).
27. OECD. OECD Principles for the Validation, for Regulatory Purposes, of (Quantitative) Structure-Activity Relationship Models. 2004. Available online: https://www.oecd.org/chemicalsafety/risk-assessment/37849783.pdf (accessed on 24 January 2023).
28. ECHA. Practical Guide–How to Use and Report (Q)SARs. ECHA-16-B-09-EN. 2016. Available online: https://echa.europa.eu/documents/10162/13655/pg_report_qsars_en.pdf/407dff11-aa4a-4eef-a1ce-9300f8460099 (accessed on 1 February 2023).
29. Benfenati, E.; Manganaro, A.; Gini, G. VEGA-QSAR: AI inside a platform for predictive toxicology. In Proceedings of the Workshop "Popularize Artificial Intelligence 2013", Turin, Italy, 5 December 2013; CEUR Workshop: Turin, Italy, 2013; Volume 1107.
30. Dimitrov, S.D.; Diderich, R.; Sobanski, T.; Pavlov, T.S.; Chankov, G.V.; Chapkanov, A.S.; Karakolev, Y.H.; Temelkov, S.G.; Vasilev, R.A.; Gerova, K.D.; et al. QSAR Toolbox–Workflow and major functionalities. *SAR QSAR Environ. Res.* **2016**, *27*, 203–219. [CrossRef]
31. Benigni, R.; Bossa, C. Mechanisms of chemical carcinogenicity and mutagenicity: A review with implications for predictive toxicology. *Chem. Rev.* **2011**, *111*, 2507–2536. [CrossRef]
32. EFSA Panel on Contaminants in the Food Chain (CONTAM); Schrenk, D.; Bodin, L.; Chipman, J.K.; del Mazo, J.; Grasl-Kraupp, B.; Hogstrand, C.; Hoogenboom, L.; Leblanc, J.C.; Nebbia, C.S.; et al. Risk assessment of ochratoxin A in food. *EFSA J.* **2020**, *18*, 6113. [CrossRef]
33. Nieto, C.H.D.; Granero, A.M.; Zon, M.A.; Fernández, H. Sterigmatocystin: A mycotoxin to be seriously considered. *Food Chem. Toxicol.* **2018**, *118*, 460–470. [CrossRef]
34. Knasmüller, S.; Parzefall, W.; Helma, C.; Kassie, F.; Ecker, S.; Schulte-Hermann, R. Toxic effects of griseofulvin: Disease models, mechanisms, and risk assessment. *Crit. Rev. Toxicol.* **1997**, *27*, 495–537. [CrossRef]
35. EFSA Panel on Contaminants in the Food Chain (CONTAM). Scientific Opinion on the risks to human and animal health related to the presence of beauvericin and enniatins in food and feed. *EFSA J.* **2014**, *12*, 3802. [CrossRef]
36. Ostry, V.; Malir, F.; Toman, J.; Grosse, Y. Mycotoxins as human carcinogens—The IARC Monographs classification. *Mycotoxin Res.* **2017**, *33*, 65–73. [CrossRef]
37. Delucca, A.J.; Dunn, J.J.; Lee, L.S.; Ciegler, A. Toxicity, mutagenicity and teratogenicity of brevianamide, viomellein and xanthomegnin; secondary metabolites of *Penicillium viridicatum*. *J. Food Saf.* **1982**, *4*, 165–168. [CrossRef]
38. Cocchi, V.; Gasperini, S.; Lenzi, M. Anthraquinones: Genotoxic until Proven Otherwise? A Study on a Substance-Based Medical Device to Implement Available Data for a Correct Risk Assessment. *Toxics* **2022**, *10*, 142. [CrossRef] [PubMed]
39. Chen, A.; Mao, X.; Sun, Q.; Wei, Z.; Li, J.; You, Y.; Zhao, J.; Jiang, G.; Wu, Y.; Wang, L.; et al. *Alternaria* mycotoxins: An overview of toxicity, metabolism, and analysis in food. *J. Agric. Food Chem.* **2021**, *69*, 7817–7830. [CrossRef] [PubMed]
40. EFSA Panel on Contaminants in the Food Chain (CONTAM). Scientific Opinion on the risks for animal and public health related to the presence of *Alternaria* toxins in feed and food. *EFSA J.* **2011**, *9*, 2407. [CrossRef]
41. European Commission. Commission Recommendation (EU) 2022/553 of 5 April 2022 on Monitoring the Presence of Alternaria Toxins in Food. O.J.E.U., L 107/90. Available online: https://eur-lex.europa.eu/legal-content/EN/TXT/PDF/?uri=CELEX:32022H0553 (accessed on 2 March 2023).
42. Garofalo, M.; Payros, D.; Penary, M.; Oswald, E.; Nougayrède, J.P.; Oswald, I.P. A novel toxic effect of foodborne trichothecenes: The exacerbation of genotoxicity. *Environ. Pollut.* **2023**, *317*, 120625. [CrossRef] [PubMed]
43. Yang, W.; Yu, M.; Fu, J.; Bao, W.; Wang, D.; Hao, L.; Yao, P.; Nüssler, A.K.; Yan, H.; Liu, L. Deoxynivalenol induced oxidative stress and genotoxicity in human peripheral blood lymphocytes. *Food Chem. Toxicol.* **2014**, *64*, 383–396. [CrossRef]
44. Abdel-Wahhab, M.A.; El-Nekeety, A.A.; Salman, A.S.; Abdel-Aziem, S.H.; Mehaya, F.M.; Hassan, N.S. Protective capabilities of silymarin and inulin nanoparticles against hepatic oxidative stress, genotoxicity, and cytotoxicity of Deoxynivalenol in rats. *Toxicon* **2018**, *142*, 1–13. [CrossRef]
45. Dönmez-Altuntas, H.; Dumlupinar, G.; Imamoglu, N.; Hamurcu, Z.; Liman, B.C. Effects of the mycotoxin citrinin on micronucleus formation in a cytokinesis-block genotoxicity assay in cultured human lymphocytes. *J. Appl. Toxicol.* **2007**, *27*, 337–341. [CrossRef] [PubMed]
46. Ayed, Y.; Ayed-Boussema, I.; Ouanes, Z.; Bacha, H. In vitro and in vivo induction of chromosome aberrations by alpha-and beta zearalenols: Comparison with zearalenone. *Mutat. Res.* **2011**, *726*, 42–46. [CrossRef]
47. Belgacem, H.; Salah-Abbeès, J.B.; Ezzdini, K.; Abdel-Wahhab, M.A.; Zinedine, A.; Abbès, S. *Lactobacillus plantarum* Mon03 counteracts zearalenone genotoxicty in mice: Chromosome aberrations, micronuclei, DNA fragmentation and apoptotique gene expression. *Mutat. Res./Genet. Toxicol. Environ. Mutagen.* **2019**, *840*, 11–19. [CrossRef]
48. Khan, R.B.; Phulukdaree, A.; Chuturgoon, A.A. Fumonisin B1 induces oxidative stress in oesophageal (SNO) cancer cells. *Toxicon* **2018**, *141*, 104–111. [CrossRef]
49. Radić, S.; Domijan, A.M.; Ljubimir, K.G.; Maldini, K.; Ivešić, M.; Štefanić, P.P.; Krivohlavek, A. Toxicity of nanosilver and fumonisin B1 and their interactions on duckweed (*Lemna minor* L.). *Chemosphere* **2019**, *229*, 86–93. [CrossRef]
50. Rumora, L.; Kovačić, S.; Rozgaj, R.; Čepelak, I.; Pepeljnjak, S.; Žanić Grubišić, T. Cytotoxic and genotoxic effects of fumonisin B1 on rabbit kidney RK13 cell line. *Arch. Toxicol.* **2002**, *76*, 55–61. [CrossRef] [PubMed]
51. Ehrlich, V.; Darroudi, F.; Uhl, M.; Steinkellner, H.; Zsivkovits, M.; Knasmueller, S. Fumonisin B1 is genotoxic in human derived hepatoma (HepG2) cells. *Mutagenesis* **2002**, *17*, 257–260. [CrossRef]

52. EFSA Panel on Contaminants in the Food Chain (CONTAM). Scientific Opinion on Ergot alkaloids in food and feed. *EFSA J.* **2012**, *10*, 2798. [CrossRef]
53. Roberts, G.; Rand, M.J. Chromosomal damage induced by some ergot derivatives in vitro. *Mutat. Res./Fundam. Mol. Mech. Mutagen.* **1977**, *48*, 205–214. [CrossRef]
54. Seifried, H.E.; Seifried, R.M.; Clarke, J.J.; Junghans, T.B.; San, R.H.C. A compilation of two decades of mutagenicity test results with the Ames *Salmonella typhimurium* and L5178Y mouse lymphoma cell mutation assays. *Chem. Res. Toxicol.* **2006**, *19*, 627–644. [CrossRef] [PubMed]
55. Dighe, R.; Vaidya, V.G. Induction of sister chromatid exchanges by ergot compounds in Chinese hamster ovary cells in vitro. *Teratog. Carcinog. Mutagen.* **1988**, *8*, 169–174. [CrossRef]
56. Elliger, C.A.; Henika, P.R.; MacGregor, J.T. Mutagenicity of flavones, chromones and acetophenones in *Salmonella typhimurium*: New structure—Activity relationships. *Mutat. Res./Genet. Toxicol.* **1984**, *135*, 77–86. [CrossRef] [PubMed]
57. Hansen, T.J. Ames mutagenicity tests on purified 3-nitropropionic acid. *Food Chem. Toxicol.* **1984**, *22*, 399–401. [CrossRef] [PubMed]
58. Lemée, P.; Fessard, V.; Habauzit, D. Prioritization of mycotoxins based on mutagenicity and carcinogenicity evaluation using combined in silico QSAR methods. *Environ. Pollut.* **2023**, *323*, 121284. [CrossRef]
59. Alonso-Jauregui, M.; Font, M.; González-Peñas, E.; López de Cerain, A.; Vettorazzi, A. Prioritization of Mycotoxins Based on Their Genotoxic Potential with an In Silico-In Vitro Strategy. *Toxins* **2021**, *13*, 734. [CrossRef]
60. Glück, J.; Buhrke, T.; Frenzel, F.; Braeuning, A.; Lampen, A. In silico genotoxicity and carcinogenicity prediction for food-relevant secondary plant metabolites. *Food Chem. Toxicol.* **2018**, *116*, 298–306. [CrossRef]
61. Lou, J.; Fu, L.; Peng, Y.; Zhou, L. Metabolites from *Alternaria* fungi and their bioactivities. *Molecules* **2013**, *18*, 5891–5935. [CrossRef] [PubMed]
62. Qileng, A.; Wei, J.; Lu, N.; Liu, W.; Cai, Y.; Chen, M.; Lei, H.; Liu, Y. Broad-specificity photoelectrochemical immunoassay for the simultaneous detection of ochratoxin A, ochratoxin B and ochratoxin C. *Biosens. Bioelectron.* **2018**, *106*, 219–226. [CrossRef] [PubMed]
63. Lerda, D. Mycotoxins factsheet. In *JRC Technical Notes*, 4th ed.; JRC: Geel, Belgium, 2011; Volume 66956, p. 2011. Available online: https://www.bezpecnostpotravin.cz/UserFiles/File/Kvasnickova2/JRC_mykotoxiny.pdf (accessed on 30 October 2022).
64. Marin, S.; Ramos, A.J.; Cano-Sancho, G.; Sanchis, V. Mycotoxins: Occurrence, toxicology, and exposure assessment. *Food Chem. Toxicol.* **2013**, *60*, 218–237. [CrossRef] [PubMed]
65. Benfenati, E.; Golbamaki, A.; Raitano, G.; Roncaglioni, A.; Manganelli, S.; Lemke, F.; Norinder, U.; Lo Piparo, E.; Honma, M.; Manganaro, A.; et al. A large comparison of integrated SAR/QSAR models of the Ames test for mutagenicity. *SAR QSAR Environ. Res.* **2018**, *29*, 591–611. [CrossRef]
66. Pereira, V.L.; Fernandes, J.O.; Cunha, S.C. Mycotoxins in cereals and related foodstuffs: A review on occurrence and recent methods of analysis. *Trends Food Sci. Technol.* **2014**, *36*, 96–136. [CrossRef]
67. Honma, M.; Kitazawa, A.; Cayley, A.; Williams, R.V.; Barber, C.; Hanser, T.; Saiakhov, R.; Chakravarti, S.; Myatt, G.J.; Cross, K.P.; et al. Improvement of quantitative structure–Activity relationship (QSAR) tools for predicting Ames mutagenicity: Outcomes of the Ames/QSAR International Challenge Project. *Mutagenesis* **2019**, *34*, 3–16. [CrossRef]
68. Aydin, M.; Rencüzoğullari, E. Genotoxic and Mutagenic Effects of Mycotoxins: A Review. *Commagene J. Biol.* **2019**, *3*, 132–161. [CrossRef]
69. Juan-García, A.; Taroncher, M.; Font, G.; Ruiz, M.J. Micronucleus induction and cell cycle alterations produced by deoxynivalenol and its acetylated derivatives in individual and combined exposure on HepG2 cells. *Food Chem. Toxicol.* **2018**, *118*, 719–725. [CrossRef] [PubMed]
70. Mamur, S.; Yuzbasioglu, D.; Yılmaz, S.; Erikel, E.; Unal, F. Assessment of cytotoxic and genotoxic effects of enniatin-A in vitro. *Food Addit. Contam. A* **2018**, *35*, 1633–1644. [CrossRef] [PubMed]
71. Madia, F.; Corvi, R. EURL ECVAM Genotoxicity and Carcinogenicity Consolidated Database of Ames Negative Chemicals. European Commission, Joint Research Centre (JRC) [Dataset] PID. 2020. Available online: http://data.europa.eu/89h/38701804-bc00-43c1-8af1-fe2d5265e8d7 (accessed on 5 October 2022).

toxins

MDPI

Article

Potential Toxicity and Mechanisms of T-2 and HT-2 Individually or in Combination on the Intestinal Barrier Function of Porcine Small Intestinal Epithelial Cells

Weihua He [1,†], Jianhua Wang [2,†], Mengyi Han [2], Lihua Wang [1], Ling Li [1], Jiahui Zhang [1], Siqi Chen [1], Jiayi Guo [1], Xiaohu Zhai [1,*] and Junhua Yang [2,*]

1 Institute of Pet Science and Technology, Jiangsu Agri-Animal Husbandry Vocational College, Taizhou 225300, China; 2008020102@jsahvc.edu.cn (W.H.); 1997010187@jsahvc.edu.cn (L.W.); 2007020099@jsahvc.edu.cn (L.L.); 202210101@jsahvc.edu.cn (J.Z.); 202210137@jsahvc.edu.cn (S.C.); 202214904@jsahvc.edu.cn (J.G.)

2 Institute for Agri-Food Standards and Testing Technology, Shanghai Academy of Agricultural Sciences, Shanghai 201403, China; wangjianhua@saas.sh.cn (J.W.); h2054426758@163.com (M.H.)

* Correspondence: zhaixiaohu@jsahvc.edu.cn (X.Z.); yangjunhua@saas.sh.cn (J.Y.); Tel.: +86-21-37191637 (J.Y.)

† These authors contributed equally to this work.

Abstract: Under natural conditions, T-2 toxin can be easily metabolized to HT-2 toxin by deacetylation, and T-2 and HT-2 are usually co-contaminated in grain and feed at a high detected rate. Our previous information indicated that T-2 toxin could injure the function of the intestinal barrier, but the combined toxicity and mechanism of T-2 and HT-2 on the intestinal cells of porcines are still unknown. Therefore, we aimed to explore T-2 and HT-2 individually and combined on cellular viability, cell membrane integrity, the expression of tight junction-related proteins, and the generation of inflammatory factors in porcine intestinal epithelial cells (IPEC-J2). The results showed that T-2 and HT-2, individually or in combination, could induce a decrease in cell viability, an increase in LDH release and IL-1, IL-6, and TNF-α generation, and a decrease in the anti-inflammatory factor IL-10. Based on the analysis of immunofluorescence staining, real-time PCR, and western blotting, the tight junction protein expressions of Claudin-1, Occludin, and ZO-1 were significantly decreased in the T-2 and HT-2 individual or combination treated groups compared with the control. Furthermore, all the parameter changes in the T-2 + HT-2 combination group were much more serious than those in the individual dose groups. These results suggest that T-2 and HT-2, individually and in combination, could induce an intestinal function injury related to an inflammatory response and damage to the intestinal barrier function in porcine intestinal epithelial cells. Additionally, T-2 and HT-2 in combination showed a synergistic toxic effect, which will provide a theoretical basis to assess the risk of T-2 + HT-2 co-contamination in porcine feed.

Keywords: T-2 toxin; HT-2 toxin; porcine intestinal epithelial cells; inflammation; intestinal barrier

Key Contribution: The exposure of T-2 and HT-2 individually and in combination to IPEC-J2 cells caused damage to intestinal barrier function, which contributed to the inflammatory response and decreased expression of tight junction proteins, and the combination of T-2 and HT-2 presented a synergistic toxic effect.

1. Introduction

T-2 toxin, a sesquigerene compound generated by *F. poae* and *F. sporotrichioides*, is characterized by the highest toxicity in type A trichohecene [1], which is found in cereals and unprocessed foods around the world, including Africa, Europe, Southeast Asia, South America, and China [2]. T-2 toxin is difficult to volatilize and is insoluble in water, but it could be resiliently degraded through hydrolysis, hydroxylation, deepoxidation, and conjugation reactions in diverse animal and plant organisms [3]. Generally, the most common metabolic pathway of T-2 is fast deacetylation of the C-4 site, which is then converted into HT-2 toxin [4].

According to the European Union survey of mycotoxins contamination in food, 57% of 11,022 samples tested positive for DON, 16% of 4166 samples tested positive for NIV, 20% of 3490 samples tested positive for T-2, and 14% of 3032 samples tested positive for HT-2 [5,6]. In addition, DON contamination was more commonly observed in wheat and corn, while NIV, T-2 toxin, and HT-2 toxin are all common contaminants in corn, wheat, and oats [6]. In China, T-2 and HT-2 were detected in grain from Sichuan province; the detection rate and average dose of T-2 were 11.64% and 0.565 g/kg, and that of HT-2 were 49.74% and 3.746 μg/kg, and the highest concentrations of T-2 toxin and HT-2 toxins were 3.332 μg/kg and 34.510 μg/kg. Furthermore, the EU is also considering the allowable limits for the T-2 + HT-2 toxin combination [7]. Following the widespread contamination of mycotoxins, the food chain contaminants group of the European Food Safety Authority defined the tolerable daily intake of T-2 toxin and HT-2 toxin at 100 ng/kg [8]. In 2016 and 2017, the mycotoxins contamination in maize was investigated in Croatia, and fumonisin B1 (85.7%), HT-2 (73.8%), deoxynivalenol (65.5%), fumonisin B2 (64.3%), zearalenone (54.8%), T-2 toxin (28.6%), aflatoxin B1 (2.4%), aflatoxin B2, and aflatoxin G1 (1.2%) were discovered in the mixture contamination, and the combined detection rate of T-2 toxin and HT-2 toxin was 26.2% in the total samples. According to the above information, it can be summarized that the levels of T-2 toxin and HT-2 toxin were wide-ranging, and the two toxins often appear in combination.

A large number of studies have shown that the mycotoxins T-2 and HT-2 have extensive toxic effects on humans and animals [9], which could inhibit protein synthesis and cell proliferation and cause toxicity and damage to the skin, immunological system, liver, reproductive system, and digestive system in vitro and in vivo. Because T-2 toxin and HT-2 toxin share the same chemical structure as epoxysesquiterpenes, their toxicity is relatively similar [10]. However, it is not possible to assess the combined toxicity of mycotoxins based on the individual components, as they may produce additive, synergistic, and antagonistic interactions [11]. In earlier studies, T-2 + HT-2 toxins showed synergistic effects at low doses and antagonistic effects at high doses in porcine Leydig cells, but the reversed effect was found in the ternary combination [12]. The gastrointestinal organ is the first target for possible hazardous effects of mycotoxins after ingestion of contaminated food [13]. At present, most studies on the toxic mechanism of mycotoxins on porcine intestinal epithelial cells focus on individual toxicity, but the mechanism of T-2 toxin and HT-2 toxin individually or combined in treatment on the damage of porcine intestinal epithelial cells is still not clear. On the basis of the previous study, cell viability, the permeability of the cell membrane, the expression and distribution of tight junction proteins, and the secretion of inflammatory factors were observed after T-2 and HT-2 individual or combined exposure. The results of this experiment will further enrich the combined toxic mechanism of mycotoxins and provide a new idea for the prevention and control of contamination.

2. Results

2.1. *Effect of T-2 and HT-2 on Cell Viability and Lactate Dehydrogenase (LDH) Activity*

As shown in Figure 1A, cell viability was dramatically reduced in response to T-2 toxin and HT-2 toxin individually or in combination groups compared to the controls ($p < 0.05$). Additionally, cell viability in the T-2 + HT-2 combination group was decreased compared with that in the T-2 toxin and HT-2 toxin single-treated groups ($p < 0.05$).

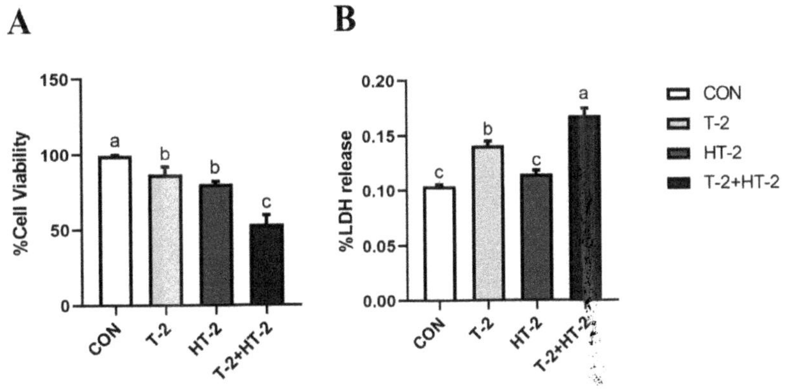

Figure 1. Changes in cell viability and LDH release of IPEC-J2 cells after exposure to T-2 (3.125 nmol/L), HT-2 (6.25 nmol/L), and T-2 (3.125 nmol/L) + HT-2 (6.25 nmol/L) for 24 h. (**A**) Cell viability; (**B**) lactate dehydrogenase activity. All values are given as the mean ± SD of four independent experiments. Values with different superscript letters are significantly different ($p < 0.05$).

The LDH released from cells could be quantitatively detected and was considered an important indicator of cell membrane integrity. As shown in Figure 1B, the LDH from IPEC-J2 cells gradually increased when treated with T-2 toxin and HT-2 toxin individually and in combination ($p > 0.05$ or $p < 0.05$), and the LDH in the combination group was obviously higher compared with that in the T-2 toxin and HT-2 toxin individual groups ($p < 0.05$).

2.2. The Expression of Tight Junction Protein by Immunofluorescence Assay

Following the treatment with T-2 toxin and HT-2 toxin, the distribution of tight junction proteins Claudin-1, Occludin, and ZO-1 in IPEC-J2 cells was observed by a laser scanning confocal microscope. As shown in Figures 2–4, the red fluorescence intensity was proportional to the expression of the three proteins. In the control group, the expression of the three proteins was high and distributed evenly, but the distribution of Claudin-1, Occludin, and ZO-1 in the combined toxin-treated group decreased, and the expression in the individual T-2 toxin and HT-2 toxin groups decreased significantly. Additionally, T-2 toxin and HT-2 toxin caused some morphological changes in tight junction proteins, which led to the barrier function damage of cell junctions.

2.3. Expression of Tight Junction-Related Protein mRNA by Real-Time PCR

The expression levels of *Claudin-1*, *Occludin*, and *ZO-1* genes were investigated by qRT-PCR, as shown in Figure 5. Compared with the control, the levels of the three tight junction protein mRNA were dramatically reduced in all the treated groups ($p > 0.05$ or $p < 0.05$), and the mRNA relative expression of *Claudin-1*, *Occludin*, and *ZO-1* genes in the T-2 + HT-2 combination group was obviously decreased compared with that in the control, T-2 toxin, and HT-2 toxin individual groups ($p < 0.05$).

2.4. Expression of Tight Junction-Related Proteins by Western-Blot

Following treatment with T-2 and HT-2 for 24 h, the expression of Claudin-1, Occludin, and ZO-1 proteins was investigated by western blotting. The protein levels of Occludin and ZO-1 were significantly lower in the T-2 toxin group than those in the control group ($p < 0.05$). The expression levels of the three proteins were dramatically reduced in the HT-2 individual group and the T-2 + HT-2 combination group compared with those in the control and T-2 individual groups ($p < 0.05$). In addition, the expression levels of Claudin-1, Occludin, and ZO-1 in the T-2 + HT-2 combined group were also higher than the HT-2 individual group ($p < 0.05$, Figure 6).

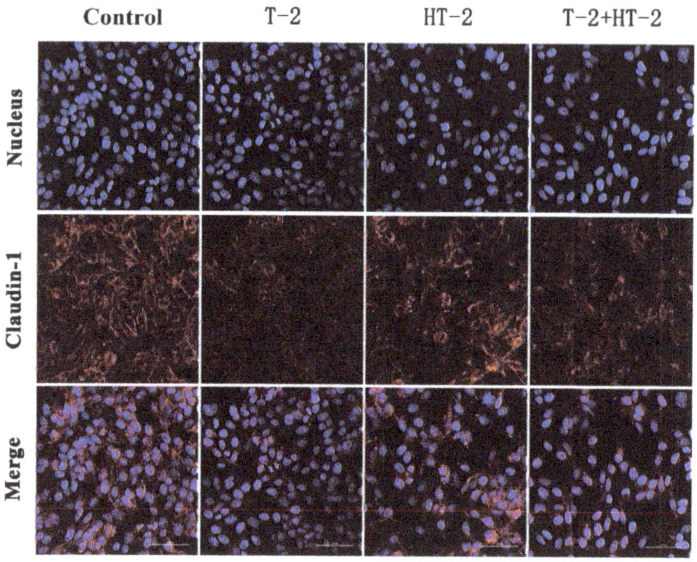

Figure 2. Changes in expression and distribution of Claudin-1 protein in IPEC-J2 cells after exposure to T-2 (3.125 nmol/L), HT-2 (6.25 nmol/L), and T-2 (3.125 nmol/L) + HT-2 (6.25 nmol/L) for 24 h (400×).

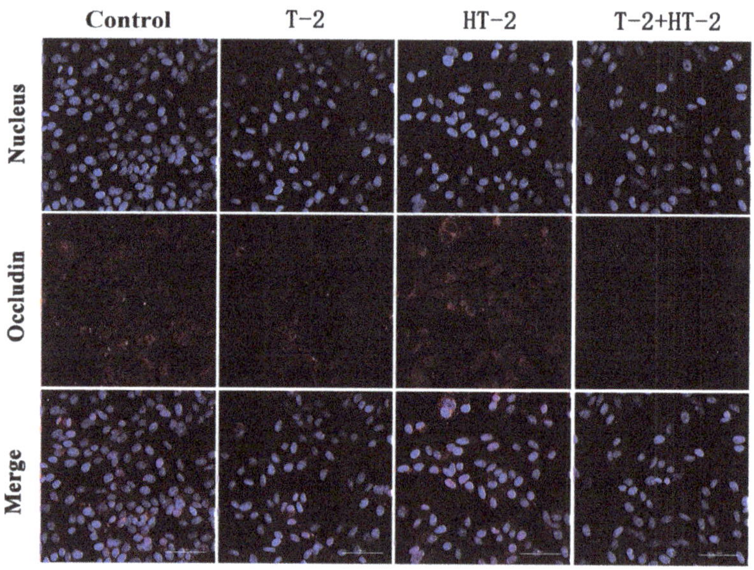

Figure 3. Changes in expression and distribution of Occludin protein in IPEC-J2 cells after exposure to T-2 (3.125 nmol/L), HT-2 (6.25 nmol/L), and T-2 (3.125 nmol/L) + HT-2 (6.25 nmol/L) for 24 h (400×).

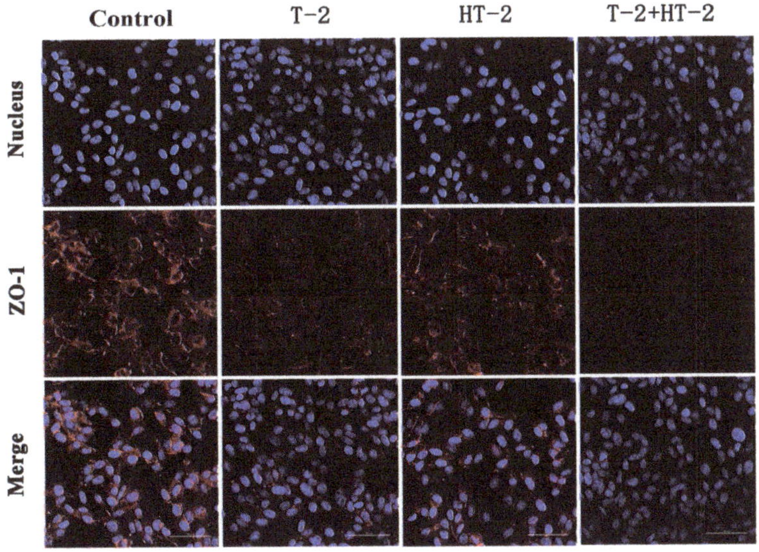

Figure 4. Changes in expression and distribution of ZO-1 protein in IPEC-J2 cells after exposure to T-2 (3.125 nmol/L), HT-2 (6.25 nmol/L), and T-2 (3.125 nmol/L) + HT-2 (6.25 nmol/L) for 24 h (400×).

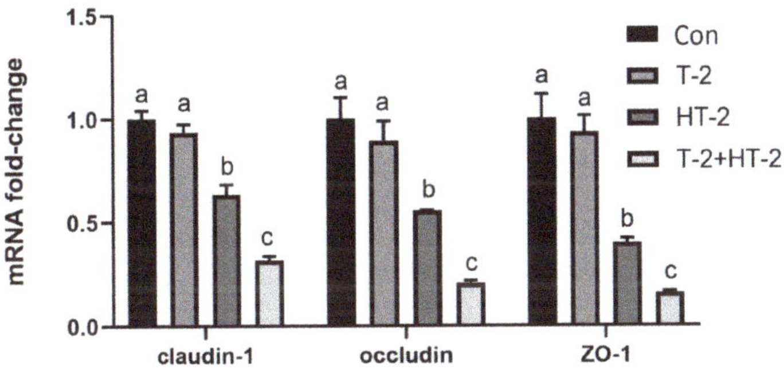

Figure 5. Changes in expression level of tight junction-related genes in IPEC-J2 cells after exposure to T-2 (3.125 nmol/L), HT-2 (6.25 nmol/L), and T-2 (3.125 nmol/L) + HT-2 (6.25 nmol/L) for 24 h. The control value was set at 1.0. All values are given as the mean ± SD of four independent experiments. Values with different superscript letters are significantly different ($p < 0.05$).

2.5. Effect of T-2 and HT-2 on the Content of Inflammatory Factors

The concentration change in T-2 and HT-2 on inflammatory factors is shown in Figure 7. Compared with the control, the contents of IL-1, IL-6, and TNF-α in IPEC-J2 cells were significantly increased when T-2 toxin and HT-2 toxin were exposed individually or in combination ($p < 0.05$), and obvious differences were found between the different groups ($p < 0.05$). However, the content of IL-10 in the HT-2 toxin and T-2 + HT-2 combination groups was dramatically decreased compared with the control and T-2 toxin groups ($p < 0.05$). Additionally, the level of IL-10 in the combination group was also obviously lower compared with the individual dose groups of T-2 toxin and HT-2 toxin ($p < 0.05$).

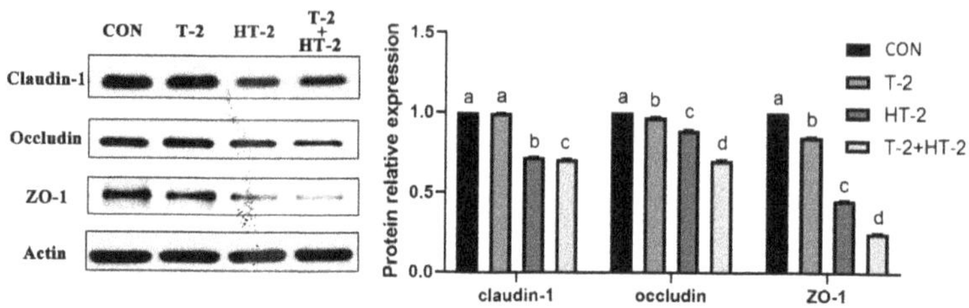

Figure 6. Changes in relative expression of tight junction-related proteins in IPEC-J2 cells after exposure to T-2 (3.125 nmol/L), HT-2 (6.25 nmol/L), and T-2 (3.125 nmol/L) + HT-2 (6.25 nmol/L) for 24 h. The control value was set at 1.0. All values are given as the mean ± SD of four independent experiments. Values with different superscript letters are significantly different ($p < 0.05$).

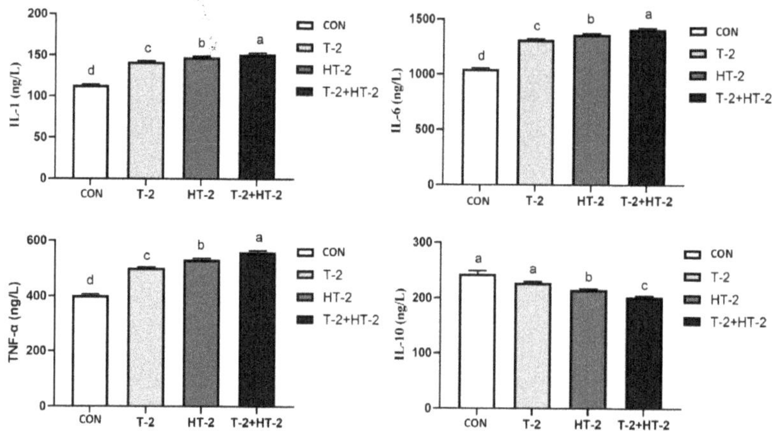

Figure 7. Changes in content of IL-1, IL-6, TNF-a, and IL-10 in IPEC-J2 cells after exposure to T-2 (3.125 nmol/L), HT-2 (6.25 nmol/L), and T-2 (3.125 nmol/L) + HT-2 (6.25 nmol/L) for 24 h. All values are given as the mean ± SD of four independent experiments. Values with different superscript letters are significantly different ($p < 0.05$).

3. Discussion

Our study results support the conclusion that exposure to T-2 toxin and HT-2 toxin individually or in combination by IPEC-J2 could induce LDH release, decrease tight junction protein expression, and generate pro-inflammatory factors. The toxicological effect of the combination group was more obvious than that of the individual groups, presenting a toxic synergy.

Generally, a basic barrier is composed of the intestinal epithelium and the luminal environment, which could be used to regulate and prevent the intake of harmful substances [14–16]. Furthermore, the tight connections (TJ) in the intestine are constituted by claudins and other transmembrane proteins, which are crucial for preserving the integrity of the intestinal mucosa and intestinal health in both humans and animals [17,18]. These proteins are aggregated and stabilized by zonula occluden 1 (ZO-1) and cytoskeletal proteins at the top and Claudin 1 at the bottom of the lateral membrane [19]. Porcine cells are one of the most sensitive cells to T-2 toxin, and our previous studies proved that T-2 toxin could easily cause barrier function damage to IPEC-J2 cells. In this research, T-2 toxin and HT-2 toxin, individually and in combination, damaged the cell membrane integrity and induced the leakage of large amounts of LDH from the cytoplasm to the outside of the cell, which then destroyed the integrity of the intestinal cells. According to reports, T-2 toxin and HT-2 toxin exposed to broiler hepatocytes could induce an increase in LDH level, which was consistent with our results [20]. Similarly, T-2 toxin exposure also led to an increase in LDH levels in TM3 Leydig cells [21]. However, there was no information reported about the LDH level change when treated with T-2 and HT-2 in combination. This study demonstrated that the level of LDH when treated with T-2 + HT-2 was less than in the individual groups, which suggested that the T-2 + HT-2 combination induced a more serious injury to the barrier function in IPEC-J2 and produced a synergistic effect.

According to laser confocal observation, the localization and shape of the tight junction were all changed by treatment with T-2 toxin and HT-2 toxin for 24 h, and changes in the combined group were even more significant. The anatomical basis for preserving the intestinal epithelium and its barrier function was the mechanism of the intestinal mucosal barrier. The main functional element of the intestinal epithelial barrier is TJ proteins, including ZO-1, Claudin-1, and Occludin, which maintain the intestinal barrier's biological function by sealing the space between adherent epithelial cells [22]. In this study, the mRNA and protein expression levels of ZO-1, Occludin, and Claudin in IPEC-J2 cells were dramatically lower, showing that the T-2 toxin and HT-2 toxin, individually or combined, could injure the barrier function of IPEC-J2 cells. Similarly, the expression levels of ZO-1, Occludin, and Claudin in the intestines of mice and in human epithelial cells were all significantly decreased after T-2 toxin exposure [23]. All the studies indicated that T-2 toxin and HT-2 toxin could damage the intestinal epithelial barrier function, and the toxicological effect of the T-2 toxin and HT-2 toxin combination is much stronger than that of the individual toxins.

Inflammation plays a vitally important role in modulating the barrier function of the intestinal mucosal immune system [24,25]. Additionally, damage to the intestinal barrier function involves an increase in epithelial permeability, translocation of intraluminal allergens and pathogens, non-specific inflammation, and overstimulation of the intestinal-related immune system [23]. Earlier studies found that the IRE1/XBP1 pathway was activated by T-2 toxin exposure, which then disrupted intestinal mucins in 4-week-old BALB/C mice, increased the generation of proinflammatory cytokines, and induced an inflammation response [26]. In the current study, all the results were similar to earlier studies in that the proinflammatory factors IL-1, IL-6, and TNF-α were increased and the anti-inflammatory factor IL-10 was decreased in the IPEC-J2 cells after T-2 and HT-2 individual and combination treatment, which was consistent with the LDH result. However, there was little information about the toxicological effect of T-2 + HT-2 on the expression of inflammatory factors. Other mycotoxins studies have shown that co-exposure to DON and cevalenol could significantly increase the level of inflammatory factors in intestinal explants, with a synergistic effect promoting inflammation [27]. Additionally, both DON and ZEN exposed to intestinal cells can induce synergistic or additive deleterious effects on the expression of inflammatory factors [28,29]. Our experimental results indicate that the influence of inflammation induced by the T-2 toxin and HT-2 toxin combination was significantly greater than that of the individual treatments. Therefore, we were correct in our hypothesis that T-2 toxin and HT-2 toxin, individually or in combination, could increase the inflammation of IPEC-J2 cells and that the combined group has synergistic effects on intestinal barrier damage in pigs. In general, the major metabolic pathway of T-2 toxin in mammals is rapid deacetylation at C-4, which results in HT-2 toxin formation. Therefore, adequate attention should be paid to the combination of T-2 toxin and HT 2 toxin in animals.

4. Conclusions

In summary, our data shows that IPEC-J2 cells treated with T-2 toxin and HT-2 toxin individually or in combination had obviously adverse effects on cell viability and caused an increase in LDH release from the cell. This could damage the barrier function of intestinal epithelial cells by disrupting the expression of tight junction protein and inducing the generation of inflammatory factors. Moreover, the damage to the barrier function caused by the T-2 + HT-2 combination group was more serious than that caused by the individual toxin treatments, which suggests a synergistic effect from the T-2 + HT-2 combination. However, their potential impacts on the intestinal health of swine are still worth further study.

5. Materials and Methods

5.1. Chemicals

The T-2 toxin and HT-2 toxin powders were supplied by Pribolab (Qingdao, Shandong, China). Dulbecco's Modified Eagle's Medium (DMEM), penicillin-streptomycin, 0.25% trypsin cell digestive fluid, fetal bovine serum (FBS), dimethyl sulfoxide (DMSO), and HBSS were obtained from Gibco (Grand Island, NY, USA). The CCK-8 kit was supplied by Beyotime Biotechnology (Shanghai, China). Prime Script RT Master Mix, SYBR Green I real-time PCR, and the Luminous kit were purchased from Takara (Dalian, China). The lactate dehydrogenase (LDH) kit was purchased from Applygen Technologies (Beijing, China). Claudin-1, Occludin, and ZO-1 antibodies and FITC-sheep anti-rabbit IgG were purchased from Proteintech (Rosemont, IL, USA). Β-actin was obtained from Servicebio (Wuhan, China).

5.2. Cell Culture

The cellular line of IPEC-J2 from Bio-World (Shang, China) was cultured in DMEM containing 10% FBS and 1% penicillin-streptomycin (100 units/mL–100 μg/mL). The incubator condition of the cells was 37 °C in a humidified chamber with 5% CO_2. T-2 toxin and HT-2 toxin were dissolved in DMSO to prepare a stock solution with a concentration of 1 mg/mL. The final concentration of DMSO was less than 0.1% in all detected processes.

5.3. Determination of Cell Viability

According to the CCK-8 kit, the cell viability was determined. IPEC-J2 cells were digested and collected at the logarithmic growth stage. Cell suspension concentration was calculated and then diluted, and then 5×10^3 cells/pores were seeded into a 96-well plate for 24 h. The cells were incubated with T-2 (3.125 nmol/L), HT-2 (6.25 nmol/L), and T-2 (3.125 nmol/L) + HT-2 (6.25 nmol/L) for 24 h, which was based on our previous experiment (unpublished). In the control group, the cells were cultured in 0.1% DMSO in DMEM, and inverted light microscopy was used to analyze the cell development and morphology. Subsequently, the 100 μL complete culture liquor containing different concentrations of T-2 and HT-2 toxins was discarded, and a 10 μL solution from the CCK-8 kit with 100 μL DMEM was added into each well. The cells were dyed and cultured at 37 °C for 2 h. The absorbance was recorded using the microplate reader iMark (Multiskan Sky), and the data were adjusted to reflect the cells that were not treated. Each group underwent six replicates, and a blank control group (no cells) was set up. Cell activity % = (treatment group OD value − blank group OD value)/(control group OD value − blank group OD value).

5.4. Determination of Lactate Dehydrogenase (LDH) Activity

IPEC-J2 cells with a density of 1×10^5 cells were inoculated into 96-well plates, and the cells were exposed to different concentrations of T-2 toxin and HT-2 toxin individually or combined for 24 h, and six replicates were performed for each treatment group. Subsequently, the culture medium of 96 well plates in each group was collected, and LDH release activity was detected according to the instructions of the LDH kit. The OD values were observed by a photometer, and the activity units were calculated according to the standard curve values. LDH release rate (%) = enzyme activity unit measured in cellular culture medium/(enzyme activity unit measured in cell lysate + enzyme activity unit measured in cell culture medium) × 100%.

5.5. Detection of ZO-1, Occludin, and Claudin-1 by Immunofluorescence Staining

IPEC-J2 cells were inoculated into 12-well plates with a density of 1.5×10^5. During the logarithmic growth period, different levels of T-2 (3.125 nM), HT-2 (6.25 nM), and T-2 + HT-2 (3.125 nM + 6.25 nM) were added to the culture medium for 24 h, and three replicates were performed for each treatment group. Subsequently, a 4% polyoxymethylene solution was immobilized for 10 min, and then 5% goat serum at room temperature was added to seal for 1 h. Finally, the primary antibodies of Claudin-1 (Proteintech13050-1-AP, 1:500), Occludin (Proteintech27260-1-AP, 1:300), and ZO-1 (Proteintech21773-1-AP, 1:500) were added and incubated at 4 °C overnight, respectively. Furthermore, the cells were incubated with biotinylated goat anti-rabbit IgG (1:1000) for 45 min at room temperature. Finally, 4′, 6-diamidino-2-phenylindole (DAPI) (100 ng/mL) and anti-fluorescence quenchant were used to reveal the immunoreaction, and the images were captured and recorded by a fluorescence microscope (Leica Microsystems Inc., Wetzlar, Germany).

5.6. RNA Extraction and Real-Time Quantitative PCR

IPEC-J2 cells were inoculated into 6-well plates with a density of 1×10^6 cells in each well. After 24 h of incubation, the cells were collected from each group of T-2 toxin and HT-2 toxin, individually or in combination, and three replicates were performed for each treatment group. The mRNA expression levels for tight junction-associated genes were determined by qPCR. TRIZOL was used to extract the total RNA, and then a Prime Script RT Master Mix kit was used to perform the cDNAs. According to the nucleotide sequences of *Claudin-1*, *Occludin*, and *ZO-1* to synthesize primers (shown in Table 1), *β-actin* was selected as the reference gene. A real-time qPCR reaction was performed according to the manufacturer's instructions for the SYBR Green I kit, and the $2^{-\Delta\Delta CT}$ method was used to analyze the relative expression of each gene [30].

Table 1. Parameters of primer for tight junction-related genes and β-actin genes.

Gene	Accession Number	Sequences (5′→3′)	Product
Claudin-1	NC_010455.5	F: GGCAGATCCAGTGCAAAGTC R: CCCAGCAGGATGCCAATTAC	94 bp
Occludin	NC_010458.4	F: CATTATGCACCCAGCAACGA R: GCACATCACGATAACGAGCA	168 bp
ZO-1	NC_010443.5	F: GGGCTCTTGGCTTGCTATTC R: AAGGCCTCGGAATCTCCAAA	160 bp
β-actin	424396	F: CTGGACTTCGAGCAGGAGATGG R: TTCGTGGATGCCGCAGGATTC	168 bp

Note: qPCR procedures were used as follows: pre-denaturation (95 °C 30 s), PCR reaction (95 °C 5 s, 60 °C 34 s, 40 cycles), 95 °C 15 s, and 60 °C 45 s. Each sample gene and *β-actin* were amplified under the same conditions and repeated 3 times.

5.7. Detection of Tight Junction-Associated Proteins

After washing the cell, RIPA protein lysate was used to extract the total protein, and the BCA kit was performed to determine the concentration of protein. Subsequently, 15 μg of protein was separated by 12.5% sodium dodecyl sulfate polyacrylamide gel electrophoresis (SDS-PAGE). The separated proteins were transferred onto a nitrocellulose (NC) membrane, and then 5% skim milk powder was used to block the membrane. After washing with TBS, membranes were probed with anti-Claudin-1 (Proteintech13050-1-AP, 1:750), anti-Occludin (Proteintech27260-1-AP, 1:750), anti-ZO-1 (Proteintech21773-1-AP, 1:750), and anti-β-actin (ServicebioGB12001, 1:3000) antibodies at 4 °C overnight. After that, the blots were incubated with horseradish peroxidase (HRP)-conjugated secondary antibody (Thermo Scientific, Waltham, MA, USA, 1:10,000) at room temperature for 1 h. Finally, the signals were captured, and the intensities of proteins on the bands were quantified using ImageJ Software (Java 1.8.0_112, National Institute of Health, Bethesda, MD, USA), and the expression of the target protein was analyzed using β-actin as an internal reference.

5.8. Effect of T-2 and HT-2 on Inflammation Factors

The levels of IL-1, IL-6, IL-10, and TNF-α were detected according to the ELISA kit from MLBIO (Shanghai Enzyme-linked Biotechnology Co., Ltd., Shanghai, China). IPEC-J2 cells were seeded at a density of 1 × 10^5 cells per hole on a 24-well plate. After incubation with T-2 (3.125 nmol/L), HT-2 (6.25 nmol/L), and T-2 (3.125 nmol/L) + HT-2 (6.25 nmol/L) for 24 h, the supernatant was centrifuged and collected to remove impurities and cell fragments. Based on the competitive ELISA, a standard curve was set, and the samples meeting calibration standards were assayed according to the manufacturer's instructions. The levels of inflammation factors were calculated by comparing the OD values of the samples to their respective standard curves at 450 nm within 30 min. Six replicates were performed for each treatment group.

5.9. Statistical Analysis

SPSS 17.0 (SPSS Inc., Chicago, IL, USA) was used for all analyses. Statistical analyses were performed by one-way ANOVA and LSD's post-hoc test. The data were presented as mean ± SEM, and $p < 0.05$ was considered statistically significant.

Author Contributions: J.Y., J.W. and X.Z. designed the research and made the revisions; L.W., L.L. and J.Z. performed the research; M.H., S.C. and J.G. contributed to data curation and formal analysis; W.H. and J.Y. wrote the paper. All authors have read and agreed to the published version of the manuscript.

Funding: This work was supported by the Social Development Science and Technology Research Projects from the Science and Technology Commission of Shanghai Municipality (No. 21DZ1201300) and the JiangSu Agri-Animal Husbandry Vocational College Academy Research Project (Grant No. NSF2023ZR06).

Informed Consent Statement: Not applicable.

Data Availability Statement: Data are contained within the article.

Conflicts of Interest: The authors declare no conflict of interest.

References

1. Janik, E.; Niemcewicz, M.; Podogrocki, M.; Ceremuga, M.; Stela, M.; Bijak, M. T-2 toxin–the most toxic trichothecene mycotoxin: Metabolism, toxicity, and decontamination strategies. *Molecules* **2021**, *26*, 6868. [CrossRef]
2. Wu, J.; Yang, C.; Liu, J.; Chen, J.; Huang, C.; Wang, J.; Liang, Z.; Wen, L.; Yi, J.; Yuan, Z. Betulinic acid attenuates T-2 toxin-induced cytotoxicity in porcine kidney cells by blocking oxidative stress and endoplasmic reticulum stress. *Comp. Biochem. Physiol. C Toxicol. Pharmacol.* **2021**, *249*, 109124. [CrossRef]
3. Kuchenbuch, H.; Becker, S.; Schulz, M.; Cramer, B.; Humpf, H. Thermal stability of T-2 and HT-2 toxins during biscuit- and crunchy muesli-making and roasting. *Food Addit. Contam. Part. A Chem. Anal. Control Expo. Risk Assess.* **2018**, *35*, 2158–2167. [CrossRef] [PubMed]
4. Wachowska, U.; Packa, D.; Wiwart, M. Microbial inhibition of fusarium pathogens and biological modification of trichothecenes in cereal grains. *Toxins* **2017**, *9*, 408. [CrossRef] [PubMed]
5. Pernica, M.; Kyralová, B.; Svoboda, Z.; Boško, R.; Brožková, I.; Česlová, L.; Benešová, K.; Červenka, L.; Běláková, S. Levels of T-2 toxin and its metabolites, and the occurrence of Fusarium fungi in spring barley in the Czech Republic. *Food Microbiol.* **2022**, *102*, 103875. [CrossRef] [PubMed]
6. Lattanzio, V.; Pascale, M.; Visconti, A. Current analytical methods for trichothecene mycotoxins in cereals. *Trends Analyt. Chem.* **2009**, *28*, 758–768. [CrossRef]
7. Wang, X.; Liu, X.; Liu, J.; Wang, G.; Wan, K. Contamination level of T-2 and HT-2 toxin in cereal crops from Aba area in Sichuan Province, China. *Bull. Environ. Contam. Toxicol.* **2012**, *88*, 396–400. [CrossRef]
8. Oplatowska-Stachowiak, M.; Kleintjens, T.; Sajic, N.; Haasnoot, W.; Campbell, K.; Elliott, C.; Salden, M. T-2 toxin/HT-2 toxin and Ochratoxin A ELISAs development and in-house validation in food in accordance with commission regulation (EU) No 519/2014. *Toxins* **2017**, *9*, 388. [CrossRef]
9. Adhikari, M.; Negi, B.; Kaushik, N.; Adhikari, A.; Al-Khedhairy, A.; Kaushik, N.; Choi, E. T-2 mycotoxin: Toxicological effects and decontamination strategies. *Oncotarget* **2017**, *8*, 33933–33952. [CrossRef]
10. Busman, M.; Poling, S.; Maragos, C. Observation of T-2 Toxin and HT-2 Toxin glucosides from fusarium sporotrichioides by liquid chromatography coupled to tandem mass spectrometry (LC-MS/MS). *Toxins* **2011**, *3*, 1554–1568. [CrossRef]
11. Sobral, M.; Faria, M.; Cunha, S.; Ferreira, I. Toxicological interactions between mycotoxins from ubiquitous fungi: Impact on hepatic and intestinal human epithelial cells. *Chemosphere* **2018**, *202*, 538–548. [CrossRef]
12. Ling, A.; Sun, L.; Guo, W.; Sun, S.; Yang, J.; Zhao, Z. Individual and combined cytotoxic effects of T-2 toxin and its four metabolites on porcine Leydig cells. *Food Chem. Toxicol.* **2020**, *139*, 111277. [CrossRef]

13. Goossens, J.; Pasmans, F.; Verbrugghe, E.; Vandenbroucke, V.; De Baere, S.; Meyer, E.; Haesebrouck, F.; De Backer, P.; Croubels, S. Porcine intestinal epithelial barrier disruption by the Fusariummycotoxins deoxynivalenol and T-2 toxin promotes transepithelial passage of doxycycline and paromomycin. *BMC Vet. Res.* **2012**, *8*, 245. [CrossRef]
14. Mayr, B.; Montminy, M. Transcriptional regulation by the phosphorylation-dependent factor CREB. *Nat. Rev. Mol. Cell Biol.* **2001**, *2*, 599–609. [CrossRef] [PubMed]
15. Shen, L.; Weber, C.; Raleigh, D.; Yu, D.; Turner, J. Tight junction pore and leak pathways: A dynamic duo. *Annu. Rev. Physiol.* **2011**, *73*, 283–309. [CrossRef] [PubMed]
16. Tajik, N.; Frech, M.; Schulz, O.; Schälter, F.; Lucas, S.; Azizov, V.; Dürholz, K.; Steffen, F.; Omata, Y.; Rings, A.; et al. Targeting zonulin and intestinal epithelial barrier function to prevent onset of arthritis. *Nat. Commun.* **2020**, *24*, 1995. [CrossRef] [PubMed]
17. Mariani, V.; Palermo, S.; Fiorentini, S.; Lanubile, A.; Giuffra, E. Gene expression study of two widely used pig intestinal epithelial cell lines: IPEC-J2 and IPI-2I. *Vet. Immunol. Immunopathol.* **2009**, *131*, 278–284. [CrossRef] [PubMed]
18. Campisi, L.; Barbet, G.; Ding, Y.; Esplugues, E.; Flavell, R.; Blander, J. Apoptosis in response to microbial infection induces autoreactive TH17 cells. *Nat. Immunol.* **2016**, *17*, 1084–1092. [CrossRef] [PubMed]
19. Zhai, Z.; Ni, X.; Jin, C.; Ren, W.; Li, J.; Deng, J.; Deng, B.; Yin, Y. Cecropin A modulates tight junction-related protein expression and enhances the barrier function of porcine intestinal epithelial cells by suppressing the MEK/ERK pathway. *Int. J. Mol. Sci.* **2018**, *19*, 1941. [CrossRef]
20. Wu, J.; Zhou, Y.; Yuan, Z.; Yi, J.; Chen, J.; Wang, N.; Tian, Y. Autophagy and apoptosis interact to modulate T-2 toxin-induced toxicity in liver cells. *Toxins* **2019**, *11*, 45. [CrossRef]
21. Lee, S. Intestinal permeability regulation by tight junction: Implication on inflammatory bowel diseases. *Intest. Res.* **2015**, *13*, 11–18. [CrossRef]
22. Groschwitz, K.; Hogan, S. Intestinal barrier function: Molecular regulation and disease pathogenesis. *J. Allergy Clin. Immunol.* **2009**, *124*, 3–20. [CrossRef] [PubMed]
23. Kang, R.; Li, S.; Perveen, A.; Shen, J.; Li, C. Effects of maternal T-2 toxin exposure on microorganisms and intestinal barrier function in young mice. *Ecotoxicol. Environ. Saf.* **2022**, *247*, 114252. [CrossRef] [PubMed]
24. Wang, X.; Wang, Y.; Qiu, M.; Sun, L.; Wang, X.; Li, C.; Xu, D.; Gooneratne, R. Cytotoxicity of T-2 and modified T-2 toxins: Induction of JAK/STAT pathway in RAW264.7 cells by hepatopancreas and muscle extracts of shrimp fed with T-2 toxin. *Toxicol. Res.* **2017**, *6*, 144–151. [CrossRef] [PubMed]
25. Chen, Y.; Cui, W.; Li, X.; Yang, H. Interaction between commensal bacteria, immune response and the intestinal barrier in inflammatory bowel disease. *Front. Immunol.* **2021**, *12*, 761981. [CrossRef] [PubMed]
26. Sun, Y.; Huang, K.; Long, M.; Yang, S.; Zhang, Y. An update on immunotoxicity and mechanisms of action of six environmental mycotoxins. *Food Chem. Toxicol.* **2022**, *163*, 112895. [CrossRef] [PubMed]
27. Jia, R.; Liu, W.; Zhao, L.; Cao, L.; Shen, Z. Low doses of individual and combined deoxynivalenol and zearalenone in naturally moldy diets impair intestinal functions via inducing inflammation and disrupting epithelial barrier in the intestine of piglets. *Toxicol. Lett.* **2020**, *333*, 159–169. [CrossRef] [PubMed]
28. Nathanail, A.; Varga, E.; Meng-Reiterer, J.; Bueschl, C.; Michlmayr, H.; Malachova, A.; Fruhmann, P.; Jestoi, M.; Peltonen, K.; Adam, G.; et al. Metabolism of the Fusarium mycotoxins T-2 toxin and HT-2 toxin in wheat. *J. Agric. Food Chem.* **2015**, *63*, 7862–7872. [CrossRef] [PubMed]
29. Reddy, K.; Kim, M.; Kim, K.; Ji, S.; Baek, Y.; Chun, J.; Jung, H.; Choe, C.; Lee, H.; Kim, M.; et al. Effect of commercially purified deoxynivalenol and zearalenone mycotoxins on microbial diversity of pig cecum contents. *Anim. Biosci.* **2021**, *34*, 243–255. [CrossRef]
30. Livak, K.; Schmittgen, D. Analysis of relative gene expression data using real-time quantitative PCR and the 2(-Delta Delta C(T)) method. *Methods* **2001**, *25*, 402–408. [CrossRef]

toxins

MDPI

Review

Mycotoxin Determination and Occurrence in Pseudo-Cereals Intended for Food and Feed: A Review

María Vanessa Vila-López, Noelia Pallarés *, Emilia Ferrer * and Josefa Tolosa

Preventive Medicine and Public Health, Food Science, Toxicology and Forensic Medicine Department, Faculty of Pharmacy, University of Valencia, Avda. Vicent Andrés Estellés, s/n, 46100 Burjassot, Spain; mavilo@alumni.uv.es (M.V.V.-L.); josefa.tolosa@uv.es (J.T.)

* Correspondence: noelia.pallares@uv.es (N.P.); emilia.ferrer@uv.es (E.F.)

Abstract: Nowadays, pseudo-cereals' consumption is increasing due to their health benefits as they possess an excellent nutrient profile. Whole pseudo-cereal grains are rich in a wide range of compounds, namely flavonoids, phenolic acids, fatty acids, and vitamins with known beneficial effects on human and animal health. Mycotoxins are common contaminants in cereals and by-products; however, the study of their natural occurrence in pseudo-cereals is currently scarce. Pseudo-cereals are similar to cereal grains; thus, mycotoxin contamination is expected to occur in pseudo-cereals. Indeed, mycotoxin-producing fungi have been reported in these matrices and, consequently, mycotoxin contents have been reported too, especially in buckwheat samples, where ochratoxin A and deoxynivalenol reached levels up to 1.79 μg/kg and 580 μg/kg, respectively. In comparison to cereal contamination, mycotoxin levels detected in pseudo-cereal samples are lower; however, more studies are necessary in order to describe the mycotoxin pattern in these samples and to establish maximum levels that ensure human and animal health protection. In this review, mycotoxin occurrence in pseudo-cereal samples as well as the main extraction methods and analytical techniques to determine them are described, showing that mycotoxins can be present in pseudo-cereal samples and that the most employed techniques for their determination are liquid and gas chromatography coupled to different detectors.

Keywords: mycotoxins; pseudo-cereals; risk assessment

Key Contribution: Mycotoxins have been found to occur in pseudo-cereal samples; however, their investigation in food and feed is still scarce, especially compared to cereal matrices, although their consumption is currently increasing exponentially. This review aims to describe the mycotoxin occurrence in pseudo-cereal samples and the main extraction methods and analytical techniques to determine them.

1. Introduction

1.1. Pseudo-Cereals for Food and Feed

Cereals are considered as a staple food in the diet, standing at the base of the food pyramid. Currently, their consumption is relatively high, and they are commonly included as feed ingredients. However, nowadays, consumers are more focused on healthy lifestyles and appropriate nutritional habits, and, on the other hand, more attention is paid to feed composition, especially in feedstuffs for pets. Thus, due to the excellent nutrient profile of pseudo-cereals (see below), their intake is currently increasing as an alternative to cereal consumption to meet the actual requirements of the target population for both human and animal consumption [1].

Botanically, pseudo-cereals are neither grasses nor true cereal grains, but they produce fruits or seeds, which are used and consumed as cereal grains. More concretely, pseudo-cereals are dicotyledonous grains and since they produce starch-rich seeds such as cereals, they have been called pseudo-cereals [1,2]. The most common pseudo-cereals used extensively worldwide are buckwheat (*Fagopyrum esculentum*; family *Polygonaceae*), amaranth (*Amaranthus hypochondriacus*, *Amaranthus caudatus* and *Amaranthus cruentus*; family *Amaranthceae*), and quinoa (*Chenopodium quinoa* sub sp. quinoa; family *Chenopodiaceae*) [3]. These species are originally from the Andean region, but they can be cultivated worldwide as they are able to grow in adverse environmental conditions and they do not have special agronomic requirements and can be grown by simple methods [2–5]. Moreover, they offer food products with innovative nutritional qualities to meet global nutritional demands, thereby reaching various goals of the United Nations' (UN) Agenda 2030 (https://www.un.org/sustainabledevelopment/sustainable-development-goals/ accessed on 31 January 2023), including the eradication of hunger, achieving food security, enhancing nutrition, and supporting sustainable agriculture [6,7].

Pseudo-cereals present an exceptional nutritional, phytochemical, and phenolic profile with good quality proteins. In fact, the amino acid profile and nutritional properties of pseudo-cereals in terms of essential amino acid index, biological value, protein efficiency ratio, and nutritional index are higher as compared to conventional cereals such as wheat, rice, and maize. In addition, pseudo-cereal grains show a high protein value, with an important content of lysine, an essential amino acid that is commonly the limiting amino acid in other cereals [8]. Moreover, the contents of lipids and minerals, such as calcium, magnesium, iron, potassium, and zinc, are higher than those of traditional cereals. For this reason, the interest in and consumption of pseudo-cereals as an alternative to cereals are increasing due to their nutritional properties, promoted by their beneficial effects on human health including the prevention and reduction of many degenerative diseases [1,6].

It must be highlighted that, due to the lack of gluten, these grains can be incorporated into the diet of people suffering from celiac disease, and due to their high content of starch, pseudo-cereals can be used like cereals in the preparation of value-added food products for celiac patients [1,3]. The availability of tasty gluten-free pseudo-cereals would also represent a step forward in ensuring that people with celiac disease are acquiring enough nutrients. In addition, pseudo-cereals have the potential to be included in a variety of processed functional foods since they contain bioactive compounds. Therefore, the commercialization of pseudo-cereals as functional foods would benefit people with a variety of lifestyle disorders in addition to those suffering from celiac disease.

From the processing point of view, it has been described that both buckwheat and quinoa must undergo dehulling or polishing to get rid of the saponins that give them the bitter flavor and then washing and drying prior to consumption [1,9].

Regarding animal nutrition, the inclusion of pseudo-cereal grains in feedstuffs has been increasing in the last few years, mainly due to the health benefits for different animal species. The most employed pseudo-cereal as a feedstuff ingredient is quinoa, followed by amaranth, which is being increasingly used in animal nutrition [10].

Quinoa has been used by the natives of South America since ancient times to feed ruminants and non-ruminant animals, such as cattle, sheep, horses, and pigs [11]. Although there is little research, quinoa supplementation is thought to have a high nutritional value as feed intended for ruminants [12]. In this sense, quinoa has been included as an ingredient in feed for ruminants and it has been demonstrated that it can replace clover hay in the diet by up to 45%, according to the research recently published by Ebeid et al. [13].

According to certain studies, quinoa can be ensiled to become a valuable fodder crop for dairy farms, resulting in good milk yield and greater dietary protein. Moreover, it has been reported that quinoa hay can be utilized in beef cattle feed when combined with other roughages and some chemicals. Thus, this plant may be crucial in ensuring that animal feed contains sufficient levels of macro- and microminerals. However, there are few studies on the use of quinoa in animal nutrition, and more thorough research on its application in animal feeding is required [5,14]. On the other hand, in non-ruminant nutrition, quinoa has been used especially in poultry feeding due to its content of protein and energy, thus supposing an alternative to corn [15].

Amaranth can be utilized as a grain and as fresh, dried, or ensiled forage for different animal species, mainly cattle, chickens, pigs, and rabbits. However, compared to its grain, amaranth forage has drawn a lot less study attention. Amaranth grain is classed as a growth-inhibiting grain due to the presence of heat-labile substances such as tannins, saponins, lectins, and trypsin inhibitors. Thus, due to the presence of certain anti-nutritional elements, some species of amaranth forage may not be suitable as ruminant feed or may need to undergo special processing before they can be used by ruminants. Additionally, the acceptance and use of amaranth grain by poultry and other monogastric animals are restricted by its known antinutritional properties [16]. In diets for monogastric animals, amaranth has shown good potential as an alternative ingredient, as it has been demonstrated that amaranth can improve the performance and health status of monogastric animals because of its high nutritional value and availability of phenolic compounds without having any adverse effect on animals' productivity. Therefore, it can be said that amaranth leaves and grains can be successfully utilized in monogastric animals with the use of various processing techniques that may need to be applied in order to eliminate anti-nutritional components before use in animals [17].

In the field of companion animals, the use of pseudocereals, especially quinoa, as sources of carbohydrates is increasing due to the concern of owners for the health of their pets, comparing their dietary habits with humans on many occasions. In fact, most of the new feeds that are produced and marketed today are grain free (GF) [18,19].

Despite all this, there is hardly any research on the effect that the use of these ingredients as sources of carbohydrates has on the health of dogs and cats [20], and the few studies that have been published are limited to investigating digestibility and behavior in extrusion for the preparation of kibbles.

Regarding the possible benefits that have been observed when including pseudocereals in feed composition, it has been observed that the incorporation of pseudo-cereals such as quinoa and amaranth in amounts equal to or greater than 40% can be accepted by dogs without altering or negatively affecting their digestibility [20,21]. In fact, it seems that, in these cases, beneficial modifications can be produced in the fecal microbiota of dogs fed with pseudo-cereals such as the above-mentioned ones. In addition, with some pseudo-cereals, there have been increases in butyrate-producing bacteria at the intestinal level and a decrease in bacteria from the *Fusobacteriaceae* family when compared to feeds made with rice [20]. Regarding the behavior in extrusion, if feeds made with GF grains are compared with those made with pseudo-cereals, it seems that there are significant changes, as well as in the palatability of the feed before adding the palatizer [21].

1.2. Major Mycotoxins in Pseudo-Cereals

Like cereals and oilseeds, pseudo-cereal grains are susceptible to fungal growth and mycotoxin contamination; however, these matrices have received little attention in the literature, especially compared to cereal grains [22], although some studies have demonstrated the actual risk of pseudo-cereal contamination by mycotoxins, as indicated in Section 2.

Mycotoxins are toxic metabolites produced by different fungal species or strains within the same fungal species during secondary fungal metabolism. The main hazardous mycotoxins reported are aflatoxins (AFs) (mainly produced by *Aspergillus*), fumonisins (FBs), zearalenone (ZEA), and trichothecenes (TCs) (produced by *Fusarium*), and ochratoxins (OTs) and patulin (PAT) (mainly produced by *Penicillium*) [23,24].

Mycotoxins have been associated with a broad range of toxic effects on both humans and animals, depending on their different chemical structures, including acute toxicity, hepatotoxicity, nephrotoxicity, genotoxicity, carcinogenicity, immunotoxicity, mutagenicity, teratogenicity, and reproductive toxicity [25,26]. Regarding their carcinogenic potential, their human carcinogenicity risk has been assessed by the International Agency for Research on Cancer (IARC), classifying AFs as carcinogenic substances in category "1" while FB1 and FB2 are classified in group "2B" as possible carcinogenic substances to humans [23,27].

Due to the above-mentioned adverse effects, some mycotoxins have been recognized by European regulatory authorities as emerging risks in food safety regarding animal and human health, thus, efficient regulatory frameworks and monitoring approaches have been developed globally relying on the latest scientific knowledge based mainly on new analytical tools and approaches. In this sense, maximum levels (MLs) have been established for the main classes of mycotoxins in several core commodities intended for food and feed by the European Union (EU), including cereals, nuts, fruits, and derived products, such as milk. Those MLs are compiled in EU Commission Regulation (EC) No 1881/2006 [28] and its amendments. However, different drawbacks exist nowadays, as the current MLs do not consider the exposure to multiple mycotoxins, and they are either based on the risk assessment of a single compound or their sum and no MLs have been established for pseudo-cereal grains [29–31].

Due to the scarce information on mycotoxin occurrence in these matrices, this manuscript aims to review the existing literature on the natural occurrence of mycotoxins in pseudo-cereals intended for both food and feed and the most common extraction and analytical methods employed for their determination.

2. Occurrence and Co-Occurrence of Mycotoxins in Pseudo-Cereal Samples for Food and Feed

Although it has been reported that pseudo-cereal grains can be as susceptible to fungal growth and mycotoxin contamination as cereal grains, there is still scarce information on the contaminating fungi and mycotoxin occurrence in these matrices. Some drawbacks can be found when searching for information on mycotoxin occurrence in pseudo-cereal grains. On the one hand, the few available studies focused primarily on the investigation of the mycoflora present in pseudo-cereal grains and not on mycotoxin contamination. On the other hand, another problem when searching mycotoxin data on pseudo-cereal grains in the literature is that this information is usually mixed together with cereal data; thus, it is difficult to search these studies by the title of the manuscript or by including "pseudo-cereal" as the keyword as they are mainly referred to as cereals [32].

As previously stated, some reported surveys have been focused on the mycoflora present in pseudo-cereal grains. In this sense, some studies have reported the presence of mycotoxin-producing fungi in pseudo-cereal samples. Thus, the presence of *Ascohyta, Altenaria, Phoma, Fusarium, Bipolaris, Cladosporium,* and *Pyronochaeta* genera in quinoa seeds (*Chenopodium quinoa)* from Bolivia, Brazil, Czech Republic, and Peru have been reported, but no data on mycotoxin occurrence were provided in these studies [31]. Some of these fungal genera were also found by Krysińska-Traczyk et al. (2007) [33] in buckwheat grain and buckwheat grain dust, where *Penicillium* spp., *Mucor mucedo, Alternaria alternata,* and *Cladosporium lignicola* were the predominant genera in buckwheat grain, while *Rhodotorula rubra, Mucor mucedo, Alternaria alternata,* and *Penicillium* spp. were more abundant in buckwheat grain dust. Regarding the *Fusarium* strains, *F. culmorum* was recovered from samples of grain dust but it was not detected in grain samples.

In the survey conducted by Pappier et al. (2008) [34], the authors reported *Penicillium* and *Aspergillus* as the most frequently encountered fungal genera in quinoa samples from Argentina. Interestingly, these authors reported a reduction in *Aspergillus* incidence when processing the grains for the removal of saponins; however, the incidence of *Penicillium, Eurotium, Mucor,* and *Rhizopus* increased after the treatment. These results agree with those reported by Bresler et al. (1995) [35] who found that the most prevalent mycoflora in amaranth grains from Argentina corresponded to the mycotoxin-producing fungal species *Aspergillus flavus* (*A. flavus*), *A. parasiticus, Penicillium chrysogenum,* and *Fusarium equiseti* (*F. equiseti*). *Penicillium* spp., *Fusarium* spp., and *Aspergillus* spp. were also the predominant mycoflora reported by Ramos-Díaz et al. (2021) in pseudo-cereal samples [31].

According to the study carried out by Vásquez-Ocmín et al. (2022) [36] in pseudo-cereal samples from Peru, there were several mycotoxigenic fungi detected, namely

Aspergillus spp., *Penicillium* spp., and *Alternaria* spp.; however, *Fusarium* spp. was not detected. On the contrary, *Fusarium* spp. was detected in 92% of the amaranth grain from Peru analyzed by Ducos et al. (2021) [37], while in the study reported by Sacco et al. (2020) [24], *Aspergillus* spp. was detected in amaranth, buckwheat, and quinoa samples and *Penicillium* spp. was detected in buckwheat and quinoa samples.

Regarding the most common mycotoxigenic fungal species, *Penicillium* spp. and *Alternaria* spp. were the most frequently detected in buckwheat samples. In quinoa samples, *Penicillium* spp. and *Aspergillus* spp. were the most reported fungal species while in amaranth grains, *Aspergillus* spp., *Penicillium* spp., and *Fusarium* spp. were detected, highlighting the contamination by *Fusarium* species.

Although a very limited number of studies have measured mycotoxins in pseudo-cereal grains, especially compared to cereals, in Table 1 it can be observed that mycotoxin occurrence in these matrices has been reported in different surveys. Original research papers as well as reviews have been included in this section, regardless of the year of publication, as all searches were performed without any time restriction.

Table 1. Mycotoxin occurrence in different pseudo-cereal samples.

Pseudocereal Sample	Detected Mycotoxins	Incidence (%) *	Mycotoxin Contents Media ± SD ** (Range) μg/kg	Reference
Amaranth	ZEA AFs OTA STG	100 nd nd nd	(420.0–1980.0) nd nd nd	[38]
Amaranth, quinoa, and buckwheat	TCs ZEA, α-ZOL, and β-ZOL	nd nd	nd nd	[39]
Buckwheat flour and dried buckwheat noodles, among others	AFs OTA FBs	nd 60 nd	nd 0.51 (0.16–1.79) nd	[40]
Buckwheat grain Buckwheat grain dust	DON OTA DON OTA	33.3 50 100 100	(74.0–87.0) (0.38–1.14) (10.0–283.0) (1.03–2.42)	[33]
Buckwheat flour and dried buckwheat noodles, among others	AFs OTA	nd 80	nd 0.51 (0.16–1.79)	[41]
Quinoa	AFs CIT CPA OTA	nd nd nd nd	nd nd nd nd	[34]
Buckwheat	STG	17 20	(0.5–25) (25–200)	[42]
Quinoa and Amaranth	AFs, OTA, FBs, NIV, DON, FUS-X, T-2/HT-2, CIT, STG, and ZEA	nd	nd	[22]
Buckwheat and buckwheat products	OTA DON ZEA AFs FBs T-2/HT-2	4 17 8 nd nd nd	2.1 98.0 ± 40 (max 141.0) 13.0 ± 7 (max 18.0) nd nd nd	[32]

Table 1. *Cont.*

Pseudocereal Sample	Detected Mycotoxins	Incidence (%) *	Mycotoxin Contents Media ± SD ** (Range) μg/kg	Reference
Buckwheat grain	AFB1	100	(2.18–8.39)	[43]
	DON	100	(300.0–580.0)	
	OTA	100	traces	
	T-2	100	(24.0–38.0)	
Buckwheat groats	AFB1	100	(2.3 ± 0.6)	[44]
	OTA	100	(0.7 ± 0.2)	
	T-2	100	(4.1 ± 0.5)	
Tartary buckwheat seeds	AFB1	7	5.62	[45]
Buckwheat flour	OTA	38.5	0.40 ± 0.34 (0.15–0.96)	[46]
	AFs	nd	nd	
Quinoa Amaranth (grain) Buckwheat	AFs	nr	4.4 ± 0.8	[24]
	FB1 + FB2	nr	368.5 ± 2.5	
	AFs	nr	1.60 ± 0.70	
	FB1 + FB2	nr	111.00 ± 0.0	
	AFs	nr	4.68 ± 0.44	
	FB1 + FB2	nr	567.25 ± 138.14	
Quinoa	AFB1	nd	-	[47]
	BEA	nd	-	
	DON	nd	-	
	ENNA	***id	nr	
	ENNA1	***id	nr	
	ENNB	***id	nr	
	ENNB1	***id	nr	
	Mycophenolic acid	nd	-	
	Tentoxin	nd	-	
	Tenuazonic acid	***id	nr	
Quinoa	AFB1	nd	nd	[36]
	FB1	32	traces	
	OTA	nd	nd	
	PAT	nd	nd	
	BEA	82	7.9 ± 0.85	

AFB1: aflatoxin B1; AFs: aflatoxins; BEA: beauvericin; CIT: citrinin; CPA: cyclopiazonic acid; DON: deoxynivalenol; ENNA: enniatin A; ENNA1: enniatin A1; ENNB: enniatin B; ENNB1: enniatin B1; FB1: fumonisin B1; FB2: fumonisin B2; FBs: fumonisins; OTA: ochratoxin A; PAT: patulin; STG: sterigmatocystin; TCs: trichothecenes; T-2: T-2 toxin; HT-2: HT-2 toxin; ZEA: zearalenone; α-ZOL and β-ZOL: α—and β—zearalenol; nd: not detected; nr: not reported; * Incidence (%): positive samples from the total number of analyzed samples expressed as a percentage; ** SD: Standard deviation; ***id: compounds identified but no quantification data provided.

The most reported mycotoxins in pseudo-cereal samples are those produced by *Fusarium* spp., such as TCs, FBs, ZEA, and the so-called emerging *Fusarium* mycotoxins, such as enniatins (ENNs) and beauvericin (BEA).

Among TCs, DON and T-2 toxins have been widely reported to occur mainly in buckwheat samples, showing higher incidences and contents for DON in this matrix [32,33,43].

FBs were reported at high levels in pseudo-cereal samples from Italy [24], with the highest contents (up to 567.25 μg/kg) corresponding to buckwheat samples.

In the survey conducted by Bresler et al. (1991) [38], ZEA was determined among the analyzed mycotoxins at worrying levels of up to 1980 μg/kg and 420 μg/kg in two samples of amaranth grains (*Amaranthus cruentus*). Later, these authors showed that amaranth grains possess certain susceptibility to ZEA contamination by incubation of amaranth seeds with *F.equiseti* [48]. ZEA was also reported to occur in buckwheat products as indicated by Kirinčič et al., 2015 [32]. However, neither ZEA nor TCs were detected in amaranth, quinoa, and buckwheat samples from Germany according to Schollenberger et al., 2005 [39], although several TCs were included in the survey, namely diacetoxyscirpenol (DAS), 15-monoacetoxyscirpenol (MAS), scirpentriol (SCIRP), T-2, HT-2 toxin (HT-2), T-2 triol, T-2 tetraol, neosolaniol (NEO), DON, 3- and 15-acetyl-DON (3-, 15-acDON), nivalenol (NIV), and fusarenon-X (FUS-X). Regarding cereal contamination, wheat has been reported to contain high DON levels, with mean concentrations up to 1025.4 μg/kg, and NIV with mean levels up to 75.2 μg/kg [30].

Other *Fusarium* mycotoxins such as ENNs and BEA have been widely reported in cereal samples even at high levels [49–51]. In the study reported by Uhlig et al. (2006) [49] ENNs and BEA were reported to occur in cereal samples showing high incidences and levels. ENNB showed the highest prevalence (100%) and the highest maximum concentration (5800 μg/kg) in wheat samples. Fortunately, lower contents have been reported in pseudo-cereals. In pseudo-cereals, ENNs have been detected in quinoa samples according to the study reported by Besaire et al. (2021) [47]. Although BEA was not detected in that survey, levels up to 7.9 μg/kg were detected according to Vásquez-Ocmín et al. (2022) [36]. Regarding food safety, these authors pointed out that the natural occurrence of BEA in quinoa grains originating from the Cajamarca region represents a point of concern, due to the promotion of quinoa cultivation in this region by the local authorities [36].

Due to their adverse toxic effects, AFs have been surveyed in several studies. Although a major number of the studies did not find AFs levels in pseudo-cereal samples [32,34,36,38,40,41,46,47], some studies revealed AFs' occurrence in these matrices, and some studies reported the absence of AFs in buckwheat samples but detected AFs contents in cereal samples, such as rye (0.01 μg/kg) and rice (1.06 μg/kg). Approximately 100% of buckwheat samples were contaminated with AFB1 as reported by Keriene et al. (2016) [43,44]. It must be highlighted that raw buckwheat hulls (without steamed treatment) showed higher AFB1 contents (75.8 μg/kg) than the grain [44], showing a protective effect of the hull over the inner part as demonstrated by other researchers in nut samples [52]. Interestingly, these authors reported that the samples with higher concentrations of AFB1 had significantly ($p < 0.05$) lower concentrations of DON, thus suggesting possible interaction between fungal species producing AFs and TCs [44].

AFB1 was also detected in the Tartary buckwheat seeds in the study reported by Ren et al. (2018) [45] and in buckwheat samples as reported by Sacco et al. (2020) [24]. Similar AFs levels have been reported in quinoa samples in surveys reported by Sacco et al. (2020) [24] and by Ramos-Díaz et al. (2021) [31], while lower contents were found in amaranth seeds in those studies [10,18]. These authors reported varying mycotoxin levels depending on the type of crop, geographical location, and agricultural practices used. Differences between regions have been observed by these authors regarding mycotoxin occurrence in pseudo-cereals and cereals. In general, pseudo-cereal grains from North Europe showed higher mycotoxin contamination than those from South America, while the opposite occurred with cereal grains [31].

Regarding OTA occurrence in pseudo-cereal grain samples, only two studies did not find this compound in quinoa and amaranth samples [36,38], while other surveys found OTA as a common mycotoxin contaminating pseudo-cereal samples. In the study conducted by Sugita-Konishi et al. (2006) [40], OTA was found in buckwheat flour and dried buckwheat noodles with contents up to 1.79 μg/kg, and similar contents were detected in this study in rye (2.59 μg/kg), while lower contents were detected in wheat flour (0.48 μg/kg) and oatmeal (0.18). These results are in accordance with those reported by Krysińska-Traczyk et al. (2007) [33], Kumagai et al. (2008) [41], and Kirinčič et al. (2015) [32], as all of them reported similar OTA contents in buckwheat samples of different origin. However, lower OTA levels were found by other researchers also in buckwheat from Lithuania and Serbia [43,44,46]. It must be highlighted that these authors also found lower levels of OTA in cereal samples, such as wheat (up to 0.40 μg/kg), barley (up to 0.11 μg/kg), and oat (up to 0.09 μg/kg); however, higher contents were detected in rye samples (up to 23 μg/kg) [46].

As in cereal samples, differences have been reported regarding mycotoxin levels in pseudo-cereal grains from organic agriculture. Thus, higher mycotoxin contents were reported by Sacco et al., 2020 [24] and Keriene et al., 2016 [43] in organic buckwheat samples.

In the study carried out by Herrera et al. (2019) [53], gluten-free cereal samples for infants under six months were analyzed, some of them including quinoa and buckwheat. The results showed that ten samples were positive for at least one mycotoxin (four for AFs and six for DON). This study found that AFB1 and AFG1 co-occurred in five samples, and one sample based on rice and quinoa contained AFB2, AFG1, and AFG2 but not AFB1. Moreover, 8 samples out of 14 samples of multi-cereals for infants above 6 months were positive for AFs, while 6 samples out of these 14 were positive for DON contamination.

3. Mycotoxin Determination

Many techniques have been approved and used for the analysis of mycotoxins in food and feed since the first mycotoxins were discovered. Despite the significant advancements made in this area, there are still several obstacles and shortcomings with these analytical techniques that need to be resolved [54]. The enormous variety of mycotoxin chemical structures, the co-occurrence of mycotoxins, problems in detecting low-level mycotoxin contamination, complex food matrices where the mycotoxin contamination occurs, and difficult extraction techniques are just a few of the analytical issues [55]. Continuous improvements in the analytical methodology for mycotoxin analysis in a variety of food matrices are required to address these issues in order to support the enforcement of mycotoxin regulations, safeguard consumer health, promote the agriculture sector, and facilitate global food trade [54].

Both the extraction methods and analytical equipment used for mycotoxin determination in pseudo-cereal samples are the same or very similar to those employed for cereal matrices, as they show a similar composition. In the next sections, the most employed techniques to extract and determine different mycotoxins in pseudo-cereal samples are described.

3.1. Extraction Methods

Given that pseudo-cereals are extremely complex food matrices, extraction to isolate the target chemicals and purification or clean-up to remove impurities must always be carried out before analysis. Moreover, due to the imposed legal constraints, mycotoxins are typically found naturally in extremely low concentrations in food. In order to facilitate the subsequent detection of the analyte, it would be necessary to perform a concentration process in addition to the other steps [56,57].

The most common methodology used for mycotoxin extraction is solid–liquid extraction (SLE) followed by a purification or clean-up step when necessary, with the most widely used technique being solid-phase extraction (SPE), mainly with C18, or immunoaffinity columns (IACs), which contain specific antibodies to the analyte of interest [57,58]. However, IACs are costly and intricate purification systems, their utility in multiclass analysis is constrained by their high selectivity, and they also suffer from low recoveries for some mycotoxins. Therefore, multiclass extraction methods that are easier to use, more effective, and environmentally beneficial are needed. Among the different strategies, the so-called QuEChERS (Quick, Easy, Cheap, Effective, Rugged, and Safe), which is a type of dispersive SPE (dSPE) used for sample preparation, is commonly used for the extraction of a wide range of mycotoxins in different food matrices (Table 2) [22]. It consists of two steps: (i) extraction based on partitioning by salting-out, which involves the equilibrium between an aqueous and organic layer, and (ii) dSPE extraction for additional clean-up utilizing a mix of $MgSO_4$ and other sorbents, such as C18 or primary and secondary amine (PSA) [22,59].

Table 2. Main extraction and analytical methods for mycotoxin determination in pseudo-cereal samples.

Samples	Mycotoxins	Extraction Method	Analytical Method	Reference
Amaranth	ZEA, AFs, OTA, STG	SLE and clean-up	TLC	[38]
Amaranth, quinoa, and buckwheat	TCs ZEA, α-ZOL, and β-ZOL	SLE and clean-up by SPE SLE and IAC	GC-MS HPLC-FLD	[39]
Buckwheat flour and dried buckwheat noodles, corn, rice, peanuts, peanut butter, popcorn, cornflakes, and sesame oil	AFs OTA FBs	SLE and IAC SLE and IAC SLE and IAC	HPLC-FLD HPLC-MS	[40]
Buckwheat grain and buckwheat grain dust	DON NIV OTA	SLE and SPE clean-up SLE and IAC	GC-MS GC-MS HPLC	[33]
Corn, processed corn, buckwheat, dry buckwheat noodles, peanuts, rice, and sesame oil	AFs OTA	SLE and IAC	HPLC	[41]
Wheat, buckwheat, barley, oats, and rye	STG	SLE and SPE clean-up	LC–MS/MS	[42]
Buckwheat, quinoa, spelt, amaranth, and white rice	AFs, FBs, OTA, T-2, HT-2, STG, CIT, ZEA, NIV, DON, and FUS-X	QuEChERS	UHPLC-MS/MS	[22]
Buckwheat and buckwheat products, wheat, maize, oat, rice, rye, barley, millet, triticale, and others	OTA DON ZEA AFs FBs T-2/HT-2	SLE and IAC	HPLC-FLD HPLC-DAD HPLC-FLD LC-MS/MS LC-MS/MS GC-MS	[32]
Buckwheat groats	AFB1 OTA T-2	-	ELISA	[44]
Tartary buckwheat seeds	AFs, FBs, OTA, ZEA, DON, and T-2/HT-2	SLE	UFLC-QTrap-MS/MS	[45]
Wheat, buckwheat, rye, oat, barley, rice, millet, and corn flour	OTA AFs	SLE and IAC	HPLC-FLD HPLC-FLD	[46]
Quinoa, amaranth (grain), and buckwheat	Afs FB1 + FB2	-	ELISA	[24]
Pea protein, soy protein, red quinoa, and wheat flour	DON, AFs, ZEA, DON, FBs, ENNA, HT-2, OTA, PAT, and T-2	QuEChERS	UHPLC-MS/MS	[47]
Corn, wheat, amaranth, rice, barley, and oats	FBs, DON, and ZEA	SLE	HPLC-MS/MS	[37]
Quinoa, kăniwa, barley, oat, and wheat	101 mycotoxins	SLE	LC-MS/MS	[31]
Quinoa	AFB1, FB1, OTA, PAT, and BEA	SLE	UHPLC−HRMS	[36]

Table 2. *Cont.*

Samples	Mycotoxins	Extraction Method	Analytical Method	Reference
Amaranth	DON and ZEA	In-phase liquid–liquid microextraction based on the solidification of a floating organic drop followed by double solvent-assisted back-extraction (DLLME-SFO-SBE)	UHPLC-MS/MS	[8]

AFs: aflatoxins; CIT: citrinin; DON: deoxynivalenol; ELISA: enzyme-linked immuno sorbent assay; ENNA: enniatin A; FLD: fluorescence detector; FBs: fumonisins; GC-MS: gas chromatography/mass spectrometry; HPLC: high-performance liquid chromatography; HPLC-DAD: high-performance liquid chromatography-diode array detector; HPLC-FLD: high-performance liquid chromatography-fluorescence detection; HPLC-MS: high-performance liquid chromatography-mass spectrometry; HPTLC: high-performance thin-layer chromatography; HPLC-UV: high-performance liquid chromatography-ultraviolet detection; HT-2: HT-2 toxin; IAC: immuno affinity columns; LC-MS: liquid chromatography-mass spectrometry; MS/MS: tandem mass spectrometry; NIV: nivalenol; OTA: ochratoxin A; PAT: patulin; QuEChERS: Quick, Easy, Cheap, Effective, Rugged, and Safe; S-L extraction: solid–liquid extraction; STG: sterigmatocystin; T-2: T-2 toxin; UHPLC−HRMS: ultra-high-performance liquid chromatography high-resolution mass spectrometry; UFLC-QTrap-MS/MS: ultra-fast liquid chromatography coupled with triple quadrupole mass spectrometry; ZEA: zearalenone.

On the other hand, in a conventional SLE, a compound or a group of compounds that are a part of a solid are dissolved in a solvent with the appropriate polarity. Most of the mycotoxins in this situation can be effectively extracted using comparatively polar solvents such as acetonitrile (MeCN), acetone, or methanol (MeOH) [60]. Additionally, it is practical to use small quantities of acidified water to moisten the solid to improve the efficiency of the extraction. By doing this, the latter's contact surface with the solvent can be increased, and the pH is made more acidic, which enhances the mycotoxins' solubility [61]. Since this kind of extraction is not very specific, it is frequently followed by purification or clean-up to get rid of any interfering substances that can affect the analysis's outcomes, such as lipids, proteins, and coloring agents. Chelating chemicals (C18 and PSA) can be used to eliminate them in significant quantities [58,61].

3.2. Analytical Determination

Among the different analytical tools for mycotoxin determination, the most commonly used include thin-layer chromatography (TLC), liquid chromatography (LC), high-performance liquid chromatography (HPLC), and gas chromatography (GC) (Table 2) [62]. For quantification purposes, LC and GC can be coupled to different detectors, such as ultraviolet (UV), fluorescence (FL), or mass spectrometry (MS) [22]. Initially, UV and FL were widely used for mycotoxin determination [32,39,40,46]. These detection techniques are based on the UV absorbance and fluorescence characteristics of mycotoxins. Generally, the sensitivity of fluorescence is 10–1000 times higher than the sensitivity of the UV detector, so FL is used conventionally in the determination of specific fluorescent compounds present in cereal and pseudo-cereal samples. FL detection has been widely used in the analysis of AFs, OTA, and ZEA in cereals and pseudo-cereals matrices with good accuracy and high precision [63]. For instance, Torović (2018) [46] analyzed the presence of AFs and OTA in samples of wheat, buckwheat, rye, oat, barley, rice, millet, and corn flour employing HPLC with FL detection. The analytical method developed showed good performance, with recoveries ranging from 53.1% to 85.0% for AFB1 and from 70.3% to 88.3% for OTA and limits of detection (LODs) and quantification (LOQ) of 0.04/0.1 μg/Kg for AFB1 and 0.07/0.2 μg/Kg for OTA, respectively.

However, nowadays, LC or HPLC coupled to tandem mass spectrometry detectors (LC–MS/MS) is the most widely used techniques as it allows one to obtain an accurate and reliable determination of several mycotoxins at even low concentrations in complex matrices. Moreover, it allows the simultaneous determination of mycotoxins belonging to different chemical families [64]. In this sense, Ramos-Diaz et al. [31] developed a multi-analyte LC–MS/MS method to detect mycotoxins and fungal metabolites in pseudo-cereal grains (quinoa and kăniwa). The study documented the detection of 101 analytes at varying levels with recoveries ranging from 70 to 120% (Table 2). Moreover, Veršilovskis et al. (2008) [42] previously developed a sensitive LC-MS/MS method for the analysis of STC in buckwheat and other cereal grains. This method included sample extraction with acetonitrile/water solution followed by SPE.

Another tool for mycotoxin determination commonly employed is ELISA, which consists of an immunological assay for mycotoxin screening [24,44]. This assay provides fast and economical measurements and does not require previous clean-up procedures. ELISA formats (direct, indirect, competitive, and sandwich) are recognized as adequate for mycotoxin screening. To favor field analysis, a transduction system with appropriate molecular recognition elements has been integrated. In this sense, different analytical methods have been developed for AFs, OTA, FB1, and TCs' detection in cereals and pseudo-cereals (Table 2) [63]. However, the number of matrices tested is limited, and at low concentrations, method precision can be reduced. In addition, the presence of matrix interferents and structurally related mycotoxins can alter antibody binding and, subsequently, mycotoxin quantification. For instance, in buckwheat groats, Keriene et al. [44] reported limits of detection (LODs) for ELISA determination of 3.5 μg/Kg for T-2, 2.5 μg/Kg for OTA, and 1 μg/Kg for AFB1. According to Sacco et al. (2020) [24], who used an ELISA for AFs and FBs determination, in this case, the ELISA detection range was 2–200 μg/Kg for AFB1, 1.75–140 μg/Kg for total AFs, and 222–6000 μg/Kg for FBs.

Although ELISA and UV and FL detectors have been widely employed, nowadays, the most employed technique is ultra-high-performance liquid chromatography coupled to mass spectrometry, commonly used for quantitative multi-mycotoxin analysis. In general, a low-resolution mass spectrometer (MS) is used for analysis or simply for compound identification purposes; however, it may not be enough to distinguish between two molecules with the same molecular mass. In this sense, HRMS constitutes a more sensitive and accurate detector able to distinguish between two substances with similar masses with a high level of certainty and confidence.

In the studies carried out by Bessaire et al. (2021) [47] and Arroyo-Manzanares et al. (2014) [22], HRMS and MS/MS techniques, respectively, were combined with QuEChERS extraction, while in other surveys, they have been combined with other novel extraction methods. Thus, Bochetto et al. (2021) [8] used an extraction/preconcentration procedure that consisted of in-phase liquid–liquid microextraction based on the solidification of a floating organic drop followed by double solvent-assisted back-extraction (DLLME-SFO-SBE) followed by UHPLC-MS/MS determination, while Vásquez-Ocmín et al. (2022) [36] used biphasic microextraction. These extraction techniques allowed one to extract even small mycotoxin amounts which can be detected and quantified due to the high specificity and sensitivity of the MS/MS and HRMS detectors. Both extraction methods allow one to reduce the solvent amount used during the extraction step; thus, the future trend in mycotoxin extraction is to develop green alternatives to conventional methods.

4. Conclusions

Mycotoxins are common chemical contaminants present mainly in cereal products and their by-products. Nowadays, in both human and animal nutrition, nutritional habits are changing towards healthier ingredient consumption; thus, in some cases, cereals are being replaced by pseudo-cereal grains, such as quinoa, amaranth, and buckwheat, due to their multifarious health benefits. These beneficial effects are due to the presence in their composition of flavonoids, phenolic acids, fatty acids, vitamins, dietary fibers, minerals, and other bioactive compounds that are present in pseudo-cereals. As cereals and pseudo-cereals are similar, it is expected for mycotoxin contamination to be found in pseudo-cereal samples; however, the investigation of their natural occurrence in these matrices is still scarce. Despite this, some studies have reported the occurrence of some mycotoxins, such as AFs, OTA, FBs, TCs, and ZEA, among others, in quinoa, amaranth, and buckwheat samples. The extraction methods and analytical tools employed for mycotoxin determination in pseudo-cereal samples are the same employed for cereal analysis, with SLE followed by SPE or clean-up being the most employed extraction techniques and LC or GC coupled to different detectors, mainly UV, FL, and MS/MS, being the most employed analytical equipment.

Author Contributions: Conceptualization, J.T.; resources, J.T. and E.F.; data curation, J.T. and E.F.; writing—original draft preparation, M.V.V.-L., J.T. and N.P.; writing—review and editing, J.T. and E.F.; visualization, J.T. and E.F.; supervision, J.T. and E.F.; funding acquisition, J.T. and E.F. All authors have read and agreed to the published version of the manuscript.

Funding: This research was funded by the Ministerio de Ciencia e Innovación of Spain PID2020-115871RB-I00 and by the Conselleria d'Innovació, Universitats, Ciència i Societat Digital (AEST/2021/077).

Institutional Review Board Statement: Not applicable.

Informed Consent Statement: Not applicable.

Data Availability Statement: The data presented in this study are available in this article.

Conflicts of Interest: The authors declare no conflict of interest.

References

1. Mir, N.A.; Riar, C.S.; Singh, S. Nutritional constituents of pseudo cereals and their potential use in food systems: A review. *Trends Food Sci. Technol.* **2018**, *75*, 170–180. [CrossRef]
2. Paucar-Menacho, L.M.; Simpalo-López, W.D.; Castillo-Martínez, W.E.; Esquivel-Paredes, L.J.; Martínez-Villaluenga, C. Reformulating Bread Using Sprouted Pseudo-cereal Grains to Enhance Its Nutritional Value and Sensorial Attributes. *Foods* **2022**, *11*, 1541. [CrossRef]
3. Thakur, P.; Kumar, K.; Dhaliwal, H.S. Nutritional facts, bio-active components and processing aspects of pseudocereals: A comprehensive review. *Food Biosci.* **2021**, *42*, 101170. [CrossRef]
4. Hinojosa, L.; González, J.A.; Barrios-Masias, F.H.; Fuentes, F.; Murphy, K.M. Quinoa abiotic stress responses: A review. *Plants* **2018**, *7*, 106. [CrossRef]
5. Asher, A.; Galili, S.; Whitney, T.; Rubinovich, L. The potential of quinoa (*Chenopodium quinoa*) cultivation in Israel as a dual-purpose crop for grain production and livestock feed. *Sci. Hortic.* **2020**, *272*, 109534. [CrossRef]
6. Haros, C.M.; Sanz-Penella, J.M. Food uses of whole pseudocereals. In *Pseudocereals: Chemistry and Technology*; Wiley-Blackwell: Hoboken, NJ, USA, 2017; pp. 163–192.
7. Graziano, S.; Agrimonti, C.; Marmiroli, N.; Gullì, M. Utilisation and limitations of pseudocereals (quinoa, amaranth, and buckwheat) in food production: A review. *Trends Food Sci. Technol.* **2022**, *125*, 154–165. [CrossRef]
8. Bochetto, A.; Merino, N.; Kaplan, M.; Guiñez, M.; Cerutti, S. Design of a combined microextraction and back-extraction technique for the analysis of mycotoxins in amaranth seeds. *J. Food Compos. Anal.* **2021**, *98*, 103818. [CrossRef]
9. Ruiz, K.B.; Khakimov, B.; Engelsen, S.B.; Bak, S.; Biondi, S.; Jacobsen, S.E. Quinoa seed coats as an expanding and sustainable source of bioactive compounds: An investigation of genotypic diversity in saponin profiles. *Ind. Crops Prod.* **2017**, *104*, 156–163. [CrossRef]
10. Bilalis, D.; Roussis, I.; Kakabouki, I.; Folina, A. Quinoa (*Chenopodium quinoa Willd.*) crop under Mediterranean conditions: A review. *Cienc. Investig. Agrar.* **2019**, *46*, 51–68. [CrossRef]
11. Kakabouki, I.; Bilalis, D.; Karkanis, A.; Zervas, G.; Hela, D. Effects of fertilization and tillage system on growth and crude protein content of quinoa (*Chenopodium quinoa willd.*): An alternative forage crop. *Emir. J. Food Agric.* **2014**, *26*, 18–24. [CrossRef]
12. Angeli, V.; Silva, P.M.; Massuela, D.C.; Khan, M.W.; Hamar, A.; Khajehei, F.; Graeff-Hönninger, S.; Piatti, C. Quinoa (*Chenopodium quinoa Willd.*): An overview of the potentials of the "Golden Grain" and socio-economic and environmental aspects of its cultivation and marketization. *Foods* **2020**, *9*, 216. [CrossRef] [PubMed]
13. Ebeid, H.M.; Kholif, A.E.; El-Bordeny, N.; Chrenkova, M.; Mlynekova, Z.; Hansen, H.H. Nutritive value of quinoa (*Chenopodium quinoa*) as a feed for ruminants: In *sacco* degradability and in vitro gas production. *Environ. Sci. Pollut. Res.* **2022**, *29*, 35241–35252. [CrossRef] [PubMed]
14. Tan, M. Macro-and micromineral contents of different quinoa (*Chenopodium quinoa Willd.*) varieties used as forage by cattle. *Turk. J. Agric. For.* **2020**, *44*, 46–53. [CrossRef]
15. Gül, M.; Tekce, E. Quinoa: A new feedstuff in animal nutrition. *Yem Mag.* **2016**, *76*, 29–35.
16. Peiretti, P.G. Amaranth in animal nutrition: A review. *Livest. Res. Rural. Dev.* **2018**, *30*, 1–20.
17. Manyelo, T.G.; Sebola, N.A.; van Rensburg, E.J.; Mabelebele, M. The probable use of *Genus amaranthus* as feed material for monogastric animals. *Animals* **2020**, *10*, 1504. [CrossRef]

18. Sanderson, S.L. Pros and cons of commercial pet foods (including grain/grain free) for dogs and cats. *Vet. Clin. Small Anim. Pract.* **2021**, *51*, 529–550. [CrossRef]
19. Laflamme, D.; Izquierdo, O.; Eirmann, L.; Binder, S. Myths and misperceptions about ingredients used in commercial pet foods. *Vet. Clin. Small Anim. Pract.* **2014**, *44*, 689–698. [CrossRef]
20. Traughber, Z.T.; He, F.; Hoke, J.M.; Davenport, G.M.; Rodriguez-Zas, S.L.; Southey, B.R.; de Godoy, M.R. Ancient grains as novel dietary carbohydrate sources in canine diets. *J. Anim. Sci.* **2021**, *99*, skab080. [CrossRef]
21. Pezzali, J.G.; Aldrich, C.G. Effect of ancient grains and grain-free carbohydrate sources on extrusion parameters and nutrient utilization by dogs. *J. Anim. Sci.* **2019**, *97*, 3758–3767. [CrossRef]
22. Arroyo-Manzanares, N.; Huertas-Pérez, J.F.; García-Campaña, A.M.; Gámiz-Gracia, L. Simple methodology for the determination of mycotoxins in pseudocereals, spelt and rice. *Food Control* **2014**, *36*, 94–101. [CrossRef]
23. Carballo, D.; Tolosa, J.; Ferrer, E.; Berrada, H. Dietary exposure assessment to mycotoxins through total diet studies. A review. *Food Chem. Toxicol.* **2019**, *128*, 8–20. [CrossRef] [PubMed]
24. Sacco, C.; Donato, R.; Zanella, B.; Pini, G.; Pettini, L.; Marino, M.F.; Rookmin, A.D.; Marvasi, M. Mycotoxins and flours: Effect of type of crop, organic production, packaging type on the recovery of fungal genus and mycotoxins. *Int. J. Food Microbiol.* **2020**, *334*, 108808. [CrossRef]
25. Marin, S.; Ramos, A.J.; Cano-Sancho, G.; Sanchis, V. Mycotoxins: Occurrence, toxicology, and exposure assessment. *Food Chem. Toxicol.* **2013**, *60*, 218–237. [CrossRef] [PubMed]
26. Bhat, R.; Rai, R.V.; Karim, A.A. Mycotoxins in food and feed: Present status and future concerns. *Compr. Rev. Food Sci. Food Saf.* **2010**, *9*, 57–81. [CrossRef]
27. Ostry, V.; Malir, F.; Toman, J.; Grosse, Y. Mycotoxins as human carcinogens—The IARC monographs classification. *Mycotoxin Res.* **2017**, *33*, 65–73. [CrossRef]
28. European Commission. Commission Regulation (EC) No. 1881/2006 Setting Maximum Levels for Certain Contaminants in Foodstuffs. Brussels, Belgium, 2006. Available online: https://www.fsai.ie/uploadedFiles/Consol_Reg1881_2006.pdf (accessed on 29 December 2022).
29. More, S.J.; Hardy, A.; Bampidis, V.; Benford, D.; Hougaard Bennekou, S.; Bragard, C.; Boesten, J.; Halldorsson, T.I.; Hernández-Jerez, A.F.; Jeger, M.J.; et al. Guidance on harmonised methodologies for human health, animal health and ecological risk assessment of combined exposure to multiple chemicals. *EFSA J.* **2009**, *17*, 5634. [CrossRef]
30. Palumbo, R.; Crisci, A.; Venâncio, A.; Cortiñas Abrahantes, J.; Dorne, J.L.; Battilani, P.; Toscano, P. Occurrence and co-occurrence of mycotoxins in cereal-based feed and food. *Microorganisms* **2020**, *8*, 74. [CrossRef]
31. Ramos-Diaz, J.M.; Sulyok, M.; Jacobsen, S.E.; Jouppila, K.; Nathanail, A.V. Comparative study of mycotoxin occurrence in Andean and cereal grains cultivated in South America and North Europe. *Food Control* **2021**, *130*, 108260. [CrossRef]
32. Kirinčič, S.; Škrjanc, B.; Kos, N.; Kozolc, B.; Pirnat, N.; Tavčar-Kalcher, G. Mycotoxins in cereals and cereal products in Slovenia–Official control of foods in the years 2008–2012. *Food Control* **2015**, *50*, 157–165. [CrossRef]
33. Krysinska-Traczyk, E.; Perkowski, J.; Dutkiewicz, J. Levels of fungi and mycotoxins in the samples of grain and grain dust collected from five various cereal crops in eastern Poland. *Ann. Agric. Environ. Med.* **2007**, *14*, 159–167.
34. Pappier, U.; Fernandez Pinto, V.; Larumbe, G.; Vaamonde, G. Effect of processing for saponin removal on fungal contamination of quinoa seeds (*Chenopodium quinoa Willd.*). *Int. J. Food Microbiol.* **2008**, *125*, 153–157. [CrossRef]
35. Bresler, G.; Brizzio, S.B.; Vaamonde, G. Mycotoxin-producing potential of fungi isolated from amaranth seeds in Argentina. *Int. J. Food Microbiol.* **1995**, *25*, 101–108. [CrossRef] [PubMed]
36. Vásquez-Ocmín, P.G.; Marti, G.; Gadea, A.; Cabanac, G.; Vásquez-Briones, J.A.; Casavilca-Zambrano, S.; Ponts, N.; Jargeatj, P.; Haddada, M.; Bertani, S. Metabotyping of Andean pseudocereals and characterization of emerging mycotoxins. *Food Chem.* **2023**, *407*, 135134. [CrossRef]
37. Ducos, C.; Pinson-Gadais, L.; Chereau, S.; Richard-Forget, F.; Vásquez-Ocmín, P.; Cerapio, J.P.; Casavilca-Zambrano, S.; Ruiz, E.; Pineau, P.; Bertani, S.; et al. Natural occurrence of mycotoxin-producing fusaria in market-bought peruvian cereals: A food safety threat for Andean populations. *Toxins* **2021**, *13*, 172. [CrossRef]
38. Bresler, G.; Vaamonde, G.; Brizzio, S. Natural occurrence of zearalenone and toxicogenic fungi in amaranth grain. *Int. J. Food Microbiol.* **1991**, *13*, 75–80. [CrossRef] [PubMed]
39. Schollenberger, M.; Müller, H.M.; Rüfle, M.; Suchy, S.; Planck, S.; Drochner, W. Survey of *Fusarium* toxins in foodstuffs of plant origin marketed in Germany. *Int. J. Food Microbiol.* **2005**, *97*, 317–326. [CrossRef]
40. Sugita-Konishi, Y.; Nakajima, M.; Tabata, S.; Ishikuro, E.; Tanaka, T.; Norizuki, H.; Itoh, Y.; Aoyama, K.; Fujita, K.; Kai, S.; et al. Occurrence of aflatoxins, ochratoxin A, and fumonisins in retail foods in Japan. *J. Food Prot.* **2006**, *69*, 1365–1370. [CrossRef] [PubMed]
41. Kumagai, S.; Nakajima, M.; Tabata, S.; Ishikuro, E.; Tanaka, T.; Norizuki, H.; Itoh, Y.; Aoyama, K.; Fujita, K.; Kai, S.; et al. Aflatoxin and ochratoxin A contamination of retail foods and intake of these mycotoxins in Japan. *Food Addit. Contam.* **2008**, *25*, 1101–1106. [CrossRef] [PubMed]
42. Veršilovskis, A.; Bartkevičs, V.; Miķelsone, V. Sterigmatocystin presence in typical Latvian grains. *Food Chem.* **2008**, *109*, 243–248. [CrossRef]

43. Keriene, I.; Mankeviciene, A.; Cesnuleviciene, R.; Janaviciene, S. Mycotoxins in Lithuanian buckwheat grain grown under sustainable and organic production systems. Proceedings book. In Proceedings of the 3rd International Conference on Sustainable Agriculture and Environment (3rd ICSAE), Warsaw, Poland, 26–28 September 2016; ISBN 978-605-9831-95-6.
44. Keriene, I.; Mankeviciene, A.; Bliznikas, S.; Cesnuleviciene, R.; Janaviciene, S.; Jablonskyte-Rasce, D.; Maiksteniene, S. The effect of buckwheat groats processing on the content of mycotoxins and phenolic compounds. *CyTA-J. Food* **2016**, *14*, 565–571. [CrossRef]
45. Ren, G.; Hu, Y.; Zhang, J.; Zou, L.; Zhao, G. Determination of multi-class mycotoxins in tartary buckwheat by ultra-fast liquid chromatography coupled with triple quadrupole mass spectrometry. *Toxins* **2018**, *10*, 28. [CrossRef]
46. Torović, L. Aflatoxins and ochratoxin A in flour: A survey of the Serbian retail market. *Food Addit. Contam. Part B* **2018**, *11*, 26–32. [CrossRef] [PubMed]
47. Bessaire, T.; Ernest, M.; Christinat, N.; Carrères, B.; Panchaud, A.; Badoud, F. High resolution mass spectrometry workflow for the analysis of food contaminants: Application to plant toxins, mycotoxins and phytoestrogens in plant-based ingredients. *Food Addit. Contam. Part A* **2021**, *38*, 978–996. [CrossRef] [PubMed]
48. Bresler, G.; Vaamonde, G.; Degrossi, C.; Pinto, V.F. Amaranth grain as substrate for aflatoxin and zearalenone production at different water activity levels. *Int. J. Food Microbiol.* **1998**, *42*, 57–61. [CrossRef]
49. Uhlig, S.; Torp, M.; Heier, B.T. Beauvericin and enniatins A, A1, B and B1 in Norwegian grain: A survey. *Food Chem.* **2006**, *94*, 193–201. [CrossRef]
50. Uhlig, S.; Jestoi, M.; Parikka, P. *Fusarium avenaceum*—The North European situation. *Int. J. Food Microbiol.* **2007**, *119*, 17–24. [CrossRef]
51. Lindblad, M.; Gidlund, A.; Sulyok, M.; Borjesson, T.; Krska, R.; Olsen, M.; Fredlund, E. Deoxynivalenol and other selected *Fusarium* toxins in Swedish wheat—Occurrence and correlation to specific *Fusarium* species. *Int. J. Food Microbiol.* **2013**, *167*, 284–291. [CrossRef]
52. Tolosa, J.; Font, G.; Mañes, J.; Ferrer, E. Nuts and dried fruits: Natural occurrence of emerging *Fusarium* mycotoxins. *Food Control* **2013**, *33*, 215–220. [CrossRef]
53. Herrera, M.; Bervis, N.; Carramiñana, J.J.; Juan, T.; Herrera, A.; Ariño, A.; Lorán, S. Occurrence and exposure assessment of aflatoxins and deoxynivalenol in cereal-based baby foods for infants. *Toxins* **2019**, *11*, 150. [CrossRef]
54. Alshannaq, A.; Yu, J.H. Occurrence, toxicity, and analysis of major mycotoxins in food. *Int. J. Environ. Res. Public Health* **2017**, *14*, 632. [CrossRef] [PubMed]
55. Shephard, G.S. Current status of mycotoxin analysis: A critical review. *J. AOAC Int.* **2016**, *99*, 842–848. [CrossRef] [PubMed]
56. Reiter, E.; Zentek, J.; Razzazi, E. Review on sample preparation strategies and methods used for the analysis of aflatoxins in food and feed. *Mol. Nutr. Food Res.* **2009**, *53*, 508–524. [CrossRef]
57. Pereira, V.L.; Fernandes, J.O.; Cunha, S.C. Mycotoxins in cereals and related foodstuffs: A review on occurrence and recent methods of analysis. *Trends Food Sci. Technol.* **2014**, *36*, 96–136. [CrossRef]
58. Cigić, I.K.; Prosen, H. An overview of conventional and emerging analytical methods for the determination of mycotoxins. *Int. J. Mol. Sci.* **2009**, *10*, 62–115. [CrossRef]
59. Iqbal, S.Z. Mycotoxins in food, recent development in food analysis and future challenges; a review. *Curr. Opin. Food Sci.* **2021**, *42*, 237–247. [CrossRef]
60. Bian, Y.; Zhang, Y.; Zhou, Y.; Wei, B.; Feng, X. Recent Insights into Sample Pretreatment Methods for Mycotoxins in Different Food Matrices: A Critical Review on Novel Materials. *Toxins* **2023**, *15*, 215. [CrossRef]
61. Rahmani, A.; Jinap, S.; Soleimany, F. Qualitative and quantitative analysis of mycotoxins. *Compr. Rev. Food Sci. Food Saf.* **2009**, *8*, 202–251. [CrossRef]
62. Zheng, M.Z.; Richard, J.L.; Binder, J. A review of rapid methods for the analysis of mycotoxins. *Mycopathologia* **2006**, *161*, 261. [CrossRef]
63. Singh, J.; Mehta, A. Rapid and sensitive detection of mycotoxins by advanced and emerging analytical methods: A review. *Food Sci. Nutr.* **2020**, *8*, 2183–2204. [CrossRef]
64. Malachová, A.; Stránská, M.; Václavíková, M.; Elliott, C.T.; Black, C.; Meneely, J.; Hajšlová, J.; Ezekiel, C.N.; Schuhmacher, R.; Krska, R. Advanced LC–MS-based methods to study the co-occurrence and metabolization of multiple mycotoxins in cereals and cereal-based food. *Anal. Bioanal. Chem.* **2018**, *410*, 801–825. [CrossRef] [PubMed]

toxins

MDPI

Review

Type A Trichothecene Metabolic Profile Differentiation, Mechanisms, Biosynthetic Pathways, and Evolution in *Fusarium* Species—A Mini Review

Jianhua Wang [1,*], Mengyuan Zhang [1,2], Junhua Yang [1], Xianli Yang [1], Jiahui Zhang [1] and Zhihui Zhao [1]

1 Institute for Agro-Food Standards and Testing Technology, Ministry of Agriculture, Shanghai Academy of Agricultural Sciences, 1000 Jinqi Road, Shanghai 201403, China; m210300781@st.shou.edu.cn (M.Z.); yangjunhua@saas.sh.cn (J.Y.); yangxianli@saas.sh.cn (X.Y.); jiahuizhang1224@163.com (J.Z.); 18918162068@163.com (Z.Z.)

2 College of Food Sciences & Technology, Shanghai Ocean University, Shanghai 201306, China

* Correspondence: wangjianhua@saas.sh.cn; Tel.: +86-21-67131637

Abstract: Trichothecenes are the most common *Fusarium* toxins detected in grains and related products. Type A trichothecenes are among the mycotoxins of greatest concern to food and feed safety due to their high toxicity. Recently, two different trichothecene genotypes within *Fusarium* species were reported. The available information showed that *Tri1* and *Tri16* genes are the key determinants of the trichothecene profiles of T-2 and DAS genotypes. In this review, polymorphisms in the *Tri1* and *Tri16* genes in the two genotypes were investigated. Meanwhile, the functions of genes involved in DAS and NEO biosynthesis are discussed. The possible biosynthetic pathways of DAS and NEO are proposed in this review, which will facilitate the understanding of the synthesis process of trichothecenes in *Fusarium* strains and may also inspire researchers to design and conduct further research. Together, the review provides insight into trichothecene profile differentiation and *Tri* gene evolutionary processes responsible for the structural diversification of trichothecene produced by *Fusarium*.

Keywords: mycotoxin; DAS; NEO; T-2; trichothecenes; *Fusarium* species

Key Contribution: Polymorphisms in the *Tri1* and *Tri16* genes in T-2 and DAS genotypes were studied, and probable DAS and NEO biosynthetic pathways are proposed in this review.

1. Introduction

Filamentous fungi within the *Fusarium* genus are the most important etiological agents of a variety of plant diseases worldwide, resulting in huge economic losses annually. Additionally, mycotoxins produced by these pathogens are also of concern due to their wide range of toxicological effects [1–4]. Among these toxic secondary metabolites, trichothecenes are the most commonly detected *Fusarium* mycotoxins, with relatively high contents compared with other ones [1–4]. Trichothecene is a large family of non-volatile sesquiterpenes that are classified into four different groups (type A, B, C, and D) according to structural variations [5,6]. To date, more than 200 trichothecenes have been identified in nature, which represent a major threat to food and feed safety [7].

Fusarium trichothecenes are divided into two major types characterized by the absence (type A trichothecenes) or presence (type B trichothecenes) of a keto group at carbon atom 8 (C-8). Type A trichothecenes contain an ester function (e.g., T-2 toxin), a hydroxyl group (e.g., neosolaniol, NEO) at C-8 of the skeleton 12,13-epoxytrichothec-9-ene (EPT) molecule, or no substituent at all at C-8 (e.g., 4,15-diacetoxyscirpenol, DAS) [5–7] (Figure 1). As illustrated in Figure 1, type B trichothecenes, such as deoxynivalenol (DON) and its acetylated derivatives, possess a keto group at C-8 [5]. Trichothecenes have been shown to be potent inhibitors of nucleotides and protein synthesis and can affect mitochondrial function, induce immunosuppression, etc., in eukaryotic organisms [8,9]. Trichothecenes can also act as virulence factors in plants, which will facilitate the colonization and spreading of the pathogens in host tissues [10–12]. The *Fusarium* trichothecenes of greatest concern are type A trichothecenes, with their high toxicity.

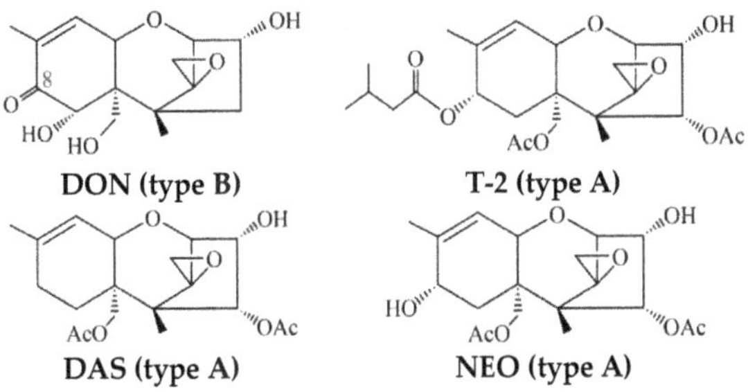

Figure 1. Chemical structures of Type A and B trichothecenes. Examples of type A trichothecenes include T-2 toxin, DAS, and NEO. DON is an example of a type B trichothecene.

This review seeks to outline recent findings on type-A-trichothecene-producing *Fusarium* species, the *Tri* gene's genetic diversity, and evolution. In addition, current gaps, advances, and potential topics for future studies on *Fusarium* and trichothecenes are also mentioned.

2. Metabolic Profile Differentiation in Type-A-Trichothecene-Producing *Fusarium*

It is well known that comprehensive research on the differentiation of type B trichothecenes has been conducted in the *Fusarium graminearum* species complex (FGSC), with the most important *Fusarium* species causing fusarium head blight in wheat around the world. In the FGSC, three different trichothecene genotypes (chemotypes) were distinguished according to their production of NIV (nivalenol), 3ADON (3-acetyldeoxynivalenol), or 15ADON (15-acetyldeoxynivalenol) [13,14]. FGSC species and trichothecene genotype diversity have been biogeographically structured worldwide. The statement suggests that the diversity of *Fusarium* species and trichothecene genotypes is not randomly distributed but instead follows geographic patterns [15–17]. This can have important implications for understanding the evolution and ecology of these organisms as well as for developing strategies for managing plant diseases caused by *Fusarium* species. Additionally, understanding the differences between different types of these mycotoxins is important for assessing their potential impact on human and animal health as well as for developing effective control and prevention strategies for reducing their occurrence in food and other agricultural products.

Several *Fusarium* species, such as *Fusarium sporotrichioides*, *Fusarium poae*, *Fusarium kyushuense*, and *Fusarium langsethiae*, are well-known type A trichothecene producers. Among these fungi, different mycotoxin productivities were also observed. For example, *F. sporotrichioides* was reported to consistently produce T-2, HT-2, DAS, and NEO and has been recognized as the main source for T-2 and HT-2 [7,18,19]. The production of mycotoxins by 109 strains of *F. langsethiae* (23 strains), *F. poae* (49 strains), *F. sporotrichioides* (35 strains), and *F. kyushuense* (2 strains) was investigated independently in different laboratories [20]. From the compiled results, it was found that *F. langsethiae* and *F. sporotrichioides* consistently produced type A trichothecene (T-2, HT-2, and NEO). However, a different profile was observed in the 49 *F. poae* strains, and 41 of them produced type B trichothecenes (NIV and 4ANIV) in addition to type A trichothecenes (DAS). For the two *F. kyushuense* strains, no type A trichothecenes were detected from either of the strains [20]. However, among the mycotoxins produced by *F. poae*, NIV, a type B trichothecene, was cited as one of the most common mycotoxins produced by this species in the studies by [21–24]. *Fusarium armeniacum* was also reported to produce type A trichothecenes, such as T-2, DAS, and NEO [25].

In the past decade, several novel *Fusarium* species that produce type A trichothecenes have been reported, such as *Fusarium sibiricum* [26], *Fusarium palustre* [27], and *Fusarium goolgardi* [28,29]. *F. sibiricum* was mainly recovered in Siberia and the Russian Far East and formally described by Yli-Mattila et al. [26] in 2011. *F. sibiricum* is phylogenetically more closely related to *F. sporotrichioides* but is morphologically more similar to *F. poae* and *F. langsethiae* [26]. Analysis of trichothecene production revealed that all the tested *F. sibiricum* isolates could produce type A trichothecene T-2 as well as DAS with mean concentrations of 17.4 ppm and 0.2 ppm, respectively [26]. *F. palustre* is a new *Fusarium* species associated with the dieback of *Spartina alterniflora* in Atlantic salt marshes [27]. Subsequently, Rocha et al. [29] proved that strains from *F. palustre* can produce type A trichothecenes, including DAS, NEO, and T-2 toxin.

Despite the discovery of various metabolic profiles in type-A-trichothecene-producing species, no particular genotypes were outlined. The identification of two distinct genotypes within type A trichothecene producers was found in *F. goolgardi* [29]. *F. goolgardi* is an emerging species identified by Laurence et al. [28] from *Xanthorrhoea glauca* in natural ecosystems in Australia. Chemical analysis revealed the production of type A trichothecenes in *F. goolgardi* cultures [29]. Among the eight *F. goolgardi* strains evaluated, four of them (RBG5411, 5417, 5419, and 5420) produced T-2 toxin, DAS, NEO, and 8-acetylneosolaniol (hereinafter referred to as T-2 genotype in this work), while the other four strains (RBG6914, 6915, 5421, and 5422) produced only DAS (hereinafter referred to as DAS genotype in this work) [29]. So, the study by Rocha et al. [29] indicated that there were at least two distinct trichothecene genotypes in *F. goolgardi* populations. It is worth noting that a novel group of type A trichothecenes (NX toxins) produced by FGSC was identified by Varga et al. [30] in 2015. In this review, we will not delve into details about FGSC. To our best knowledge, only a single genotype has been reported for *F. langsethiae*, *F. sibiricum*, and *F. sporotrichioides*, and strains from these species can produce T-2 and some other type A trichothecenes, such as DAS and NEO.

According to the present data, in general, type-A-trichothecene-producing *Fusarium* species may be indigenous and possibly endemic to their origin at a low frequency. If these strains become more abundant or are spread and exchanged widely through transportation and trade, type A trichothecenes could become a common contaminant in cereals and related products. For example, a high prevalence has been found for *F. langsethiae* on oats; however, it is now spreading even to barley cultivated in Mediterranean environments [31]. So, it will be important to monitor whether *Fusarium* species, such as *F. goolgardi*, have a selective advantage in specific ecosystems. Nevertheless, it is worth noting that the occurrence of type-A-trichothecene-producing *Fusarium* species in different geographic locations in the world suggests their wide distribution in nature. Novel type-A-trichothecene-producing species may be identified in further studies with more extensive collections in the future.

3. *Tri1* and *Tri16* Genes Are the Key Determinants of Trichothecene Profiles

The biosynthetic pathway and molecular regulation mechanism of trichothecenes are now relatively clear, and many studies have been conducted since the 1990s. Up until now, 15 trichothecene biosynthesis genes (*Tri* genes) have been identified and characterized in the *Fusarium* genome (Table 1). Molecular genetics revealed that these *Tri* genes occur at three loci. The 12-gene core locus on chromosome 2 is located within a 25 kb region as a cluster response for the synthesis of the EPT skeleton molecule and subsequent modifications at C-3, C-4, and C-15. The *Tri1*–*Tri16* locus on chromosome 1 is essential for the hydroxylation and acylation of C-8, respectively. The single-gene locus on chromosome 4, *Tri101*, is responsible for acetylation of the hydroxyl group at C-3, converting isotrichodermol to isotrichodermin. This step has been proven to serve as a mechanism for the self-protection of the trichothecene-producing organism [32], which can significantly reduce the toxicity of trichothecenes.

Table 1. Function of *Tri* genes and major phenotype of individual *Tri* gene disruption in Type A trichothecene biosynthesis in *F. sporotrichioides*.

Tri Gene	Function	Mutant Phenotype
Tri8	C-3 deacetylase	3-acetyl T-2
Tri7	C-4 acetyltransferase	HT-2
Tri3	C-15 acetyltransferase	15-decalonectrin, 3,15-didecalonectrin
Tri4	multifunctional oxygenase	trichodiene
Tri6	zinc finger transcription factor	low levels of trichodiene
Tri5	trichodiene synthase	no trichothecenes
Tri10	regulatory gene	no trichothecenes
Tri9	unknown	not determined
Tri11	C-15 hydroxylase	isotrichodermin
Tri12	trichothecene efflux pump	no trichothecenes
Tri13	C-4 hydroxylase	8-hydroxycalonectrin, 4-deoxy T-2, 8-hydroxy-3-deacetylcalonectrin
Tri14	virulence factor	T-2
Tri1	C-8 hydroxylase	DAS
Tri16	C-8 acyltransferase	NEO, DAS
Tri101	C-3 acetyltransferase	isotrichodermol

In trichothecene-producing *Fusarium* species, the *Tri1*–*Tri16* locus determines type A versus type B trichothecene production. In T-2 producers, *Tri1* and *Tri16* are responsible for the specific hydroxylation and acylation, respectively, at the C-8 position [33–35]. However, in type B trichothecene producers, the enzyme encoded by the *Tri1* gene catalyzes the hydroxylation of trichothecenes at both C-7 and C-8, but the *Tri16* gene is non-functional due to the presence of frameshifts and stop codons in its coding region [36]. Meanwhile, a non-functional *Tri16* in type-A-trichothecene-producing *Fusarium* strains should prevent the acylation of the hydroxyl of trichothecene intermediates at the C-8 position, and a non-functional *Tri1* gene should equally prevent hydroxylation of the C-8 and, of course, prevent the later acylation reaction catalyzed by the Tri16 enzyme.

Sequence analysis of the *Tri1*–*Tri16* locus in four *F. sibiricum*, seven *F. langsethiae*, and six *F. sporotrichioides* revealed that the orientation and order of the two genes were the same as previously characterized for *F. sporotrichioides*, although the length of the *Tri1*–*Tri16* intergenic region varied among and within species [26]. According to the phylogenetic analysis of the *Tri1* and *Tri16* gene coding sequences, the two genes in *F. sibiricum* strains are more closely related to those of *F. sporotrichioides*. In *F. sibiricum*, the *Tri1*–*Tri16* locus is more similar in organization and sequence to those of *F. langsethiae* and *F. sporotrichioides* than to that in the species of *F. poae* [26].

Recently, two different trichothecene metabolic profiles were identified in *F. goolgardi* strains. To reveal the reason why this phenomenon exists, the nucleotide sequences of different *Tri* genes from several type-A-trichothecene-producing *Fusarium* species were analyzed and compared with the two *F. goolgardi* groups. The results showed that no major differences were observed in the coding regions of the core cluster genes (including *Tri3*–*Tri8*, *Tri11*, *Tri13*, and *Tri101*) among these strains [29]. However, significant differences were identified in the *Tri1* and *Tri16* sequences. As shown in Figure 2, there is a transition (C-to-T) in the coding region of the *Tri1* gene, which resulted in a premature stop codon in the gene of the *F. goolgardi* strain with the DAS genotype [29]. According to previous studies, in T-2-producing *Fusarium* species, *Tri16* is an intronless gene [26]. In comparison with *F. sporotrichioides*, the *Tri16* coding region of DAS-genotype strains exhibited a single-nucleotide deletion, which introduced a frameshift mutation and caused two premature stop codons in the gene of the *F. goolgardi* strain with the DAS genotype. However, *Tri1* and *Tri16* orthologs from all the T-2 producers, including the *F. goolgardi* T-2 genotype strains, did not contain the same or any other similar nonsense or frameshift mutations in the coding regions [29]. Overall, the results showed that the *Tri1* gene is essential for the hydroxylation of type A trichothecene at C-8, and this gene determines the production of DAS, NEO, and T-2 toxins in *F. goolgardi* [29].

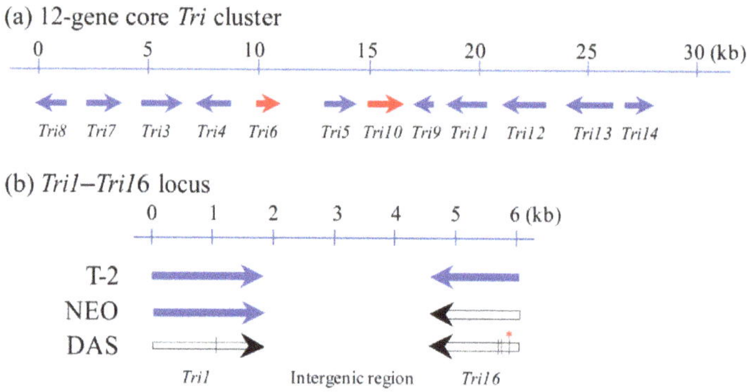

Figure 2. Type A trichothecene biosynthetic loci in *Fusarium* species. (**a**) The 12-gene core *Tri* cluster. (**b**) Comparison of the *Tri1*–*Tri16* locus in *Fusarium* species with T-2, NEO, and DAS genotypes, respectively. *Tri6* and *Tri10*, the two transcription factors, are shown in red. Arrows indicate genes and the direction of transcription. Filled arrows indicate that the *Tri* genes are functional, while the non-functional genes are indicated with empty arrows. Premature stop codons are indicated by vertical lines on the arrows; the frameshift that occurred in the *Tri16* gene in the DAS genotype is indicated above the panel by * and together with a vertical line on the arrow.

In the previous studies by Brown et al. [34], Peplow et al. [35], and Proctor et al. [36], the *Tri16* gene was found to be truncated in the *F. poae* strains examined, which would explain why some *F. poae* strains cannot produce T-2 toxin. Moreover, the contradictory reports about the ability of different geographically originated *F. poae* strains to produce type A trichothecenes may also be the cause of the misidentification of *Fusarium* species due to their high morphological similarity [18,19,26].

The organization and genomic context of the trichothecene biosynthetic locus *Tri1*–*Tri16* are similar in *F. langsethiae*, *F. sibiricum*, *F. sporotrichioides*, and *F. goolgardi* (including both the T-2 and DAS genotypes), but significantly different from those described for some of *F. poae* [26,29,36]. On the other hand, the occurrences of nonsense mutation and frameshift mutation in the coding region of *Tri1* and *Tri16* genes [29], respectively, led to the loss of functions of the two genes in type A trichothecene producers, such as *F. goolgardi* strains that possess a strict DAS genotype (Figure 2). These results indicate that we still have a lot to do about trichothecene biosynthesis in the *Fusarium* genus and also suggest the necessity of re-explaining the diversity of trichothecene production in these complicated fungi. Based on the research of the past, we can hypothesize that *Fusarium* strains that produce only trichothecene with a hydroxyl at C-8 may be identified in the future.

4. Proposed Biosynthetic Pathways of DAS and NEO and Comparisons with T-2

The biosynthesis of trichothecenes begins with the cyclization of the precursor substance trans-farnesyl pyrophosphate (FPP) to form trichodiene, followed by oxygenation, isomerization, cyclization, and esterification to finally form trichothecene toxins with various structures. The types and chemical structures of trichothecene toxins are mainly determined by the metabolic pathways involved and the genetic differences in *Tri* genes. According to the chemical structures of trichothecenes and even some newly identified trichothecene orthologs produced by *Fusarium* species, the biosynthetic pathways and gene functions can probably be predicted based on our existing understanding of trichothecene biosynthesis [37]. Comprehensive studies have been conducted to reveal the biosynthesis of T-2 since the 1990s, and its biosynthetic pathway and molecular regulation mechanism are now relatively clear. To our best knowledge, however, limited information is available for the other type A trichothecenes, such as DAS and NEO.

Structurally, there are two hydrogen atoms at C-8 in the DAS molecule, while one of the two hydrogen atoms is replaced by a hydroxyl group and an isovalerate group, respectively, in NEO and T-2. That means the DAS genotype strains do not have the ability to synthesize the enzymes that can catalyze the hydroxylation and isovalerate addition to C-8. On the other hand, co-occurrence of DAS, NEO, and T-2 toxins in a single strain demonstrated that all the 3-acetylneosolaniol, 3,4,15-triacetoxyscirpenol, and 3-acetyl T-2 can be served as substrates of Tri8 (an esterase) by which the C-3 acetyl group is replaced by a hydroxyl [38]. The differential activity of Tri8, as defined by the DNA sequence, determines the production of either 3ADON or 15ADON in FGSC [39]. However, so far, it is still unclear whether strains with a type A trichothecene genotype that produces only NEO but no trichothecene with isovalerate function at C-8 naturally exist in the genus *Fusarium*.

As shown in Table 1, the specific functions of most *Tri* genes in trichothecene biosynthesis have been studied in *Fusarium* species [33,35,40–48], which makes it possible to predict the biosynthetic pathway of different trichothecene metabolites, such as DAS and NEO. For example, the *Tri1* genes in T-2 producers are responsible for oxidation at C-8 of the trichothecene scaffold [33]. Target gene disruption of the *Tri1* gene blocks production of C-8-oxygenated trichothecenes and leads to the accumulation of DAS in *F. sporotrichioides* [33]. The recently identified DAS producer that carries a non-functional *Tri1* gene in *F. goolgardi* species further confirmed our thinking [29]. The data by Peplow et al. [35] indicate that *Tri16* encodes an acyltransferase that catalyzes the formation of ester side groups at C-8 during T-2 biosynthesis in *F. sporotrichioides*. Similarly, *Fusarium* strains with type A trichothecene genotypes containing a non-functional *Tri16* gene may produce NEO.

Based on the findings of trichothecene biosynthesis in *Fusarium*, as shown in Figure 3, we proposed the biosynthetic pathways of DAS and NEO and made a comparison with the T-2 biosynthetic pathway. As reviewed by Chen et al. [37] and Chen et al. [49], in trichothecene biosynthetic pathways, the reaction steps catalyzing FPP to calonectrin (CAL) are shared among *Fusarium* species. For detailed information on type A and type B trichothecene biosynthesis, please refer to previous publications [6,7,50–52]. In T-2 producers, intermediate metabolite CAL, which is eventually converted to T-2 toxin, undergoes a series of steps catalyzed by Tri13-Tri7-Tri1-Tri16-Tri8 sequentially. As we predicted, the same reactions occurred in the immediate two following steps, catalyzed by Tri13 and Tri7, respectively, after CAL during the biosynthesis of DAS, NEO, and T-2. CAL is hydroxylated by Tri13 at the C-4 position to produce the intermediate metabolite 3,15-diacetoxyscirpenol, and the hydroxyl group is subsequently converted to an acetyl group by the enzyme of Tri7 to produce 3,4,15-triacetoxyscirpenol (Figure 3). As mentioned above, the DAS strains have a pseudo-*Tri1* gene, so differences arise in the later steps after 3,4,15-triacetoxyscirpenol during the biosynthesis of DAS, NEO, and T-2.

As shown in Figure 3, in DAS producers, 3,4,15-triacetoxyscirpenol is deacetylated by esterase encoded by *Tri8*, leading to the formation of DAS. In NEO and T-2 producers, 3,4,15-triacetoxyscirpenol is further converted to 3-acetylenosolaniol through the activity of the Tri1 enzyme. The product 3-acetylenosolaniol is deacetylated by the enzyme of Tri8 at C-3 to produce NEO. In light of this, we draw the conclusion that CAL is catalyzed via the Tri13-Tri7-Tri8 and Tri13-Tri7-Tri1-Tri8 pathways, respectively, during the biosynthesis of DAS and NEO. It is easy to understand that T-2 producers can also produce portions of DAS and NEO which are the intermediates of T-2 biosynthesis, since the T-2 strains contain all the functional *Tri* genes required for DAS and NEO biosynthesis. However, the definite biosynthetic pathway and detailed regulation mechanisms of DAS and NEO are unclear, and systematic studies still should be conducted, especially using the strict DAS genotype strains identified, such as the DAS-genotype *F. goolgardi* strains. Moreover, the proposed biosynthetic pathways of DAS and NEO in this work will provide new insights into trichothecene biosynthesis and guide researchers to carry out more extensive studies on this topic.

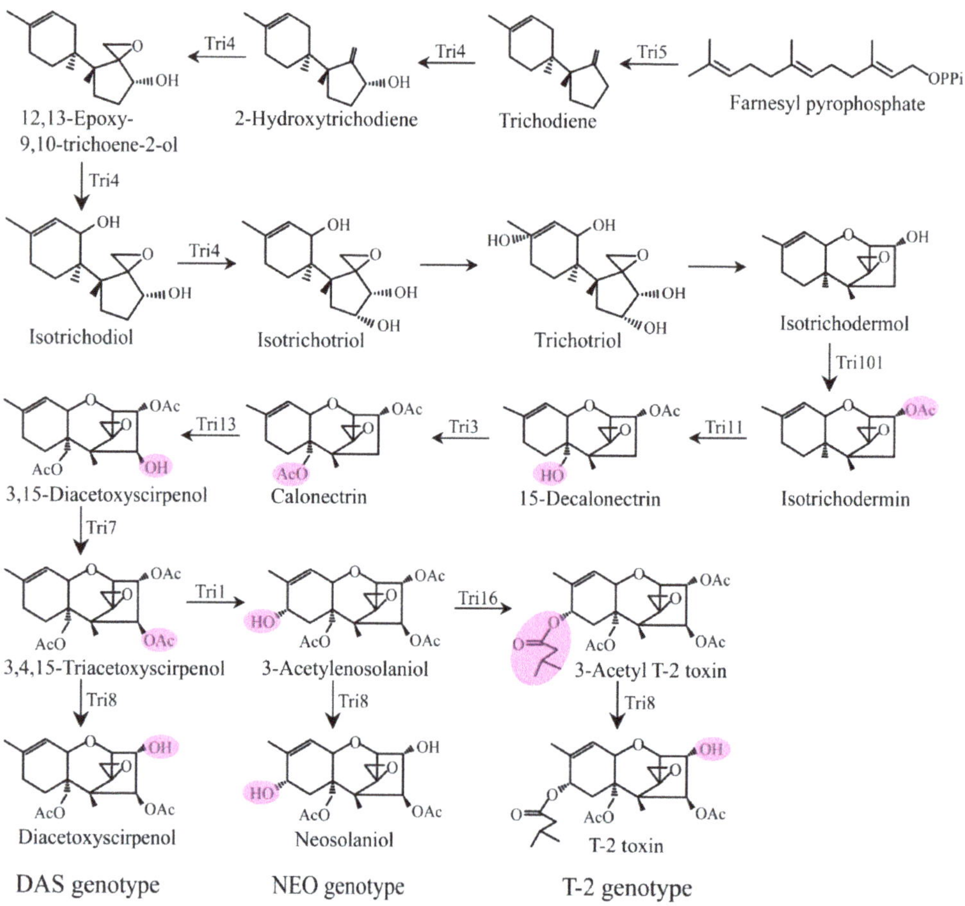

Figure 3. The proposed biosynthetic pathways of DAS, NEO, and their comparison with T-2 in *Fusarium*. Steps catalyzed by Tri enzymes are identified near the arrow showing the step. Unlabeled arrows indicate steps for which the specific genes or enzymes are unknown.

5. Evolution Potential of Type A Trichothecene Metabolic Profile Differentiation in *Fusarium*

Studies of trichothecene-producing *Fusarium* species indicate that the evolutionary process of the *Tri* loci is complex in fusaria and suggest that gain or loss functions, mutations, translocations, and non-functionalization occurred within and between *Tri* loci [23,36,39,53]. The structure diversity of trichothecenes is the cause of genetic polymorphism in the *Tri* genes. It was found that the evolution of *Tri* genes does not always correlate with the evolutionary process of *Fusarium* species, which has been maintained through balancing selection and accompanied by the evolution process of the fungi [54]. The studies by Proctor et al. [36] and Kelly et al. [55] reported inconsistencies between species phylogenies and *Tri1*–*Tri16*-based phylogenies. Specifically, trans-species evolution and genomic translocations of the *Tri1* gene have been identified, and this gene is found in at least four genomic contexts [36]. Recently, Kelly et al. [55] revealed that the evolution of a novel trichothecene-producing population in FGSC was accompanied by a marked change in selective pressure on *Tri1*. However, the genomic context and evolutionary affinities of the *Tri1* variants from type-A-trichothecene-producing strains have not been investigated. A wide range of sequencing and phylogenetic analyses of *Tri1* from diverse *Fusarium* strains is warranted to further reveal the origins and evolutionary processes of the type-A-trichothecene-producing strains with different genotypes.

Proctor et al. [36] have also suggested that the *Tri1–Tri16* locus was the ancestral character state in the ancestral trichothecene-producing *Fusarium* species, and the gene was probably functional in the ancestral strains, as it is more likely for a gene to lose functionality than for a non-functional gene (such as due to deletions and nonsense mutations, etc.) to become functional [29,36]. So, we hypothesize that the two genotypes within *F. goolgardi* evolved from the same ancestor. The *Tri1* gene in *F. goolgardi* strains with the T-2 genotype is probably ancestral to the allele in strains with the DAS genotype.

Nevertheless, it is worth noting whether strains that primarily produce NEO or co-occurrences of NEO and DAS without T-2 exist in nature or not. If this is the case, three genotypes will be classified within type A trichothecene strains. To simplify the description of type A trichothecene genotypes, we recommend using DAS, NEO, and T-2, respectively, for the strains, which will facilitate the implementation of scientific research and academic exchanges.

The results of previous studies provide evidence for a complex evolutionary process of *Tri* loci and specific *Tri* genes that included gain, loss, functional changes, rearrangement, and trans-species polymorphism [23,36,39,53]. The structural diversity of trichothecenes potentially reflects differences in selection pressure experienced by the fungi that produce the analogs [54]. Ward et al. [54] concluded that trichothecene structural diversity in the FGSC has been maintained through balancing selection. Thus, further investigations are required to reveal the important evolutionary event that has given rise to type A trichothecene structural differences through comparative analyses of different *Fusarium* species. These results will provide new insights into genetic basis changes or biochemical alterations that occurred in trichothecene biosynthesis and regulation as fungi with the pathways adapt to various environmental conditions. Multispecies comparisons of *Tri* loci and *Tri* genes may also provide key insights into the evolution process of trichothecene metabolism in *Fusarium*.

6. Conclusions and Future Prospects

Two major type-A-trichothecene-producing *Fusarium* groups were identified in nature: one group can produce trichothecene containing an ester function at the C-8 position and is represented by T-2; the other group produces trichothecene without a substituent at C-7 and C-8 but not T-2 and is represented by DAS. The phylogenetic relationship assessment of *Tri* genes provided important evidence for the genetic basis of chemotype differentiation within this species. The *Tri1–Tri16* locus is responsible for the chemical structure variation of these two genotypes; both *Tri1* and *Tri16* are functional in the T-2 genotype but non-functional in *Fusarium* strains with the DAS genotype due to the occurrence of premature stop codons caused by a point mutation within their coding regions [29]. The apparent genetic changes within type-A-trichothecene-producing *Fusarium* species highlight the need for monitoring and more phenotypic characterization of trichothecene-producing populations.

As previously reviewed, the *Fusarium* genus and trichothecene genotype diversity vary significantly among different hosts and geographic locations [15–17]. Further investigations are required to track the spread of different trichothecene genotypes and to elucidate potential differences in their competitive abilities, including environmental adaptability and aggressiveness in different plant hosts. The environmental drivers of trichothecene metabolic profile differentiation in *Fusarium* are waiting to be further revealed, and continuous studies will be required to elucidate the ethology, host preference, economic loss caused, forecast and prediction, and control methods of different *Fusarium* populations. Most importantly, the molecular mechanisms of DAS and NEO biosynthesis should be comprehensively clarified.

Author Contributions: Ideation, J.W. and M.Z.; literature search J.W., M.Z., J.Y., X.Y., J.Z. and Z.Z.; writing—original draft preparation, J.W., M.Z., J.Y. and X.Y.; writing—review and editing, J.W.; supervision, J.W.; project administration, J.W.; funding acquisition, J.W., J.Y. and X.Y. All authors have read and agreed to the published version of the manuscript.

Funding: This research was funded by the Science and Technology Commission of Shanghai Municipality (grant number 21DZ1201300) and Natural Science Foundation of Shanghai (grant numbers 20ZR1437200 and 23ZR1455700).

Institutional Review Board Statement: Not applicable.

Informed Consent Statement: Not applicable.

Data Availability Statement: Not applicable.

Conflicts of Interest: The authors declare no conflict of interest.

References

1. D'Mello, J.P.F.; Placinta, C.M.; Macdonald, A.M.C. *Fusarium* mycotoxins: A review of global implications for animal health, welfare and productivity. *Anim. Feed. Sci. Technol.* **1999**, *80*, 183–205.
2. Yazar, S.; Omurtag, G.Z. Fumonisins, trichothecenes and zearalenone in cereals. *Int. J. Mol. Sci.* **2008**, *9*, 2062–2090. [PubMed]
3. Arunachalam, C.; Doohan, F.M. Trichothecene toxicity in eukaryotes: Cellular and molecular mechanisms in plants and animals. *Mol. Toxicol. Lett.* **2013**, *217*, 149–158.
4. Wu, Q.; Wang, X.; Nepovimova, E.; Miron, A.; Liu, Q.; Wang, Y.; Su, D.; Yang, H.; Li, L.; Kuca, K. Trichothecenes: Immunomodulatory effects, mechanisms, and anticancer potential. *Arch. Toxicol.* **2017**, *91*, 3737–3785. [CrossRef] [PubMed]
5. Wang, J.; Zhao, Z.; Yang, X.; Yang, J.; Gong, A.; Zhang, J.; Chen, L.; Zhou, C. *Fusarium graminearum* species complex and trichothecene genotype. In *Mycotoxins and Food Safety*; Sabuncuoglu, S., Ed.; IntechOpen: London, UK, 2019.
6. Kimura, M.; Tokai, T.; Takahashi-Ando, N.; Ohsato, S.; Fujimura, M. Molecular and genetic studies of *Fusarium* trichothecene biosynthesis: Pathways, genes, and evolution. *Biosci. Biotechnol. Biochem.* **2007**, *71*, 2105–2123.
7. McCormick, S.P.; Stanley, A.M.; Stover, N.A.; Alexander, N.J. Trichothecenes: From simple to complex mycotoxins. *Toxins* **2011**, *3*, 802–814.
8. Wu, J.; Jing, L.; Yuan, H.; Peng, S. T-2 toxin induces apoptosis in ovarian granulosa cells of rats through reactive oxygen species-mediated mitochondrial pathway. *Toxicol. Lett.* **2011**, *202*, 168–177.
9. Rocha, O.; Ansari, K.; Doohan, F.M. Effects of trichothecene mycotoxins on eukaryotic cells: A review. *Food Addit. Contam.* **2015**, *22*, 369–378.
10. Proctor, R.H.; Hohn, T.M.; McCormick, S.P. Reduced virulence of *Gibberella zeae* caused by disruption of a trichothecene toxin biosynthetic gene. *Mol. Plant Microbe Interact.* **1995**, *8*, 593–601. [CrossRef]
11. Desjardins, A.E.; Proctor, R.H.; Bai, G.; McCormick, S.P.; Shaner, G.; Buechely, G.; Hohn, T.M. Reduced virulence of trichothecene-nonproducing mutants of *Gibberella zeae* in wheat field tests. *Mol. Plant Microbe Interact.* **1996**, *9*, 775–781.
12. Jansen, C.; Von Wettstein, D.; Schafer, W.; Kogel, K.H.; Felk, A.; Maier, F.J. Infection patterns in barley and wheat spikes inoculated with wild-type and trichodiene synthase gene disrupted *Fusarium graminearum*. *Pro. Natl. Acad. Sci. USA* **2005**, *102*, 16892–16897. [CrossRef] [PubMed]
13. Miller, J.D.; Greenhalgh, R.; Wang, Y.; Lu, M. Trichothecene chemotypes of three *Fusarium* species. *Mycologia* **1991**, *83*, 121–130. [CrossRef]
14. Wang, J.; Li, H.; Qu, B.; Zhang, J.; Huang, T.; Chen, F.; Liao, Y. Development of a generic PCR detection of 3-acetyldeoxynivalenol-, 15-acetyldeoxynivalenol- and nivalenol-chemotypes of *Fusarium graminearum* clade. *Int. J. Mol. Sci.* **2008**, *9*, 2495–2504. [PubMed]
15. Zhang, J.; Li, H.; Dang, F.; Qu, B.; Xu, Y.; Zhao, C.; Liao, Y. Determination of the trichothecene mycotoxin chemotypes and associated geographical distribution and phylogenetic species of the *Fusarium graminearum* clade from China. *Mycol. Res.* **2007**, *111*, 967–975.
16. Wang, J.H.; Ndoye, M.; Zhang, J.B.; Li, H.P.; Liao, Y.C. Population structure and genetic diversity of the *Fusarium graminearum* species complex. *Toxins* **2011**, *3*, 1020–1037.
17. Edwards, S.G.; Imathiu, S.M.; Ray, R.V.; Back, M.; Hare, M.C. Molecular studies to identify the *Fusarium* species responsible for HT-2 and T-2 mycotoxins in UK oats. *Int. J. Food Microbiol.* **2012**, *156*, 168–175.
18. Torp, M.; Nirenberg, H.I. *Fusarium langsethiae* sp. nov. on cereals in Europe. *Int. J. Food Microbiol.* **2004**, *95*, 247–256.
19. Torp, M.; Langseth, W. Production of T-2 toxin by a *Fusarium* resembling *Fusarium poae*. *Mycopathologia* **1999**, *147*, 89–96. [CrossRef]
20. Thrane, U.; Adler, A.; Clasen, P.; Galvano, F.; Langseth, W.; Lew, H.; Logrieco, A.; Nielsen, K.F.; Ritieni, A. Diversity in metabolite production by *Fusarium langsethiae*, *Fusarium poae*, and *Fusarium sporotrichioides*. *Int. J. Food Microbiol.* **2004**, *95*, 257–266.
21. Covarelli, L.; Beccari, G.; Prodi, A.; Generotti, S.; Etruschi, F.; Juan, C.; Ferrer, E.; Mañes, J. *Fusarium* species, chemotypes characterization and trichothecene contamination of durum and soft wheat in an area of central Italy. *J. Sci. Food Agr.* **2015**, *95*, 540–551.
22. Stenglein, S.A.; Dinolfo, M.I.; Barros, G.; Bongiorno, F.; Chulze, S.N.; Moreno, M.V. *Fusarium poae* pathogenicity and mycotoxin accumulation on selected wheat and barley genotypes at a single location in Argentina. *Plant Dis.* **2014**, *98*, 1733–1738. [CrossRef]
23. Vogelgsang, S.; Sulyok, M.; Hecker, A.; Jenny, E.; Krska, R.; Schuhmacher, R.; Forrer, H.R. Toxigenicity and pathogenicity of *Fusarium poae* and *Fusarium avenaceum* on wheat. *Eur. J. Plant Pathol.* **2008**, *122*, 265–276.

24. Vogelgsang, S.; Sulyok, M.; Bänziger, I.; Krska, R.; Schuhmacher, R.; Forrer, H.R. Effect of fungal strain and cereal substrate on in vitro mycotoxin production by *Fusarium poae* and *Fusarium avenaceum*. *Food Addit. Contam. Part A* **2008**, *25*, 745–757. [CrossRef] [PubMed]
25. Nichea, M.J.; Cendoya, E.; Zachetti, V.G.L.; Chiacchiera, S.M.; Sulyok, M.; Krska, R.; Torres, A.M.; Chulze, S.N.; Ramirez, M.L. Mycotoxin profile of *Fusarium armeniacum* isolated from natural grasses intended for cattle feed. *World Mycotoxin J.* **2014**, *8*, 451–457. [CrossRef]
26. Yli-Mattila, T.; Ward, T.J.; O'Donnell, K.; Proctor, R.H.; Burkin, A.A.; Kononenko, G.P.; Gavrilova, O.P.; Aoki, T.; McCormick, S.P.; Gagkaeva, T.Y. *Fusarium sibiricum* sp. nov, a novel type A trichothecene-producing *Fusarium* from northern Asia closely related to *F. sporotrichioides* and *F. langsethiae*. *Int. J. Food Microbiol.* **2011**, *147*, 58–68.
27. Elmer, W.H.; Marra, R.E. New species of *Fusarium* associated with dieback of *Spartina alterniflora* in Atlantic salt marshes. *Mycologia* **2011**, *103*, 806–819. [CrossRef]
28. Laurence, M.H.; Walsh, J.L.; Shuttleworth, L.A.; Robinson, D.M.; Johansen, R.M.; Petrovic, T.; Vu, T.T.H.; Burgess, L.W.; Summerell, B.A.; Liew, E.C.Y. Six novel species of *Fusarium* from natural ecosystems in Australia. *Fungal Divers.* **2016**, *77*, 349–366.
29. Rocha, L.O.; Laurence, M.H.; Proctor, R.H.; McCormick, S.P.; Summerell, B.A.; Liew, E.C.Y. Variation in type A trichothecene production and trichothecene biosynthetic genes in *Fusarium goolgardi* from natural ecosystems of Australia. *Toxins* **2015**, *7*, 4577–4594.
30. Varga, E.; Wiesenberger, G.; Hametner, C.; Ward, T.J.; Dong, Y.; Schöfeck, D.; McCormick, S.; Broz, K.; Stückler, R.; Schuhmacher, R.; et al. New tricks of an old enemy: Isolates of *Fusarium graminearum* produce a type A trichothecene mycotoxin. *Environ. Microbiol.* **2015**, *17*, 2588–2600. [PubMed]
31. Lattanzio, V.M.T.; Ciasca, B.; Terzi, V.; Ghizzoni, R.; McCormick, S.P.; Pascale, M. Study of the natural occurrence of T-2 and HT-2 toxins and their glucosyl derivatives from field barley to malt by high-resolution Orbitrap mass spectrometry. *Food Addit. Contam. Part A* **2015**, *32*, 1647–1655. [CrossRef] [PubMed]
32. Kimura, M.; Kaneko, I.; Komiyama, M.; Takatsuki, A.; Koshino, H.; Yoneyama, K.; Yamaguchi, I. Trichothecene 3-*O*-acetyltransferase protects both the producing organism and transformed yeast from related mycotoxins. *J. Biol. Chem.* **1998**, *273*, 1654–1661. [PubMed]
33. Meek, I.B.; Peplow, A.W.; Ake, C., Jr.; Phillips, T.D.; Beremand, M.N. *Tri1* encodes the cytochrome P450 monooxygenase for C-8 hydroxylation during trichothecene biosynthesis in *Fusarium sporotrichioides* and resides upstream of another new *Tri* gene. *Appl. Environ. Microbiol.* **2003**, *69*, 1607–1613.
34. Brown, D.W.; Proctor, R.H.; Dyer, R.B.; Plattner, R.D. Characterization of a *Fusarium* 2-gene cluster involved in trichothecene C-8 modification. *J. Agric. Food Chem.* **2003**, *51*, 7936–7944. [CrossRef] [PubMed]
35. Peplow, A.W.; Meek, I.B.; Wiles, M.C.; Phillips, T.D.; Beremand, M.N. *Tri16* is required for esterification of position C-8 during trichothecene mycotoxin production by *Fusarium sporotrichioides*. *Appl. Environ. Microbiol.* **2003**, *69*, 5935–5940. [CrossRef]
36. Proctor, R.H.; McCormick, S.P.; Alexander, N.J.; Desjardins, A.E. Evidence that a secondary metabolic biosynthetic gene cluster has grown by gene relocation during evolution of the filamentous fungus *Fusarium*. *Mol. Microbiol.* **2009**, *74*, 1128–1142. [PubMed]
37. Chen, L.; Yang, J.; Wang, H.; Yang, X.; Zhang, C.; Zhao, Z.; Wang, J. NX toxins: New threat posed by *Fusarium graminearum* species complex. *Trends Food Sci. Technol.* **2022**, *119*, 179–191.
38. McCormic, S.P.; Alexander, N.J. *Fusarium Tri8* encodes a trichothecene C-3 esterase. *Appl. Environ. Microbiol.* **2002**, *68*, 2959–2964. [CrossRef]
39. Alexander, N.J.; McCormick, S.P.; Waalwijk, C.; van der Lee, T.; Proctor, R.H. The genetic basis for 3-ADON and 15-ADON trichothecene chemotypes in *Fusarium graminearum*. *Fungal Genet. Biol.* **2011**, *48*, 485–495.
40. Brown, D.W.; McCormic, S.P.; Alexander, N.J.; Proctor, R.H.; Desjardins, A.E. A genetic and biochemical approach to study trichothecene diversity in *Fusarium sporotrichioides* and *Fusarium graminearum*. *Fungal Genet. Biol.* **2001**, *32*, 121–133.
41. McCormic, S.P.; Hohn, T.M.; Desjardins, A.E. Isolation and characterization of *Tri3*, a gene encoding 15-*O*-acetyltransferase from *Fusarium sporotrichioides*. *Appl. Environ. Microbiol.* **1996**, *62*, 353–359. [CrossRef]
42. Hohn, T.M.; Desjardins, A.E.; McCormic, S.P. The *Tri4* gene of *Fusarium sporotrichioides* encodes a cytochrome P450 monooxygenase involved in trichothecene biosynthesis. *Mol. Gen. Genet.* **1995**, *248*, 95–102. [PubMed]
43. Proctor, R.H.; Hohn, T.M.; McCormic, S.P.; Desjardins, A.E. *Tri6* encodes an unusual zinc finger protein involved in regulation of trichothecene biosynthesis in *Fusarium sporotrichioides*. *Appl. Environ. Microbiol.* **1995**, *61*, 1923–1930. [PubMed]
44. Hohn, T.M.; Vanmiddlesworth, F. Purification and characterization of the sesquiterpene cyclase trichodiene synthetase from *Fusarium sporotrichioides*. *Arch. Biochem. Biophys.* **1986**, *251*, 756–761. [PubMed]
45. Tag, A.G.; Garifullina, G.F.; Peplow, A.W.; Ake, C., Jr.; Phillips, T.D.; Hohn, T.M.; Beremand, M.N. A novel regulatory gene, *Tri10*, controls trichothecene toxin production and gene expression. *Appl. Environ. Microbiol.* **2001**, *67*, 5294–5302. [CrossRef]
46. Alexander, N.J.; Hohn, T.M.; McCormic, S.P. The *Tri11* gene of *Fusarium sporotrichioides* encodes a cytochrome P-450 monooxygenase required for C-15 hydroxylation in trichothecene biosynthesis. *Appl. Environ. Microbiol.* **1998**, *64*, 221–225. [CrossRef]
47. Alexander, N.J.; McCormic, S.P.; Hohn, T.M. *Tri12*, a trichothecene efflux pump from *Fusarium sporotrichioides*: Gene isolation and expression in yeast. *Mol. Gen. Genet.* **1999**, *261*, 977–984. [CrossRef]
48. McCormic, S.P.; Alexander, N.J.; Trapp, S.E.; Hohn, T.M. Disruption of *Tri101*, the gene encoding trichothecene 3-*O*-acetyltransferase, from *Fusarium sporotrichioides*. *Appl. Environ. Microbiol.* **1999**, *65*, 5252–5256.

49. Chen, Y.; Kistler, H.C.; Ma, Z. *Fusarium graminerum* trichothecene mycotoxins: Biosynthesis, regulation, and management. *Annu. Rev. Phytopathol.* **2019**, *57*, 15–39. [CrossRef]
50. Alexander, N.J.; Proctor, R.H.; McCormick, S.P. Gens, gene clusters, and biosynthesis of trichothecenes and fumonisins in *Fusarium*. *Toxin Rev.* **2009**, *28*, 198–215. [CrossRef]
51. Desjardins, A.E. From yellow rain to green wheat: 25 years of trichothecene biosynthesis research. *J. Agr. Food Chem.* **2009**, *57*, 4478–4484. [CrossRef]
52. Foroud, N.A.; Eudes, F. Trichothecenes in cereal grains. *Int. J. Mol. Sci.* **2009**, *10*, 147–173. [PubMed]
53. Proctor, R.H.; McCormick, S.P.; Kim, H.; Cardoza, R.E.; Stanley, A.M.; Lindo, L.; Kelly, A.; Brown, D.W.; Lee, T.; Vaughan, M.M.; et al. Evolution of structural diversity of trichothecenes, a family of toxins produced by plant pathogenic and entomopathogenic fungi. *PLoS Pathog.* **2018**, *14*, e1006946.
54. Ward, T.J.; Bielawski, J.P.; Kistler, H.C.; Sullivan, E.; O'Donnell, K. Ancestral polymorphism and adaptive evolution in the trichothecene mycotoxin gene cluster of phytopathogenic *Fusarium*. *Proc. Natl. Acad. Sci. USA* **2002**, *99*, 9278–9283. [CrossRef] [PubMed]
55. Kelly, A.; Proctor, R.H.; Belzile, F.; Chulze, S.N.; Clear, R.M.; Cowger, C.; Elmer, W.; Lee, T.; Obanor, F.; Waalwijk, C.; et al. The geographic distribution and complex evolutionary history of the NX-2 trichothecene chemotype from *Fusarium graminearum*. *Fungal Genet. Biol.* **2016**, *95*, 39–48. [PubMed]

MDPI AG
Grosspeteranlage 5
4052 Basel
Switzerland
Tel.: +41 61 683 77 34

Toxins Editorial Office
E-mail: toxins@mdpi.com
www.mdpi.com/journal/toxins

www.ingramcontent.com/pod-product-compliance
Lightning Source LLC
Chambersburg PA
CBHW061522210926
51850CB00008B/2003
9783725851010